Blood Cells and Redox Homeostasis in Health and Disease

Guest Editors

Angelo D'Alessandro
Alkmini T. Anastasiadi

MDPI

Basel • Beijing • Wuhan • Barcelona • Belgrade • Novi Sad • Cluj • Manchester

Guest Editors
Angelo D'Alessandro
Department of Biochemistry and Molecular Genetics
University of Colorado Denver
Aurora, CO
United States

Alkmini T. Anastasiadi
Department of Biochemistry
University of Patras
Patras
Greece

Editorial Office
MDPI AG
Grosspeteranlage 5
4052 Basel, Switzerland

This is a reprint of the Special Issue, published open access by the journal *Antioxidants* (ISSN 2076-3921), freely accessible at: https://www.mdpi.com/journal/antioxidants/special_issues/blood_cells.

For citation purposes, cite each article independently as indicated on the article page online and as indicated below:

Lastname, A.A.; Lastname, B.B. Article Title. *Journal Name* **Year**, *Volume Number*, Page Range.

ISBN 978-3-7258-3629-1 (Hbk)
ISBN 978-3-7258-3630-7 (PDF)
https://doi.org/10.3390/books978-3-7258-3630-7

Contents

About the Editors . . . . **vii**

Alkmini T. Anastasiadi and Angelo D'Alessandro
E Pluribus, Unum: Emergent Redox Harmony from the Chaos of Blood Cells
Reprinted from: *Antioxidants* **2024**, *13*, 1151, https://doi.org/10.3390/antiox13091151 . . . . . . **1**

Serge Cedrick Toya Mbiandjeu, Angela Siciliano, Alessandro Mattè, Enrica Federti, Massimiliano Perduca, Davide Melisi, et al.
Nrf2 Plays a Key Role in Erythropoiesis during Aging
Reprinted from: *Antioxidants* **2024**, *13*, 454, https://doi.org/10.3390/antiox13040454 . . . . **4**

Tomoki Bo, Hidekazu Nohara, Ken-ichi Yamada, Satoshi Miyata and Junichi Fujii
Ascorbic Acid Protects Bone Marrow from Oxidative Stress and Transient Elevation of Corticosterone Caused by X-ray Exposure in *Akr1a*-Knockout Mice
Reprinted from: *Antioxidants* **2024**, *13*, 152, https://doi.org/10.3390/antiox13020152 . . . . **25**

Xia Pan, Martin Köberle and Mehrdad Ghashghaeinia
Vitamin C-Dependent Uptake of Non-Heme Iron by Enterocytes, Its Impact on Erythropoiesis and Redox Capacity of Human Erythrocytes
Reprinted from: *Antioxidants* **2024**, *13*, 968, https://doi.org/10.3390/antiox13080968 . . . . **39**

Payal Chakraborty, Hajnalka Orvos and Edit Hermesz
Molecular Study on Twin Cohort with Discordant Birth Weight
Reprinted from: *Antioxidants* **2023**, *12*, 1370, https://doi.org/10.3390/antiox12071370 . . . . . . **55**

Tomas Jasenovec, Dominika Radosinska, Katarina Jansakova, Maria Kopcikova, Aleksandra Tomova, Denisa Snurikova, et al.
Alterations in Antioxidant Status and Erythrocyte Properties in Children with Autism Spectrum Disorder
Reprinted from: *Antioxidants* **2023**, *12*, 2054, https://doi.org/10.3390/antiox12122054 . . . . . . **67**

Julie A. Reisz, Monika Dzieciatkowska, Daniel Stephenson, Fabia Gamboni, D. Holmes Morton and Angelo D'Alessandro
Red Blood Cells from Individuals with Lesch–Nyhan Syndrome: Multi-Omics Insights into a Novel S162N Mutation Causing Hypoxanthine-Guanine Phosphoribosyltransferase Deficiency
Reprinted from: *Antioxidants* **2023**, *12*, 1699, https://doi.org/10.3390/antiox12091699 . . . . . . **81**

Beata Szlachta, Anna Birková, Beáta Čižmárová, Anna Głogowska-Gruszka, Paulina Zalejska-Fiolka, Maria Dydoń and Jolanta Zalejska-Fiolka
Erythrocyte Oxidative Status in People with Obesity: Relation to Tissue Losses, Glucose Levels, and Weight Reduction
Reprinted from: *Antioxidants* **2024**, *13*, 960, https://doi.org/10.3390/antiox13080960 . . . . **101**

Aimilia Giannaki, Hara T. Georgatzakou, Sotirios P. Fortis, Alkmini T. Anastasiadi, Efthimia G. Pavlou, Efrosyni G. Nomikou, et al.
Stratification of $\beta^S\beta^+$ Compound Heterozygotes Based on L-Glutamine Administration and RDW: Focusing on Disease Severity
Reprinted from: *Antioxidants* **2023**, *12*, 1982, https://doi.org/10.3390/antiox12111982 . . . . . . **120**

Tiffany A. Thomas, Richard O. Francis, James C. Zimring, Joseph P. Kao, Travis Nemkov and Steven L. Spitalnik
The Role of Ergothioneine in Red Blood Cell Biology: A Review and Perspective
Reprinted from: *Antioxidants* **2024**, *13*, 717, https://doi.org/10.3390/antiox13060717 . . . . **135**

Moua Yang and Roy L. Silverstein
Targeting Cysteine Oxidation in Thrombotic Disorders
Reprinted from: *Antioxidants* **2024**, *13*, 83, https://doi.org/10.3390/antiox13010083 . . . . **153**

About the Editors

Angelo D'Alessandro

Angelo D'Alessandro is a tenured full professor at the Department of Biochemistry and Molecular Genetics. His research focuses on understanding the shared molecular mechanisms that drive systemic responses to trauma/hemorrhagic shock, I/R injury, sickle cell disease, aging and inflammation, mammalian hibernation, and pulmonary hypertension, using high throughput omics technologies. He is the founder and director of the Metabolomics Core of CU Anschutz School of Medicine, the director of the Mass Spectrometry Shared Resource for the University of Colorado Cancer Center and is also the founder and CSO of Omix Technologies Inc, Altis Biosciences, and serves as an advisory board member for Hemanext Inc, and Macopharma. D'Alessandro has published over 600 papers, with research sponsored by NHLBI, NIDDK, and NIGMS. He is a Boettcher Investigator, AABB Hall of Fame, Jean Julliard (ISBT), and RISE (AABB) awardee. Also, he is an affiliate investigator to the Vitalant Research Institute, the Linda Crnic Institute for Down syndrome, and the Gates Grubstake Center for Regenerative Medicine.

Alkmini T. Anastasiadi

Alkmini Anastasiadi is a postdoctoral researcher at the Department of Biochemistry, School of Medicine, University of Patras. Her research focuses on the potential of red blood cell characteristics to serve as biomarkers in several disease states, as well as the personalization of transfusion therapy. She specializes in biochemical and physiological analyses to explore red cell integrity, proteostasis, and redox status. She has contributed to numerous international publications and is the Principal Investigator of a grant awarded by the European Hematology Association aiming to identify novel erythrocytic drug targets in sickle cell disease.

antioxidants

MDPI

Editorial

E Pluribus, Unum: Emergent Redox Harmony from the Chaos of Blood Cells

Alkmini T. Anastasiadi [1,*] and Angelo D'Alessandro [2,*]

1 Department of Biochemistry, School of Medicine, University of Patras, 26504 Patras, Greece
2 Department of Biochemistry and Molecular Genetics, Anschutz Medical Campus, University of Colorado Denver, Aurora, CO 80045, USA
* Correspondence: aanastasiadi@upatras.gr (A.T.A.); angelo.dalessandro@cuanschutz.edu (A.D.); Tel.: +30-2610996154 (A.T.A.); +1-303-7240096 (A.D.)

Blood cells play a fundamental role in maintaining systemic homeostasis, by responding dynamically to various physiological and environmental stimuli. Nevertheless, they are continuously exposed to stressors such as hypoxia, inflammation, and oxidative agents, which influence not only their function and survival but also their interactions with other cell types and their contributions to systemic processes. The articles amassed in this Special Issue explore some of the myriad ways in which oxidative stress influences blood cell functionality, starting a journey through the labyrinth of redox balance, from the microscopic to the systemic, and back again.

Redox equilibrium is essential from the dawn of life of blood cells until their final moments. The unobstructed progress of hematopoiesis—from stem cell precursors to mature erythrocytes—relies, among other things, on the redox status of the organism [1,2]. When this equilibrium is disrupted, and the scales tilt toward the accumulation of reactive oxygen species (ROS), the ability of hematopoietic stem cells to thrive, proliferate, and differentiate falters beneath the weight of oxidative stress [3]. Mbiandjeu et al. [Contribution 1] shine a light on this delicate balance by presenting evidence that the nuclear factor erythroid-derived 2 (Nrf2) regulates the antioxidant response during erythropoiesis, while dysfunction in this pathway leads to severe oxidative lesions, premature death of erythroblasts and overall ineffective erythropoiesis. The authors show that this perturbation is attenuated upon antioxidant supplementation, a finding analogous to that of Bo et al. [Contribution 2], who revealed that ascorbic acid seems to protect bone marrow cells from excessive oxidative stress generated by X-ray irradiation. The ability of this simple molecule to restore white blood cell counts upon supplementation in knockout rodent models underscores the profound interdependence between antioxidant systems and hematopoiesis. This relationship even extends to iron metabolism—a key player in hemoglobin synthesis. More specifically, ascorbic acid indirectly regulates erythropoiesis, by facilitating the uptake of iron from enterocytes [Contribution 3]. The acquired iron is then exported, binds to plasma transferrin, and is supplied to erythroid precursors to proceed with their differentiation.

Oxidative stress is a sword of Damocles for circulating blood cells, whose lifespan in circulation is regulated by the balance between the production of ROS and the body's ability to neutralize them [4], a process that ultimately leads to cellular damage and dysfunction. A case in point is presented in the study by Chakraborty et al. [Contribution 4], who examined RBCs in twins with discordant birth weights. It was shown that the impaired nitric oxide synthesis aids in the generation of reactive species, which oxidize the RBC membrane and induce morphological alterations in the lower-weight sibling cohort; thus, they potentially exert a significant impact on the physiological status of the vascular system. Autism spectrum disorder (ASD) is another condition in which high systemic oxidative stress is mirrored in the RBC redox status and membrane properties. Given the close association

Citation: Anastasiadi, A.T.; D'Alessandro, A. E Pluribus, Unum: Emergent Redox Harmony from the Chaos of Blood Cells. *Antioxidants* **2024**, *13*, 1151. https://doi.org/10.3390/antiox13091151

Received: 19 September 2024
Accepted: 22 September 2024
Published: 23 September 2024

between RBC physiology and function and the high oxygen needs of nerve cells, the deterioration of RBCs could be linked to ASD clinical manifestations [Contribution 5]. Both findings highlight that oxidative stress not only affects immediate cellular function but can also cast long shadows on long-term health outcomes. Of course, oxidative dysregulation might be a result of specific mutations, as in the case of Lesch–Nyhan syndrome, where the deficiency in hypoxanthine–guanine phosphoribosyltransferase fuels the generation of urate, which is accompanied by excessive production of hydrogen peroxide. As thoroughly discussed in this Special Issue [Contribution 6], the RBCs of such subjects are characterized by altered metabolism, elevated oxidative stress markers, and significantly compromised integrity. Notably, treatment with allopurinol only partially alleviates some of the observed imbalances, without fully restoring normal RBC function.

In the face of such bleak prospects, the antioxidant defenses of blood cells are invaluable for the maintenance of redox balance. The cellular network of antioxidant systems includes enzymatic antioxidants such as superoxide dismutase or peroxiredoxin and non-enzymatic antioxidants like glutathione [5]. However, this defense system can be compromised in pathophysiological conditions. For instance, in the context of obesity, when the organism is in a state of constant oxidative stress, RBCs are expected to address increased oxidative insults. As revealed in the study by Szlachta et al. [Contribution 7], oxidative markers are reduced upon weight loss, while antioxidant enzyme activities are differentially affected based on the glycemic status of each participant, highlighting the complex connection between metabolic and antioxidant networks within RBCs, as well as the adaptability of the latter depending on their environment. Based on the observation that redox equilibrium is disrupted in disease states, the promise of redemption lies in therapeutic strategies that can restore redox balance and ultimately improve cellular function [6]. For instance, the high-dose administration of the antioxidant L-glutamine to patients with sickle cell disease alleviates several hallmarks of the disease, including inflammation and coagulation, probably through its redox-restorative potential [Contribution 8]. A similar effect is highlighted with ergothioneine, a dietary molecule that can inactivate several ROS. The comprehensive analysis by Thomas et al. [Contribution 9] pointed to the potential benefits of ergothioneine supplementation for patients with hematological conditions characterized by redox imbalance, due to the ability of ergothioneine to enter erythroid progenitors and exert its antioxidant properties. Moreover, ergothioneine levels could be also associated with the superior storability of RBCs or their post-transfusion outcomes. A similar study [Contribution 10], which focuses on another formed element of blood, namely platelets, further highlights the significance of antioxidant therapies. Cysteine oxoforms are involved in oxidative stress-related thrombotic phenomena via their interaction with platelet receptors and thiol isomerases. Therefore, antioxidant strategies that target oxidative cysteine modification have the potential to prevent thrombosis, as is the case of nucleophile ligands, which can target oxidized cysteines.

In the tale of blood cells, the script of redox balance is all but a linear one. From their birth in the marrow to their final moments in circulation, the crosstalk between 25 trillion blood cells—coming together as a sort of circulating organ [7]—and the remaining 5 trillion cells in the human body, is based upon a stride between the opposing forces of oxidative stress and antioxidant defense. The studies presented in this Special Issue further our understanding of the relationship between oxidative stress and blood cell functionality, offering new insights into the systemic importance of maintaining redox balance and pointing to new frontiers in both research and therapy.

Conflicts of Interest: The authors declare no conflicts of interest.

List of Contributions:

1. Mbiandjeu, S.C.T.; Siciliano, A.; Mattè, A.; Federti, E.; Perduca, M.; Melisi, D.; Andolfo, I.; Amoresano, A.; Iolascon, A.; Valenti, M.T.; et al. Nrf2 Plays a Key Role in Erythropoiesis during Aging. *Antioxidants* **2024**, *13*, 454. https://doi.org/10.3390/antiox13040454

2. Bo, T.; Nohara, H.; Yamada, K.-i.; Miyata, S.; Fujii, J. Ascorbic Acid Protects Bone Marrow from Oxidative Stress and Transient Elevation of Corticosterone Caused by X-ray Exposure in *Akr1a*-Knockout Mice. *Antioxidants* **2024**, *13*, 152. https://doi.org/10.3390/antiox13020152
3. Pan, X.; Köberle, M.; Ghashghaeinia, M. Vitamin C-Dependent Uptake of Non-Heme Iron by Enterocytes, Its Impact on Erythropoiesis and Redox Capacity of Human Erythrocytes. *Antioxidants* **2024**, *13*, 968. https://doi.org/10.3390/antiox13080968
4. Chakraborty, P.; Orvos, H.; Hermesz, E. Molecular Study on Twin Cohort with Discordant Birth Weight. *Antioxidants* **2023**, *12*, 1370. https://doi.org/10.3390/antiox12071370
5. Jasenovec, T.; Radosinska, D.; Jansakova, K.; Kopcikova, M.; Tomova, A.; Snurikova, D.; Vrbjar, N.; Radosinska, J. Alterations in Antioxidant Status and Erythrocyte Properties in Children with Autism Spectrum Disorder. *Antioxidants* **2023**, *12*, 2054. https://doi.org/10.3390/antiox12122054
6. Reisz, J.A.; Dzieciatkowska, M.; Stephenson, D.; Gamboni, F.; Morton, D.H.; D'Alessandro, A. Red Blood Cells from Individuals with Lesch–Nyhan Syndrome: Multi-Omics Insights into a Novel S162N Mutation Causing Hypoxanthine-Guanine Phosphoribosyltransferase Deficiency. *Antioxidants* **2023**, *12*, 1699. https://doi.org/10.3390/antiox12091699
7. Szlachta, B.; Birková, A.; Čižmárová, B.; Głogowska-Gruszka, A.; Zalejska-Fiolka, P.; Dydoń, M.; Zalejska-Fiolka, J. Erythrocyte Oxidative Status in People with Obesity: Relation to Tissue Losses, Glucose Levels, and Weight Reduction. *Antioxidants* **2024**, *13*, 960. https://doi.org/10.3390/antiox13080960
8. Giannaki, A.; Georgatzakou, H.T.; Fortis, S.P.; Anastasiadi, A.T.; Pavlou, E.G.; Nomikou, E.G.; Drandaki, M.P.; Kotsiafti, A.; Xydaki, A.; Fountzoula, C.; et al. Stratification of $\beta^S\beta^+$ Compound Heterozygotes Based on L-Glutamine Administration and RDW: Focusing on Disease Severity. *Antioxidants* **2023**, *12*, 1982. https://doi.org/10.3390/antiox12111982
9. Thomas, T.A.; Francis, R.O.; Zimring, J.C.; Kao, J.P.; Nemkov, T.; Spitalnik, S.L. The Role of Ergothioneine in Red Blood Cell Biology: A Review and Perspective. *Antioxidants* **2024**, *13*, 717. https://doi.org/10.3390/antiox13060717
10. Yang, M.; Silverstein, R.L. Targeting Cysteine Oxidation in Thrombotic Disorders. *Antioxidants* **2024**, *13*, 83. https://doi.org/10.3390/antiox13010083

References

1. Pimkova, K.; Jassinskaja, M.; Munita, R.; Ciesla, M.; Guzzi, N.; Cao Thi Ngoc, P.; Vajrychova, M.; Johansson, E.; Bellodi, C.; Hansson, J. Quantitative analysis of redox proteome reveals oxidation-sensitive protein thiols acting in fundamental processes of developmental hematopoiesis. *Redox Biol.* **2022**, *53*, 102343. [CrossRef] [PubMed]
2. Sattler, M.; Winkler, T.; Verma, S.; Byrne, C.H.; Shrikhande, G.; Salgia, R.; Griffin, J.D. Hematopoietic growth factors signal through the formation of reactive oxygen species. *Blood* **1999**, *93*, 2928–2935. [CrossRef] [PubMed]
3. Ludin, A.; Gur-Cohen, S.; Golan, K.; Kaufmann, K.B.; Itkin, T.; Medaglia, C.; Lu, X.J.; Ledergor, G.; Kollet, O.; Lapidot, T. Reactive oxygen species regulate hematopoietic stem cell self-renewal, migration and development, as well as their bone marrow microenvironment. *Antioxid. Redox Signal.* **2014**, *21*, 1605–1619. [CrossRef]
4. Sies, H. Oxidative stress: A concept in redox biology and medicine. *Redox Biol.* **2015**, *4*, 180–183. [CrossRef]
5. He, L.; He, T.; Farrar, S.; Ji, L.; Liu, T.; Ma, X. Antioxidants Maintain Cellular Redox Homeostasis by Elimination of Reactive Oxygen Species. *Cell. Physiol. Biochem. Int. J. Exp. Cell. Physiol. Biochem. Pharmacol.* **2017**, *44*, 532–553. [CrossRef]
6. Forman, H.J.; Zhang, H. Targeting oxidative stress in disease: Promise and limitations of antioxidant therapy. *Nat. Rev. Drug Discov.* **2021**, *20*, 689–709. [CrossRef] [PubMed]
7. Nemkov, T.; Reisz, J.A.; Xia, Y.; Zimring, J.C.; D'Alessandro, A. Red blood cells as an organ? How deep omics characterization of the most abundant cell in the human body highlights other systemic metabolic functions beyond oxygen transport. *Expert Rev. Proteom.* **2018**, *15*, 855–864. [CrossRef] [PubMed]

antioxidants

MDPI

Article

Nrf2 Plays a Key Role in Erythropoiesis during Aging

Serge Cedrick Toya Mbiandjeu [1], Angela Siciliano [2,3], Alessandro Mattè [1], Enrica Federti [2,3], Massimiliano Perduca [4], Davide Melisi [1], Immacolata Andolfo [5,6], Angela Amoresano [7], Achille Iolascon [5,6], Maria Teresa Valenti [8], Francesco Turrini [9], Michele Bovi [4], Arianna Pisani [4], Antonio Recchiuti [10], Domenico Mattoscio [10], Veronica Riccardi [2], Luca Dalle Carbonare [2,3], Carlo Brugnara [11,12], Narla Mohandas [13] and Lucia De Franceschi [2,3,*]

1 Department of Medicine, University of Verona, 37134 Verona, Italy; sergecedrick@gmail.com (S.C.T.M.); alessandro.matte@gmail.com (A.M.); davide.melisi@univr.it (D.M.)
2 Dipartimento Ingegneria per la Medicina di Innovazione—DIMI, University of Verona, 37134 Verona, Italy; angela.siciliano@univr.it (A.S.); federti.enrica@gmail.com (E.F.); veronica.riccardi@univr.it (V.R.); luca.dallecarbonare@univr.it (L.D.C.)
3 Department of Medicine, AOUI Verona, 37134 Verona, Italy
4 Department of Biotechnology, University of Verona, 37134 Verona, Italy; massimiliano.perduca@univr.it (M.P.); michelebovi78@gmail.com (M.B.); ariannapisani92@gmail.com (A.P.)
5 Dipartimento di Medicina Molecolare e Biotecnologie Mediche, Università degli Studi di Napoli Federico II, 80131 Naples, Italy; immandolfo@gmail.com (I.A.); achille.iolascon@unina.it (A.I.)
6 CEINGE Biotecnologie Avanzate, 80131 Naples, Italy
7 Department of Chimical Sciences, University Federico II, 80138 Naples, Italy; angela.amoresano@unina.it
8 Department of Neuroscience, University of Verona, 37134 Verona, Italy; mariateresa.valenti@univr.it
9 Department of Oncology, University of Torino, 10124 Torino, Italy; francesco.turrini@unito.it
10 Department of Medical, Oral, and Biotechnology Science, "G. d'Annunzio" University Chieti–Pescara, 66013 Chieti, Italy; antonio.recchiuti@unich.it (A.R.); d.mattoscio@unich.it (D.M.)
11 Department of Laboratory Medicine, Boston Children's Hospital, Boston, MA 02114, USA; carlo.brugnara@childrens.harvard.edu
12 Department of Pathology, Harvard Medical School, Boston, MA 02114, USA
13 New York Blood Center Enterprises, New York, NY 10065, USA; mnarla@nybc.org
* Correspondence: lucia.defranceschi@univr.it

Abstract: Aging is characterized by increased oxidation and reduced efficiency of cytoprotective mechanisms. Nuclear factor erythroid-2-related factor (Nrf2) is a key transcription factor, controlling the expression of multiple antioxidant proteins. Here, we show that Nrf2$^{-/-}$ mice displayed an age-dependent anemia, due to the combined contributions of reduced red cell lifespan and ineffective erythropoiesis, suggesting a role of Nrf2 in erythroid biology during aging. Mechanistically, we found that the expression of antioxidants during aging is mediated by activation of Nrf2 function by peroxiredoxin-2. The absence of Nrf2 resulted in persistent oxidation and overactivation of adaptive systems such as the unfolded protein response (UPR) system and autophagy in Nrf2$^{-/-}$ mouse erythroblasts. As Nrf2 is involved in the expression of autophagy-related proteins such as autophagy-related protein (Atg) 4-5 and p62, we found impairment of late phase of autophagy in Nrf2$^{-/-}$ mouse erythroblasts. The overactivation of the UPR system and impaired autophagy drove apoptosis of Nrf2$^{-/-}$ mouse erythroblasts via caspase-3 activation. As a proof of concept for the role of oxidation, we treated Nrf2$^{-/-}$ mice with astaxanthin, an antioxidant, in the form of poly (lactic-co-glycolic acid) (PLGA)-loaded nanoparticles (ATS-NPs) to improve its bioavailability. ATS-NPs ameliorated the age-dependent anemia and decreased ineffective erythropoiesis in Nrf2$^{-/-}$ mice. In summary, we propose that Nrf2 plays a key role in limiting age-related oxidation, ensuring erythroid maturation and growth during aging.

Keywords: oxidation; astaxanthin; nanoparticles; PLGA; red cells; UPR system; autophagy; ATF6; GADD34; ineffective erythropoiesis

Citation: Mbiandjeu, S.C.T.; Siciliano, A.; Mattè, A.; Federti, E.; Perduca, M.; Melisi, D.; Andolfo, I.; Amoresano, A.; Iolascon, A.; Valenti, M.T.; et al. Nrf2 Plays a Key Role in Erythropoiesis during Aging. *Antioxidants* **2024**, *13*, 454. https://doi.org/10.3390/antiox13040454

Academic Editor: Consuelo Borras

Received: 15 March 2024
Revised: 6 April 2024
Accepted: 8 April 2024
Published: 12 April 2024

1. Introduction

Nuclear-factor erythroid derived 2 (Nrf2) is a redox-related transcription factor involved in acute phase response against oxidation [1,2]. The activation of Nrf2 results in the upregulation of genes carrying antioxidant-related element (ARE-) motif encoding for antioxidants and cytoprotectors such as catalase, peroxiredoxin-2 (Prdx2), NADPH dehydrogenase quinone-1 (Nqo1), heme-oxygenase-1 (Ho-1), Sestrin 2 (Srxn2) or thioredoxin (Trxn) [1–8]. Mice genetically lacking Nfr2 (Nrf2$^{-/-}$) were noted to develop age-dependent immune-mediated anemia, which has been linked to the presence of naturally occurring antibody, promoting erythrophagocytosis [2,9]. However, mechanistic studies to further characterize the origin of the anemia in Nrf2$^{-/-}$ mice are lacking. Recent studies in a mouse model for β-thalassemia demonstrate a persistent activation of Nrf2, which results in intracellular accumulation of inactive antioxidant systems [10,11]. This sustains cell engulfment and overactivation and impairment of autophagy, leading to apoptosis of erythroblasts and ineffective erythropoiesis. In murine β-thalassemia, we recently documented that functional collaboration between Nrf2 and Prdx2 regulates the extent of ineffective erythropoiesis [11]. Although progress has been made regarding Nrf2 function in response to oxidation, much remains to be defined regarding its role in normal and discorded erythropoiesis.

Aging is a physiologic phenomenon characterized by increased oxidation and progressive decline in cell functions, with impairment of adaptive mechanisms such as the unfolded protein system (UPR) and autophagy [12,13]. Nrf2 is sensitive to overactivation of the UPR system and regulates the expression of autophagy-related proteins (Atgs) and key cargo proteins such as p62 [14,15]. Previous studies in various cell and animal models have shown that age-dependent dysregulation of these systems induces apoptosis by (i) cell engulfment with damaged proteins blocking autophagic flux and (ii) upregulation of UPR-system-related pro-apoptotic pathways such as the DNA damage protein inducing growth arrest (GADD34) and the caspase 3-dependent cascade [14]. In normal and stress erythropoiesis, increasing evidence indicates an important role of autophagy in supporting cell growth and differentiation [16–18]. However, limited insights are presently available regarding the contribution of the UPR system, except for the observation that Activating Transcription Factor 4 (ATF4) is involved in the cellular response to heme-mediated oxidation during erythroid maturation processes [17,18].

Here, we studied erythropoiesis in Nrf2$^{-/-}$ mice. We found an age-dependent anemia due to decreased red cell lifespan in conjunction with ineffective erythropoiesis. We also documented the importance of Nrf2 activation and the upregulation of Prdx2 expression during erythropoiesis in old wild-type mice. Furthermore, in sorted Nrf2$^{-/-}$ erythroid precursors, we documented a decline in cytoprotective mechanisms that included the UPR system and autophagy compared with wild-type animals, promoting ineffective erythropoiesis. As a proof of concept for defining the contribution of oxidation to ineffective erythropoiesis of Nrf2$^{-/-}$ aged mice, we used lyophilized PLGA nanoparticles loaded with astaxanthin (ATS-NPs), a nontoxic and organic carotenoid with strong antioxidative properties [19] but without any pro-oxidative properties [20]. ATS-NP administration significantly decreased ineffective erythropoiesis and anemia in old Nrf2$^{-/-}$ mice, mitigating endoplasmic reticulum (ER) stress and improving autophagy. Similar findings were noted with β-thalassemic mice treated with ATS-NPs. Collectively, our data indicate that Nrf2 is activated to support erythropoiesis during aging, mitigating age-related oxidation and modulating cytoprotective mechanisms such as the UPR system and autophagy.

2. Materials and Methods

2.1. Design of the Study

The present study was performed using C57BL/6J as control (wild-type; WT), Nrf2$^{-/-}$ and Prdx2$^{-/-}$ mouse strains at 4 and 12 months of age. We used female mice since erythropoietin responsiveness and iron homeostasis are affected by gender [21,22]. Mouse blood was collected by retro-orbital venipuncture in anesthetized mice using heparinized capillaries according to the general guidelines of the animal facility (CIRSAL) at the University of Verona. Where indicated, severe anemia was induced by intraperitoneal injection of doxorubicin (0.25 mg/kg body weight) [23]. Cells from the spleen and bone marrow were collected at day 9 following doxorubicin injection. Where indicated, mice were treated through intraperitoneal injection with astaxanthin-loaded PLGA nanoparticles (see also Section 2.6) at a dose of 2 mg/kg or vehicle every two days for four weeks. Hematological parameters, red cell indices and reticulocyte count were evaluated on an ADVIA 120 Hematology System (Siemens Healthcare GmbH, Henkestr, Germany) as previously described [23]. Hematocrit (Hct) and hemoglobin (Hb) were also manually determined. Red cells from wild-type or Nrf2$^{-/-}$ mice were labeled with carboxyfluorescein succinimidyl ester dye (CFSE: 10 μM; Molecular probe, Invitrogen, Waltham, MA, USA) in PBS and 0.5% BSA for 20 min at 37 °C to assess red cell lifespan without any toxicity. CFSE covalently binds to intracellular and cell surface proteins with its succinidyl group [24,25]. Hemichromes bound to the red cell membrane and the extent of binding of naturally occurring anti-band 3 antibodies were assessed as previously reported [26].

2.2. *Flow Cytometric Analysis of Erythropoiesis*

Flow cytometric analysis of erythroid precursors from the bone marrow and spleen of mice was carried out using the CD44-Ter119 gating strategy as previously described [11]. Briefly, cells were centrifuged at 1500 rpm for 5 min at 4 °C and resuspended in BEPS (PBS 1X, BSA 1%, EDTA 2 mM, NaCl 25 mM). Cells were incubated first with CD16/32 (MA5-29707 eBioscience, San Diego, CA, USA) to block the Fc receptor for 15 min at 4 °C in the dark and then later incubated with CD45-APC-Cy7 (47-9459-42 Invitrogen, Waltham, MA, USA), CD44-FITC (SAB4700182 Sigma Aldrich, Merck Group, Darmstadt, Germany), CD71-PE and Ter119-APC (12-0711-82, A18447 eBioscience, San Diego, CA, USA) antibodies for 45 min at 4 °C in the dark. Cells were washed and centrifuged at 1500 rpm for 5 min at 4 °C and resuspended in BEPS, and 7AAD (A1310 Invitrogen, Waltham, MA, USA) was added immediately before the analysis to assess cell viability. Apoptotic erythroblasts were analyzed on CD44-Ter119-gated populations using the Annexin-V PE Apoptosis detection kit (88-8102-72 eBioscience, San Diego, CA, USA), following the manufacturer's instructions [11]. ROS levels of the erythroid precursors were determined using the General Oxidative Stress Indicator, CM-H2DCFDA (C6827 LifeTechnologies, Carlsbad, CA, USA) on CD44-Ter119-gated populations as previously described [11].

Oxidative DNA damage of erythroid precursors was determined by 8OHdG flow cytometric analysis as previously described with some modifications. Briefly, erythroid precursors from mouse bone marrow were stained with CD44-FITC (SAB4700182 Sigma Aldrich, Merck Group, Darmstadt, Germany), CD71-PE and Ter119-BV480 (12-0711-82, 414-5921-82 eBiosciences, CA, USA) and fixed and permeabilized with BD Cytofix/Cytoperm and the BD Cytoperm Plus permeabilization reagent (554714 BD Biosciences, CA, USA), respectively. Cells were then stained with primary anti-8OHdG (sc-393871 Santa Cruz Biotechnology, Santa Cruz, CA, USA) and secondary anti-mouse eFluor647 (A-31571 Invitrogen, Waltham, MA, USA). All the analyses were performed with a FACS Canto-II^{TM} flow cytometer (Becton Dickinson, San Jose, CA, USA), and data were analyzed with the FlowJo v10 software (Tree Star, Ashland, OR, USA).

2.3. *Analysis of Red Cells and Sorted Erythroid Precursors*

Red cell membrane ghost and cytosol fractions were prepared as previously described [27]. For Prdx2 evaluation by Western blot, 100 mM of NEM was added to the lysis buffer to avoid possible artifacts related to Prdx2 oxidation during preparation.

Erythroblasts (CD44^{+}Ter119^{+}FSC^{high}) were sorted from WT and Nrf2$^{-/-}$ mouse bone marrow using a FACS Aria-III^{TM} cell sorter (Becton Dickinson, San Jose, CA, USA) as previously reported [11]. Sorted cells were used for (i) immunoblot analysis; (ii) immunofluorescence assay; (iii) molecular analysis performed using QRT-PCR; (iv) CPP32/Caspase-3 Fluorometric protease assay (K105-25 BioVision, Milpitas, CA, USA; following manufacturer's instructions); and (v) nuclear protein isolation, using the Q-proteome Nuclear Protein Kit (37582 Qiagen, Hilden, Germany) according to the manufacturer's instructions).

2.3.1. Molecular Analysis of Erythroid Precursors

The protocols used for RNA isolation, cDNA preparation and qRT-PCR have been previously described [11]. Quantitative RT-PCR (qRT-PCR) was performed by the SYBR-green method, following standard protocols with an Applied Biosystems ABI PRISM 7900HT Sequence Detection system. Relative gene expression was calculated using the 2-DCt method, where DCt indicates the differences in the mean Ct between selected genes and the normalization control (Gapdh) [28]. The qRT-PCR primers for each gene were designed using Primer Express software version 2.0 (Life Technologies, Carlsbad, CA, USA). The primers used are reported in Table 1.

Table 1. Primers.

Gene	Forward Primer Sequence (5′→ 3′)	Reverse Primer Sequence (5′→ 3′)
Gapdh	CCACATCGCTCAGACACCAT	AGTTAAAAGCAGCCCTGGTGAC
Erfe	ATGGGGCTGGAGAACAGC	TGGCATTGTCCAAGAAGACA
Atf6	GCCCGGTGAATGGAAAACTT	GTCTCCTCAGCACAGCGATA
P62	AGCGGTTACTCACTCCATGGAAGA	CTTGGGGAGGTTTCGTCTCT
Txn	CATGGCCAACAAAATCATGCAGCT	CGATCTGTTCAATTTTCGTTGGG
Srxn2	TTTGGCAGAGACAT	ACTTTCAGCGTGGCTGGG
Prdx2	CGCCTAGTCCAGGCCTTTCCAA	GATGGTGTCACTGCCGGG
Ho-1	GCACAGGGTGACAGAAGAG	GTCAGCATCACCTGCAGCTC

2.3.2. Immunoblot Analysis of Red Cells and Erythroid Precursors

Red cells were separated into cytosol and membrane fractions as previously described [27]. Proteins from ghosts and the cytosol fraction were solubilized in reducing or non-reducing sample buffer (50 mM Tris, pH 6.8, 2% SDS, 10% glycerol, few grains of bromphenol blue added of 5% β-mercaptoethanol for reducing conditions) and analyzed by one-dimensional SDS–polyacrylamide gel electrophoresis. Gels were either stained with colloidal Coomassie or transferred to nitrocellulose membranes for immunoblot analysis with specific antibodies: anti-Prdx2 (ab109367 Clone1E8, Abcam, Cambridge, UK); anti-catalase (ab76110 Abcam, Cambridge, UK); anti-HSP70, anti-G6PD, anti-TrxR-1 and anti-Nqo1 (sc32239, sc373886, sc365658, sc271116 Santa Cruz Biotechnology, Texas, USA); anti-HSP90 (4874 Cell signaling Technology, Leiden, NL, USA); anti-Actin (SAB4301137 clone BIII-136; Sigma-Aldrich). Protein oxidation was detected by Oxyblot (S7150 Sigma-Aldrich), which allows the detection of carbonyl groups introduced into proteins by oxidative reactions according to the manufacturer's instructions.

Sorted erythroid precursors were prepared for immunoblot analysis as previously described [11]. The following specific antibodies were used: anti-NFkB-phospho-S536 (3033 Cell Signaling Technology, Leiden, NL, USA); anti-NFkB p65 and anti-Atg5 (8242, 12994 Cell Signaling Technology, Leiden, NL, USA); anti-Nrf2-phospho-S40 (ab76026 Clone EP1809Y, Abcam, Cambridge, UK); anti-Nrf2 (ab62352 Abcam, Cambridge, UK); anti Gadd34, anti-Lamp-1, anti-SQSTM1/P62, anti-Rab5 and anti-LC3A/B (ab9869, ab24170, ab109012, ab109534, ab62721 Abcam, Cambridge, UK); anti-APG7 (Atg7) (3615 ProSci, Poway, CA, USA); anti-ATF6 (70B1413.1 Novus Biologicals, Centennial, CO, USA); anti-CHOP (ma1-250 Thermo Fisher Scientific, Waltham, MA, USA); anti-Actin (SAB4301137 clone BIII-136; Sigma-Aldrich, Saint Louis, MO, USA) and anti-GAPDH (HPA040067 Sigma-Aldrich, Saint Louis, MO, USA) as loading controls. Blots were then developed using the Luminata Forte or Luminata Classico Western chemiluminescence reagents. Images were acquired using Image Quant Las Mini 4000 Digital Imaging System (GE Healthcare Life Sciences, Little Chalfont, UK), and densitometric analysis of band intensities was carried out using the ImageQuant TL 10.2 software (GE Healthcare Life Sciences).

2.4. Preparation and Characterization of Astaxanthin-Loaded PLGA Nanoparticles

A single-emulsion solvent evaporation method under sterile conditions at 20 °C was used for the synthesis of the astaxanthin (ATS)-loaded PLGA nanoparticles (NPs) [29]. Briefly, 20 mg of PLGA (7–17 kDa PLGA 50:50 with uncapped end groups (Sigma-Aldrich, St. Louis, MO, USA) and 1 mg of astaxanthin (CNLAB NUTRITION, ASIAN GROUP) were dissolved in a 2 mL mixture of 85% acetone and 15% ethanol. The organic phase was added dropwise to 20 mL of 0.5% aqueous polyvinyl alcohol surfactant under stirring. The obtained emulsion was maintained under stirring overnight to evaporate the solvents. The nanoparticles generated were collected by centrifugation at 13,000× *g* for 20 min at 10 °C and washed several times with 0.01 M phosphate-buffered saline, pH 7.4 (PBS), to remove residual solvents. Then, 5% mannitol was added as a cryoprotectant, and the NPs were divided into aliquots and lyophilized for storage. The mean size and zeta potential of NPs were estimated using the dynamic light scattering (DLS) technique (Nano ZetaSizer ZS, ZEN3600, Malvern Instruments, Malvern, Worcestershire, UK), and the synthesized NPs were suspended in PBS buffer (137 mM NaCl, 2.7 mM KCl, 10 mM Na_2HPO_4 and 1.8 mM KH_2PO_4) at the final concentration of 5 mg/mL; the samples were maintained at 25 °C. The encapsulation efficiency was calculated from an astaxanthin UV absorbance calibration curve prepared with different amounts of the carotenoid dissolved in dimethyl sulfoxide (DMSO). The calibration curve (Figure S6a) was set up by plotting the absorbance against the astaxanthin concentration. The calibration equation line was estimated using linear regression analysis, $y = ax + b$, with the correlation coefficient R^2. Absorbance was recorded at 492 nm, the typical astaxanthin absorbance wavelength in this solvent, using empty nanoparticles' absorbance as baseline correction.

To assess the stability of the nanoparticles and the ability to retain the entrapped astaxanthin over time both at physiological and DLS analysis temperatures (37 and 25 °C, respectively), a release study in PBS was performed. Samples were collected at different time intervals. Each ATS-NP aliquot was dissolved in DMSO, and the released carotenoid was spectrophotometrically quantified at 492 nm using the same calibration curve described above.

2.5. Mass Spectrometric Analysis of Plasma and Organ ATS-NP Distribution

The calibration curve for astaxanthin was developed for quantifying astaxanthin using the external standard method. The standard was dissolved in methanol acidified by 0.1% formic acid at a concentration of 1 mg/mL. Standard solutions of varying concentrations were prepared by serial dilution from the stock solution in a range from 0.1 ppb to 100 ppb. Details on extraction from serum and organs as well as the conditions for mass spectrometric analysis are described in the Supplementary Methods.

2.6. Splenic Macrophage Analysis and Erythrophagocytosis

Spleens were gently dissociated to generate single-cell suspensions with 70 μm nylon mesh strainers; fixed (30 min, rt) with 3% formalin; and incubated (15 min, 4 °C) with fluorochrome-labeled antibodies against CD45, F4/80 (a pan macrophage marker) and CD206 (an M2 macrophage marker). Cells were permeabilized (30 min, room temperature), counterstained with anti-Ter-119 antibody (15 min, 4 °C) and analyzed with a BD FACS Canto II flow cytometer for the simultaneous detection of RBCs phagocytosed by spleen macrophages expressing surface CD206. For microRNA (miRNA) expression analysis, spleens were homogenized in a lysis buffer with 10 mM β-mercaptoethanol, and the small RNA fraction was purified using Norgen isolation kits. miRNA (100 ng) was reverse transcribed with the miScript II kit (Qiagen), and 1 ng/reaction was used as a template for real-time PCR analysis. The relative expression of miRNAs was calculated using the $2^{-\Delta Ct}$ method [30], using SNORD95 as a housekeeping miRNA for the control of equal template loading.

2.7. Statistics

A two-tailed unpaired Student *t*-test or two-way analysis of variance with Tukey's multiple comparisons was used for data analyses. Wherever indicated, we used an unpaired Student *t*-test with Sidak correction or one-way ANOVA algorithm for repeated measures. Normality was assessed with the Shapiro–Wilk test. Data show values from individual mice and are presented as mean ± SD. Differences with $p < 0.05$ were considered significant.

3. Results

3.1. Nrf2$^{-/-}$ Mice Show Accelerated Red Cell Senescence, Downregulation of Nrf2/ARE Antioxidant Systems and Age-Dependent Hyporegenerative Anemia

Nrf2$^{-/-}$ mice showed chronic mild normochromic normocytic anemia characterized by the presence of erythrocytes containing Howell–Jolly bodies (Table 2, Figure 1a).

Table 2. Hematologic parameters and red cell indices in aging wild-type and Nrf2$^{-/-}$ mice.

	Wild-Type Mice		
	4-Month-Old Mice (n = 6)	**8-Month-Old Mice (n = 6)**	**12-Month-Old Mice (n = 6)**
Hct (%)	46.1 ± 1.4	45.9 ± 0.7	44.8 ± 0.2
Hb (g/dL)	14.8 ± 0.5	15.0 ± 0.1	14.3 ± 0.4
MCV (fl)	51.3 ± 0.2	51.0 ± 0.1	52.2 ± 0.3
MCH (pg)	15.9 ± 0.7	16.5 ± 0.3	15.6 ± 0.2
RDW (%)	12.4 ± 0.08	13.5 ± 0.1	12.7 ± 0.3
Retics (10^3 cells/uL)	450 ± 22	431 ± 51	248 ± 24 °
MCVr (fl)	54.9 ± 2	56.7 ± 3	59.9 ± 1.8 °
	Nrf2$^{-/-}$ Mice		
	4-Month-Old Mice (n = 6)	**8-Month-Old Mice (n = 6)**	**12-MonthsOld Mice (n = 6)**
Hct (%)	44.3 ± 0.8	41.8 ± 1.1 °*	33.6 ± 3 °*
Hb (g/dL)	13.2 ± 0.5	12.0 ± 0.2 °*	11.0 ± 0.5 °*
MCV (fl)	51.8 ± 1.5	50.0 ± 2.0	57.2 ± 1.3 °*
MCH (pg)	16.7 ± 1.1	16.0 ± 0.3	16.1 ± 0.4
RDW (%)	13.9 ± 0.55	13.2 ± 0.4	14.1 ± 0.4 °*
Retics (10^3 cells/uL)	380 ± 20 *	190 ± 59 °*	180 ± 12 °*
MCVr (fl)	61.2 ± 1.3 *	61.0 ± 1.4 *	65.0 ± 0.2 °*

Hct: hematocrit; Hb: hemoglobin; MCV: mean corpuscular volume; MCH: mean corpuscular hemoglobin; RDW: red cell distribution width; Retics: reticulocytes; * $p < 0.05$ compared to wild-type mice; ° $p < 0.05$ compared to 4-month-old mice.

Nrf2$^{-/-}$ mice developed age-dependent anemia characterized by a progressive increase in MCV and RDW with reduced reticulocyte count compared to either younger Nrf2$^{-/-}$ mice or age-matched wild-type animals (Table 1). Nrf2$^{-/-}$ mouse erythrocytes were characterized by increased production of reactive oxygen species (ROS) and higher amounts of Annexin-V+ red cells compared to wild-type animals (Figure 1b). The expression of Nrf2/ARE-related antioxidant and cytoprotective systems such as Nqo1, catalase, Prdx2 and thioredoxin reductase 1 (TrxR1) were significantly reduced in the cytoplasmic fraction of Nrf2$^{-/-}$ mouse erythrocytes compared to wild-type red cells (Figure 1c). The expression of antioxidants was similarly reduced in young and old Nrf2$^{-/-}$ mouse red cells (Figure S1a) and was associated with increased oxidation of red cell membrane proteins as determined by Oxyblot (Figure 1d). In agreement, in Nrf2$^{-/-}$ mouse erythrocytes, we found (i) membrane translocation of classical heat shock protein-70 (HSP70) and -90 (HSP90) (Figure 1e) and (ii) reduced membrane translocation of Prdx2, which competes with hemichromes on the same band 3 binding sites (Figure 1e) [26]. We indeed found increased hemichromes bound to red cell membranes in Nrf2$^{-/-}$ mice when compared to erythrocytes from wild-type animals (Figure 1f). The increased membrane oxidative damage was associated with increased binding of naturally occurring anti-band

3 antibodies (Figure S1b). The cumulative oxidative damage in Nrf2$^{-/-}$ mouse red cells resulted in a decreased lifespan compared to that of wild-type erythrocytes (Figure 1g). No major difference in red cell survival was observed between young and old Nrf2$^{-/-}$ mice (T50: 18.5 ± 0.5 days in 4-month-old Nrf2$^{-/-}$ mice vs. 17.0 ± 0.5 days in 12-month-old Nrf2$^{-/-}$ mice).

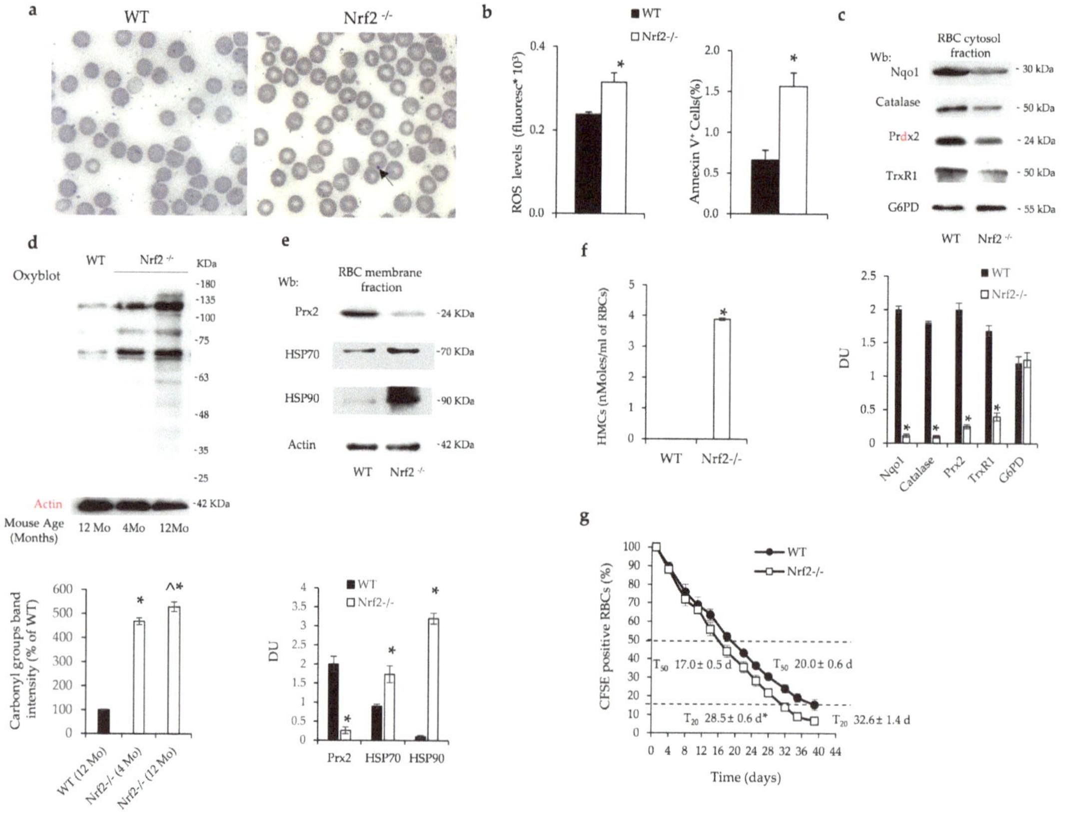

Figure 1. Nrf2$^{-/-}$ mouse red cells display increased membrane oxidation resulting in accelerated senescence (**a**) Representative May–Grunwald–Giemsa staining for the morphology of red cells from wild-type (WT) and Nrf2$^{-/-}$ mice. Arrow indicates Howell–Jolly body in Nrf2$^{-/-}$ mouse red cells. Original magnification ×100. (**b**) ROS values in red cells and Annexin-V^{+} erythrocytes from WT and Nrf2$^{-/-}$ mice. Data are presented as means ± SD (n = 4) * $p < 0.05$ compared to WT. (**c**) Western blot (Wb) analysis using specific antibodies against Nqo1, Catalase, Prdx2, TrxR1 and G6PD of the cytosolic fraction of red cells from WT and Nrf2$^{-/-}$ mice. A representative gel of these four proteins is shown (upper panel). Densitometric analysis of immunoblots shown in the lower panel. Data are presented as means ± SD (n = 4); * $p < 0.05$ compared to WT mice. DU: densitometric unit. (**d**) Oxyblot analysis of the ghost fraction of red cells from WT and Nrf2$^{-/-}$ mice at 4 and 12 months (Mo) of age. A representative gel of the other four is shown (upper panel). Actin is used as loading control. Densitometric analysis is shown in the lower panels. Data are presented as means ± SD (n = 4); * $p < 0.05$ compared to WT mice (12 Mo); ^ $p < 0.05$ compared to Nrf2$^{-/-}$ mice at 4 Mo of age. (**e**) Wb analysis using specific antibodies against Prdx2, HSP70 and HSP90 of the ghost fraction of red cells from WT and Nrf2$^{-/-}$ mice. A representative gel is shown (upper panel). Actin is used as protein loading control. Densitometric analysis of immunoblots is shown in the lower panel. Data are presented as means ± SD (n = 4); * $p < 0.05$ compared to WT mice. (**f**) Hemichrome quantification in

red cells from WT and Nrf2$^{-/-}$ mice at 12 months of age. Data are presented as means ± SD (n = 4); * $p < 0.05$ compared to WT mice. (**g**) Survival of CFSE-labeled red cells from WT and Nrf2$^{-/-}$ mice at 12 months of age (n = 6–7). Data are mean ± SD. * $p < 0.05$ compared to T20 in WT mice.

Taken together these data indicate that the absence of Nrf2 accelerates the senescence of circulating red cells, leading to the development of an age-dependent anemia. Furthermore, the progressive reduction in reticulocyte count implies perturbation of erythropoiesis during aging in mice genetically lacking Nrf2.

3.2. *Nrf2$^{-/-}$ Mice Develop Age-Dependent Ineffective Erythropoiesis Associated with Oxidative Damage Due to Lack of Nrf2/ARE-Dependent Antioxidant Systems*

Nrf2$^{-/-}$ mice displayed an age-dependent splenomegaly (Figure 2a) associated with extramedullary erythropoiesis due to increased splenic erythropoietic activity (Figure 2b). The erythroid maturation profile was similar in bone marrow from Nrf2$^{-/-}$ mice when compared to wild-type animals (Figure S2a). However, we found (i) a marked increase in ROS values in maturating erythroblasts in the bone marrow of Nrf2$^{-/-}$ mice from pop I, II and III corresponding to pro-, basophilic- and polychromatic erythroblasts, respectively (Figure 2c); (ii) increased amounts of 8-Hydroxydeoxyguanosine (8-OHdG) accumulation in total erythroblasts in Nrf2$^{-/-}$ mice (Figure 2d); and (iii) an increase in Annexin-V positivity in erythroblasts from Nrf2$^{-/-}$ mice compared to erythroblasts from wild-type animals (Figure 2e). As expected, sorted erythroblasts displayed reduced expression of Nrf2-dependent ARE genes such as *Catalase, Ho-1, Prdx2, Srxn1* and *Trxn*, with the finding confirmed by immunoblot analysis (Figures 2f and S2b). Taken together, these results imply that the absence of Nrf2 increases oxidative stress and cell apoptosis, supporting an important role of Nrf2 in erythropoiesis during aging.

3.3. *Nrf2 Is Activated during Erythropoiesis and Facilitated by Nuclear Translocation of Prdx2 during Aging*

To assess whether Nrf2 may be important during erythropoiesis in aging mice, we evaluated Nrf2's localization and activity in sorted erythroblasts from wild-type mice at 4 and 12 months of age.

As shown in Figure 3a, Nfr2 nuclear localization increased in sorted erythroid precursors from 12-month-old wild-type mice compared to younger 4-month animals (see also Figure S3a for brightfield images) in conjunction with an increase in the phospho-Nrf2 form in sorted erythroblasts from 12-month-old wild-type mice compared to younger animals (Figure 3b). No major change in the activation of NF-kB was evident in erythroid precursors from aging wild-type mice (Figure 3b). Indeed, we noted an age-dependent upregulation of Nrf2-related ARE genes such as catalase and Prdx2 in old wild-type animals compared to younger animals (Figure 3c). Since we previously described a functional collaboration between Nrf2 and Prdx2 supporting stress erythropoiesis [11], we evaluated Prdx2 distribution in erythroblasts from aging wild-type mice. This is important since in other cell models, Prdx2 has been shown to translocate to the nucleus to protect cells against oxidation-induced DNA damage [31–33]. We indeed noted nuclear translocation of Prdx2 in sorted erythroblasts from old wild-type mice compared to younger animals (Figures 3d and S3b). This finding was further validated by immunoblot analysis of nuclear fraction from aged wild-type mice compared to the nuclear fraction from age-matched sorted erythroblasts from mice genetically lacking Prdx2 (Prdx2$^{-/-}$ mice) (Figure S3c).

To define the potential differences in the binding of Prdx2 to regulatory regions of the chromatin in old wild-type mouse erythroblasts, we performed a chromatin immunoprecipitation (ChIP) assay using a specific anti-Prdx2 antibody. We analyzed the putative promoter regions of the erythroferrone (*Erfe*), *Ho-1* and *Nqo1* genes to identify Nrf2 and NF-kB binding sites by MatInspector (www.genomatix.de) and looked for enrichment of immunoprecipitated chromatin using the anti-Prdx2 antibody as compared to control IgG (Figure 3e). We could indeed demonstrate the recruitment of Prdx2 to the promoter regions of *Erfe, Ho-1* and *Nqo1*, linking Nrf2 to *Erfe* gene expression for the first time. Importantly, we observed downregulation of *Erfe* in sorted erythroblasts from Nrf2$^{-/-}$ mice (Figure S3d).

Collectively, these findings indicate that Nrf2 activation during aging in wild-type mice involves Prdx2 nuclear translocation and is key for protecting cells against oxidation-induced DNA damage.

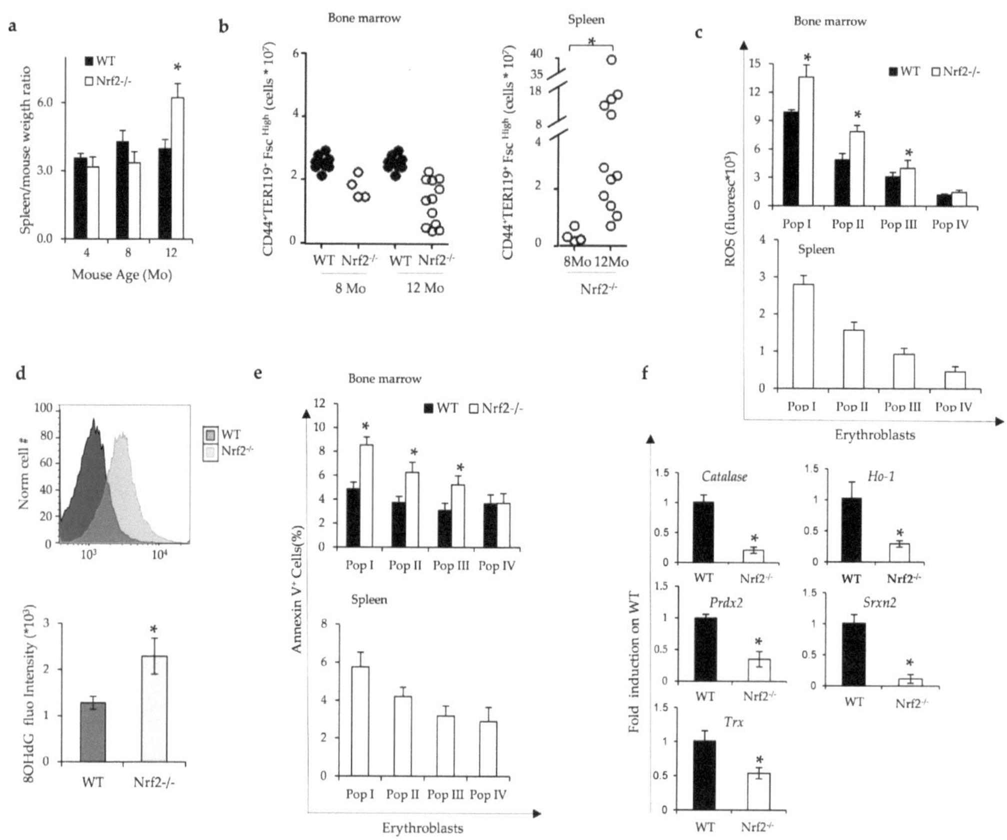

Figure 2. The absence of Nrf2 promotes ineffective erythropoiesis and increased oxidation of erythroblasts. (**a**) Spleen weight/mouse weight ratio of wild-type (WT) and Nrf2$^{-/-}$ mice at 4, 8 and 12 months (Mo) of age. Data are presented as means ± SD (n = 4–12) * $p < 0.05$ compared to WT. (**b**) Flow cytometric analysis, combining CD44-Ter119 and cell size marker strategy (CD44+/Ter119+/FSC^{high}), of the erythropoietic activity in the bone marrow and spleen from WT and Nrf2$^{-/-}$ mice at 8 and 12 months (Mo) of age. Data are presented as single dots (n = 4–12) * $p < 0.05$ compared to WT. (**c**) ROS values in bone marrow and spleen erythroblasts from WT and Nrf2$^{-/-}$ mice at 12 months of age. Data are presented as means ± SD, * $p < 0.05$ compared to WT. (**d**) DNA oxidative damage measured flow cytometrically as 8OHdG fluorescence intensity of erythroblast populations from WT and Nrf2$^{-/-}$ mice at 12 months of age. Data are presented as means ± SD, * $p < 0.05$ compared to WT. (**e**) Amount of Annexin V+ cells in bone marrow and spleen erythroid precursors from Nrf2$^{-/-}$ and WT mice as in (**c**); * $p < 0.05$ compared to WT. (**f**) mRNA expression of *catalase*, *Ho-1*, *Prdx2*, *Srxn2* and *Trx* by qRT-PCR in erythroblasts from WT and Nrf2$^{-/-}$ mice (fold on WT). Experiments were performed in triplicate. * $p < 0.01$, Nrf2$^{-/-}$ vs. WT mice. p value was calculated by t-test.

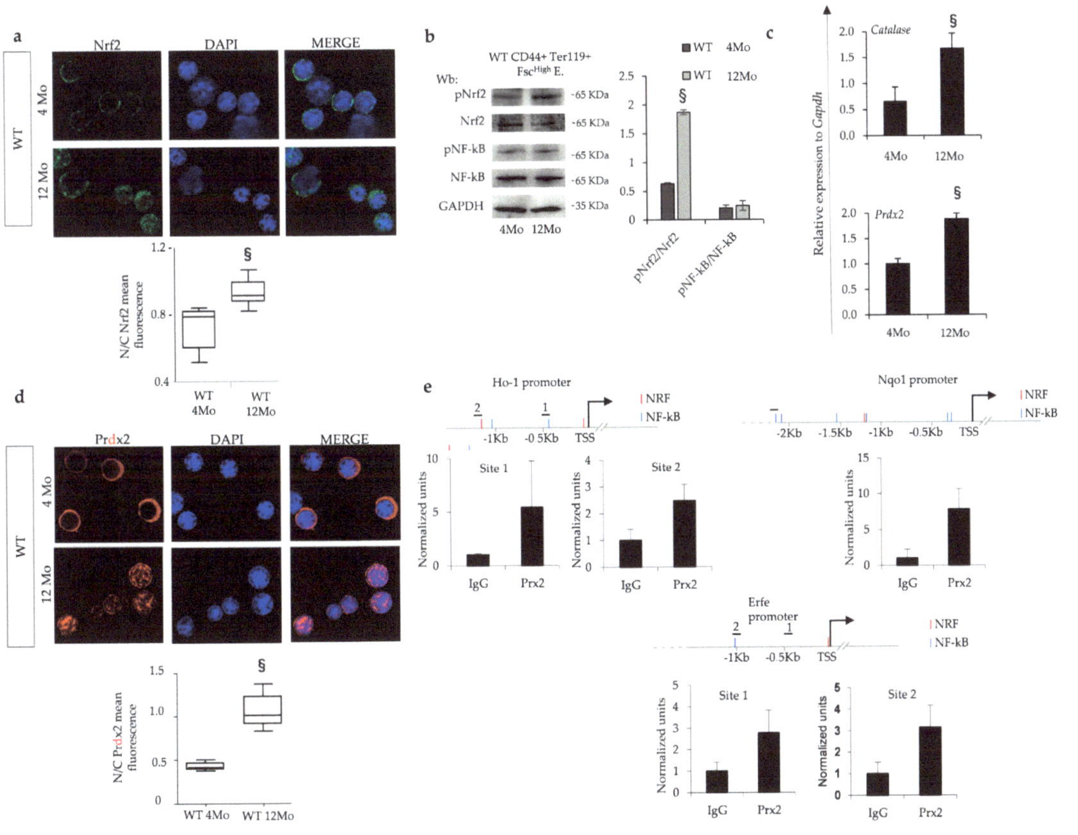

Figure 3. In aging wild-type mice, erythroblasts show age-dependent activation of Nrf2 associated with nuclear translocation of Prdx2, which is recruited in promoter regions of Nrf2. (**a**) Nrf2 immunostaining of sorted erythroid precursors from bone marrow of 4- and 12-month-old wild-type (WT) mice. DAPI was used to stain nuclei. Nrf2 mean fluorescence in the nucleus and cytoplasm was measured using ImageJ (https://imagej.net/ij/download.html). At least 25 cells were analyzed in 6 different fields of acquisition. Data are presented as median and minimum/maximum, with boxes indicating 25th–75th percentiles. § $p < 0.05$ compared to 4-month-old WT mice. Original magnification 100×. (**b**) Wb analysis with specific anti-phospho-Nrf2 (p-Nrf2), Nrf2, phospho-NF-kB(p-NF-kB) and NF-kB in sorted erythroid precursors in bone marrow of 4- and 12-month-old WT mice. GAPDH was used as protein loading control. A representative gel is shown. Densitometric analysis of immunoblots is shown in the right panel. Data are presented as means ± SD (n = 4); § $p < 0.05$ compared to 4-month-old WT mice. DU: densitometric unit (**c**) mRNA expression of *catalase* and *Prdx2* by qRT-PCR of erythroblasts from WT mice at 4 and 12 months of age. Data are mean ± SD (n = 6–8). Experiments were performed in triplicate. § $p < 0.01$, WT 4Mo vs. WT 12Mo mice. p value was calculated by t-test. (**d**) Prdx2 immunostaining of sorted erythroid precursors from bone marrow of 4- and 12-month-old wild-type (WT) mice. DAPI was used to stain nuclei. Prdx2 mean fluorescence in the nucleus and cytoplasm was measured using ImageJ. At least 40 cells were analyzed in 5 different fields of data acquisition. Data are presented as median and minimum/maximum, with boxes indicating 25th–75th percentiles; § $p < 0.01$, WT 4Mo vs. WT 12Mo mice. Original magnification 100×. (**e**) Predicted NRF and NF-kB binding sites in the indicated promoters are shown in red and blue, respectively. Horizontal bars indicate the regions amplified in the ChIP experiment. Histograms show qPCR quantification of a ChIP assay with the indicated antibodies from cells treated as above. Error bars indicate standard deviations.

3.4. $Nrf2^{-/-}$ Mice Display Increased Sensitivity to Doxorubicin-Induced Stress Erythropoiesis

Doxorubicin (Doxo) promotes severe oxidative stress and induces DNA damage, driving cells into senescence [23,34]. We evaluated the effect of Doxo treatment on erythropoiesis in $Nrf2^{-/-}$ mice compared to wild-type mice. Following administration of Doxo for 9 days, we noted a significant reduction in Hct in $Nrf2^{-/-}$ mice compared to wild-type mice (Figure S4a, upper panel) in association with a lower reticulocyte count (Figure S4a, lower panel). In agreement, we found a significant reduction in the total numbers of erythroblasts in both bone marrow and spleen in Doxo-treated $Nrf2^{-/-}$ mice compared to wild-type mice (Figure S4b). Furthermore, Annexin V^+ cells were significantly increased in polychromatic and orthochromatic erythroblasts from $Nrf2^{-/-}$ mice as compared to wild-type animals (Figure S4c). Collectively, our findings indicate that mice genetically lacking Nrf2 are characterized by a blunted response to stress erythropoiesis induced by Doxo, by accelerating cell senescence and apoptosis.

3.5. The Absence of Nrf2 Promotes Overactivation of UPR System and Impairment of Autophagy and Facilitates the Activation of Caspase-3 Pro-Apoptotic Pathway

Since the absence of Nrf2 negatively impacts erythropoiesis during aging, we hypothesized that stress-sensitive transcription factor NF-kB might be activated to support stress erythropoiesis in $Nrf2^{-/-}$ mice. Indeed, we found increased NF-kB activation in sorted erythroblasts from $Nrf2^{-/-}$ mice compared to wild-type animals (Figure 4a). Other adaptive mechanisms against sustained oxidation such as UPR and autophagy could also be involved. In sorted $Nrf2^{-/-}$ erythroid precursors, we found increased expression of (i) HSP70 and (ii) ATF6, which is one of the three UPR sensors and upregulates the expression of the pro-apoptotic protein GADD34 towards the caspase-3 cascade (Figure 4b). Indeed, we found a significantly increased activity of caspase-3 in sorted $Nrf2^{-/-}$ mouse erythroblasts compared to wild-type cells (Figure 4c).

Previous studies have shown that in the presence of severe or prolonged stress, the UPR system induces autophagy [35]. Given that efficient autophagy is required for erythroid maturation, we evaluated the expression of several key autophagy proteins in sorted erythroid precursors from the bone marrow of $Nrf2^{-/-}$ and wild-type mice. We focused our analysis on (i) LC3 as an initiator of autophagy; (ii) Atg4, involved in the fusion of autophagosomes with lysosomes; (iii) Atg5, involved in the maturation of autophagosomes together with Rab5, which contributes to recycling endosomes and thereby regulates the fusion process between endosomes; and (iv) p62 as a cargo protein [36]. As shown in Figure 4d, we found increased LC3II conversion, associated with a reduction in Atg4, and Atg5 expression and accumulation of Rab5, suggesting a possible impairment of the terminal phases of autophagy in $Nrf2^{-/-}$ erythroblasts (Figure 4d). Indeed, $Nrf2^{-/-}$ erythroblasts displayed *punctae* of Rab5 organized in large clusters more abundantly than in wild-type erythroblasts (Figures 4e and S5a). We further documented a reduced level of mRNA and protein p62 expression in $Nrf2^{-/-}$ erythroblasts compared to wild-type cells (Figures 4e and S5b).

Thus, the absence of Nrf2 promotes premature cell senescence via impaired autophagy and overactivation of the UPR systems mainly involving ATF6-GADD34 towards the pro-apoptotic caspase-3 pathway.

3.6. In $Nrf2^{-/-}$ Mice, Treatment with Astaxanthin PLGA Nanoparticles Enhances Erythropoiesis and Improves Anemia

Reduction in oxidative stress may beneficially impact $Nrf2^{-/-}$ mouse erythropoiesis. Thus, we treated $Nrf2^{-/-}$ mice with lyophilized PLGA nanoparticles loaded with astaxanthin (Figure 5a, see also stability of ATS-NP release at 25 °C and 37 °C in Supplementary Figure S6a–c). Astaxanthin (3,3′-dihydroxy-β-carotene-4,4′-dione) is a nontoxic and organic carotenoid with potent antioxidative activity with no pro-oxidative properties. Its unique molecular structure quenches singlet oxygen and scavenges free radicals, preventing lipid peroxidation [37]. Previous studies have shown that astaxanthin has a wide variety of biological effects including anti-inflammatory, antiapoptotic, neuroprotective and cardioprotective effects [37–41].

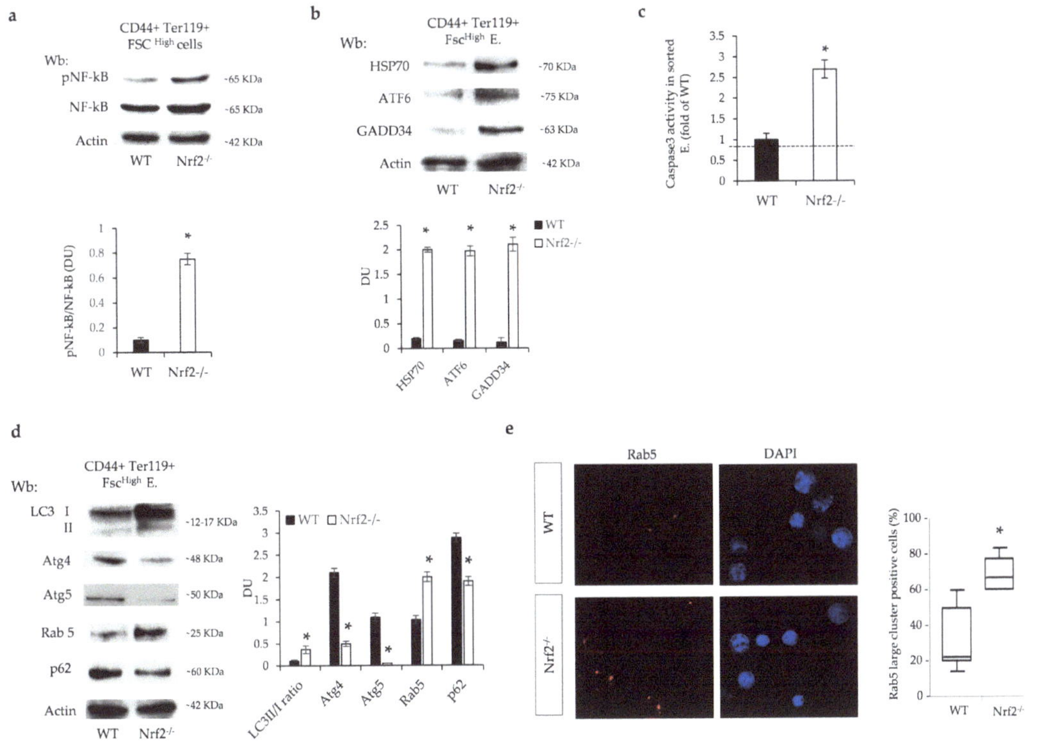

Figure 4. Sorted Nrf2$^{-/-}$ mouse erythroblasts display overactivation of the system and impaired autophagy with caspase-3 pro-apoptotic pathway activation. (**a**) Wb analysis with specific anti-phospho-NF-kB (p-NF-kB) and NF-kB in sorted erythroid precursors from bone marrow of 12-month-old wild-type (WT) and Nrf2$^{-/-}$ mice. Actin was used as protein loading control. One representative gel of the other four is shown (upper panel). Densitometric analysis of immunoblots is shown in the lower panel. Data are presented as means ± SD (n = 4); * $p < 0.05$ compared to WT mice. (**b**) Wb analysis with specific anti-HSP70, ATF6 and GADD34 in sorted erythroid precursors as in a. Densitometric analysis of immunoblots is shown in the lower panel. Data are presented as means ± SD (n = 4); * $p < 0.05$ compared to WT mice. (**c**) Caspase 3 activity determined by cleavage of a fluorescent substrate in sorted erythroid precursors from bone marrow of WT and Nrf2$^{-/-}$ mice. Data are presented as means ± SD * $p < 0.05$ compared to WT. (**d**) Wb analysis with specific anti-LC3 I/II, Atg4, Atg5, Rab5 and p62 in sorted erythroid precursors from bone marrow of 12-month-old WT and Nrf2$^{-/-}$ mice. Actin was used as protein loading control. Densitometric analysis of immunoblots is shown in the right panel. Data are presented as means ± SD (n = 4); * $p < 0.05$ compared to WT mice. (**e**) Rab5 immunostaining of sorted erythroid precursors from bone marrow of 12-month-old WT and Nrf2$^{-/-}$ mice. DAPI was used to stain nuclei. Large clusters of positive cells were measured using ImageJ. At least 40 cells were analyzed in 8 different fields of acquisition. Data are presented as median and minimum/maximum, with boxes indicating 25–75th percentiles; * $p < 0.05$ compared to WT mice.

The organ distribution of ATS was determined by mass spectrometric analysis at 24 h after ATS-NP administration (Figure 5a,b, see also Figure S6d). ATS was identified in the spleen, liver and kidney in both mouse strains (Figure 5a,b). Higher levels of ATS were found in Nrf2$^{-/-}$ mouse organs when compared to wild-type animals (Figure 5b). Previous studies have shown that PLGA-NPs are generally cleared by phagocytic uptake by (i) the spleen and liver, (ii) hepatic filtration and (iii) kidney extraction [42].

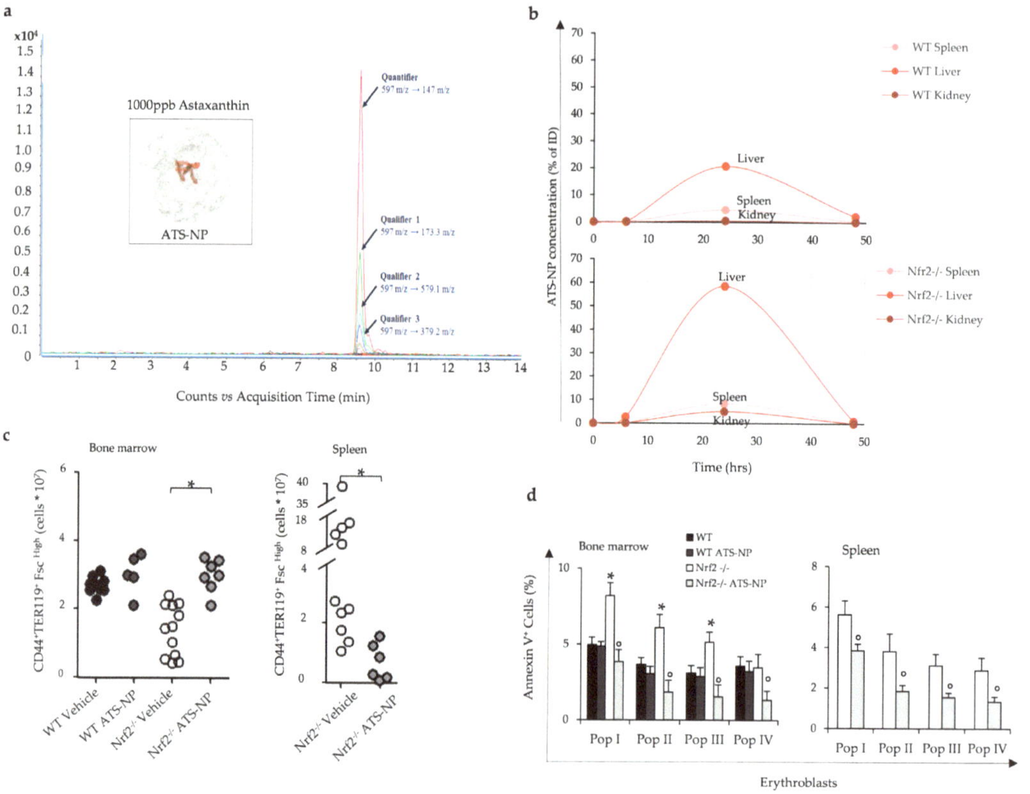

Figure 5. Astaxanthin PLGA nanoparticles improve Nrf2$^{-/-}$ mouse ineffective erythropoiesis. (**a**) Extracted ion chromatogram for astaxanthin illustrates three qualifier transitions and was used to confirm molecule identification. Inset: ATS-NP: pictorial representation of PLGA nanoparticles embedding astaxanthin; the relationship between the different molecules is illustrative. (**b**) Organ distribution of astaxanthin molecules in the two mice strains during treatment determined by mass spectrometric analysis (see also Figure S6d and Table S1). (**c**) Flow cytometric analysis, combining CD44-Ter119 and cell size marker strategy (CD44+/Ter119+/FSC^{high}), of the erythropoietic activity in the bone marrow and spleen from WT and Nrf2$^{-/-}$ mice treated with vehicle or ATS-NPs (2 mg/kg every two days for four weeks). Data are presented as single dots (n = 5–12) * $p < 0.05$ compared to vehicle-treated WT mice. (**d**) Amount of Annexin V+ cells in erythroid precursors from bone marrow and spleen of WT and Nrf2$^{-/-}$ mice treated as in (**c**). Data are presented as means ± SD; * $p < 0.05$ compared to WT mice, ° $p < 0.05$ compared to vehicle-treated mice.

ATS-NP treatment decreased the extent of anemia of Nrf2$^{-/-}$ mice by reducing ineffective erythropoiesis and improving the quality control process for red cell production. As shown in Figure 5c, ATS-NPs increase bone marrow erythropoietic activity and decrease extramedullary erythropoiesis (Figure S7a). This was associated with a significant reduction in apoptosis of erythroblasts from ATS-NP-treated Nrf2$^{-/-}$ mice compared to vehicle-treated animals (Figure 5d).

The beneficial effects of ATS-NPs on Nrf2$^{-/-}$ mouse erythropoiesis were also supported by (i) the downregulation of *Atf6* expression (Figure 6a), (ii) a reduction in caspase-3 activity (Figure 6b) and (iii) the improvement of autophagy with the clearance of large Rab5 clusters (Figure 6c lower panel).

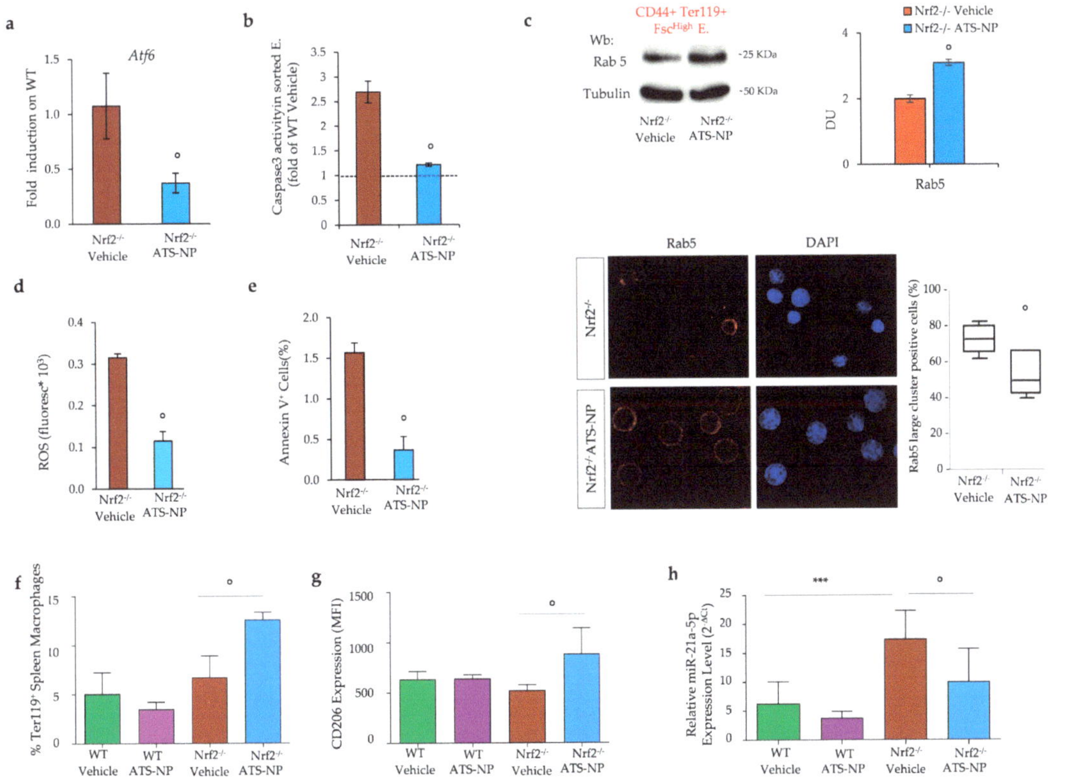

Figure 6. Astaxanthin PLGA nanoparticles reduce oxidation, prevent overactivation of adaptative systems in erythroblasts and modulate splenic macrophages towards a pro-resolving pattern in Nrf2$^{-/-}$ mice. (**a**) mRNA expression of *Atf6* by qRT-PCR on the erythroblasts from vehicle and ATS-NP treated Nrf2$^{-/-}$ mice. Data are mean ± SD (n = 6–8). Experiments were performed in triplicate. ° $p < 0.01$, Nrf2$^{-/-}$ vehicle vs. Nrf2$^{-/-}$ ATS-NP mice by *t*-test; (**b**) Caspase 3 activity determined by its cleavage of a fluorescent substrate in sorted erythroid precursors from bone marrow of Nrf2$^{-/-}$ mice treated as in (**a**); ° $p < 0.01$, Nrf2$^{-/-}$ vehicle vs. Nrf2$^{-/-}$ ATS-NP mice by *t*-test. (**c**) Upper panel: Wb analysis with specific anti-Rab5 in sorted erythroid precursors from bone marrow of Nrf2$^{-/-}$ mice treated as in (**a**). Densitometric analysis of immunoblots is shown in the right panel. Data are presented as means ± SD (n = 4); ° $p < 0.01$, Nrf2$^{-/-}$ vehicle vs. Nrf2$^{-/-}$ ATS-NP mice by *t*-test. Lower panel: Rab5 immunostaining of sorted erythroid precursors from bone marrow of Nrf2$^{-/-}$ mice treated as in (**a**). DAPI was used to stain nuclei. Large clusters of positive cells were measured using ImageJ. At least 35 cells were analyzed in 5 different fields of acquisition. Data are presented as median and minimum/maximum, with boxes indicating 25th–75th percentiles. (**d**) ROS levels in red cells and (**e**) Annexin-V^{+} erythrocytes from Nrf2$^{-/-}$ mice treated as in (**a**). Data are presented as mean ± SD (n = 4) ° $p < 0.05$ compared to vehicle-treated mice. (**f**) Flow cytometry analysis of phagocytosis of Ter-119^{+} cells, membrane expression of CD206 (**g**) and real-time PCR analysis of mir-21-5p (**h**) in splenic macrophages. Data are mean ± SD (n = 6); ***, $p < 0.001$; ° $p < 0.05$ by 1-way ANOVA.

The improvement of the quality control process during erythroid maturation promoted by ATS-NP treatment resulted in the amelioration of anemia and red cell features in Nrf2$^{-/-}$ mice (Table 3).

In ATS-NP-treated Nrf2$^{-/-}$ mice, we also found decreased (i) ROS, (ii) AnnexinV+ for red cells, (iii) red cell membrane protein oxidation and (iv) red cell membrane translocation of HSP70 and HSP90 (Figures 6d,e and S7b,c). This was associated with the optimization of erythrophagocytosis by Nrf2$^{-/-}$ mouse macrophages (Figure 6f) and a reduction in apoptotic Ly6G^{+} neutrophils (Figure S7d). Furthermore, in Nrf2$^{-/-}$ mice, ATS-NP treatment enhanced the expression of CD206, a classical surface marker of M2 polarization, and reduced the relative expression of miR-21-5p, a microRNA that promotes the acquisition of an M1 inflammatory macrophage phenotype [43], which was significantly over-expressed in Nrf2$^{-/-}$ mice compared to wild-type mice (Figure 6g,h). Taken together, our data indicate that ATS-NPs support stress erythropoiesis, reduce erythrophagocytosis and optimize spleen macrophage function skewing toward a pro-resolving, M2 phenotype in cells lacking Nrf2.

Table 3. Effects of ATS-PLGA nanoparticles on hematological parameters and red cell indices in wild-type and Nrf2$^{-/-}$ mice.

	Vehicle Wild-Type Mice (n = 6)	ATS-NP Wild-Type Mice (n = 6)	Vehicle Nrf2$^{-/-}$ Mice (n = 6)	ATS-NP Nrf2$^{-/-}$ Mice (n = 5)
Hct (%)	44.8 ± 0.2	45.9 ± 0.7	33.6 ± 3 *	43.5 ± 0.9 *^
Hb (g/dL)	14.3 ± 0.4	15.0 ± 0.1	11.0 ± 0.5 *	14.0 ± 0.4 ^
MCV (fl)	52.2 ± 0.3	51.0 ± 0.1	57.2 ± 1.3 *	53.5 ± 1.1 ^
MCH (pg)	15.6 ± 0.2	16.5 ± 0.3	16.1 ± 0.4	16.3 ± 0.2
RDW (%)	12.7 ± 0.3	13.5 ± 0.1	14.1 ± 0.4 *	12.6 ± 0.4 ^
Retics (10^3 cells/uL)	248 ± 24	431 ± 51 ^	180 ± 12 *	355 ± 65 ^
MCVr (fl)	59.9 ± 1.8	56.1 ± 2.0	65.0 ± 0.2 *	62.0 ± 1.6 ^

Hct: hematocrit; Hb: hemoglobin; MCV: mean corpuscular volume; MCH: mean corpuscular hemoglobin; RDW: red cell distribution width; Retics: reticulocytes; * $p < 0.05$ compared to wild-type mice; ^ $p < 0.05$ compared to vehicle-treated animals.

4. Discussion

In the present study, we show the key role of Nrf2 in supporting erythropoiesis and the quality control processes involved in erythroid maturation during mouse aging. We confirm that the absence of Nrf2 results in the impairment of red cell antioxidant machinery [1,2,9], promoting an accelerated removal of Nrf2$^{-/-}$ mouse erythrocytes. This was associated with an age-dependent anemia characterized by reduced reticulocyte count, suggesting a perturbation of erythropoiesis. The presence of ineffective erythropoiesis in 12-month-old Nrf2$^{-/-}$ mice was documented by (i) extramedullary erythropoiesis, (ii) increased ROS throughout erythroblast maturation, (iii) oxidative DNA damage in erythroblasts and (iv) increased cell apoptosis at all stages of erythroblast maturation. We also demonstrated that Nrf2 is activated in wild-type erythropoiesis during mouse aging, with slight but stable activation of NF-kB in sorted erythroblasts from young vs. old wild-type mice. This agrees with previous observations of primarily cytoplasmic localization of Nrf2 in maturating erythroid cells from healthy adults [44]. The activation of NF-kB is a possible compensatory mechanism for the absence of Nrf2 in erythroblasts from old Nrf2$^{-/-}$ mice when compared to younger animals, as suggested by other models [45,46]. A similar observation has been also reported in models of anemia in the terminal stage of renal disease [47,48]. Indoxyl sulfate, a uremic toxin, accelerates cell senescence towards apoptosis and reduces EPO responsiveness in both K562- and CD34+-derived cells by the inhibition of Nrf2 and modulation of NF-kB activity [47,48]. The importance of Nrf2 in cell growth and maturation of erythroblasts during aging is further supported by (i) the upregulation of Prdx2; (ii) Prdx2 nuclear translocation; and (iii) Prdx2 recruitment on the promoter region of *Erfe*, *Ho-1* and *Nqo1*, linking Nrf2 to *Erfe* gene expression in erythroblasts from wild-type mice for the first time, in addition to the reported role of the signal transducer and activator of transcription-5 (STAT5) in *Erfe* gene expression [49,50]. Collectively, these data indicate the importance of Nrf2 in erythropoiesis during aging and the collaborative action of Nrf2 and Prdx2 in limiting oxidation and assisting in cell growth and maturation in erythroblasts. The blunted response to stress erythropoiesis induced by Doxo, which promotes oxidation DNA damage and accelerates the aging phenomenon, further supports the increased susceptibility of Nrf2$^{-/-}$ mice to oxidative stress.

Perturbations of cellular adaptative mechanisms against oxidation such as the unfolded protein response (UPR) system and autophagy have been recently described in pathologic erythropoiesis such as β-thalassemia [16,18,51]. In old Nrf2$^{-/-}$ mice, prolonged oxidation resulted in the sustained activation of the UPR systems, mainly involving ATF6, promoting the upregulation of HSP70 and GADD34 as part of the EIF2a-ATF4 signaling pathway towards autophagy [16,18,52]. However, in erythroblasts from Nrf2$^{-/-}$ mice, we observed an impairment of late-phase autophagy. This might be favored by the downregulation of Nrf2-dependent Sestrin-2 expression, which contributes to the modulation of autophagy [53]. In addition, we note that Atg4, 5 and p62 genes also contain the ARE-sequence, modulated by Nrf2 [54]. Thus, we were not surprised that in sorted Nrf2$^{-/-}$ erythroblasts, Atg4, 5 and p62 expression was downregulated compared to wild-type cells [20,55–58]. This resulted in the accumulation of Rab5 organized in large clusters, supporting the impairment of autophagy in Nrf2$^{-/-}$ erythroblasts. The cumulative effect of sustained oxidation, prolonged overactivation of the UPR system and impairment of autophagy triggers the caspase-3 pro-apoptotic system, contributing to the ineffective erythropoiesis in old Nrf2$^{-/-}$ mice.

As a proof of concept, our experiments with novel ATS-NPs support this working model. We found that in Nrf2$^{-/-}$ mice, ATS-NPs ameliorate age-dependent macrocytic anemia and improve ineffective erythropoiesis. The reduction in intracellular oxidation by ATS-NPs rebalances the overwhelmed autophagy observed in Nrf2$^{-/-}$ mouse erythroblasts. This results in a significant reduction in the accumulation of large Rab5 aggregates in erythroblasts from ATS-NP-treated Nrf2$^{-/-}$ mice. Of note, we observed a difference in the amounts of ATS detected in the liver and spleen from Nrf2$^{-/-}$ vs. wild-type mice, which might be related to pro-inflammatory vs. pro-resolving profile of macrophages in the spleen and liver lacking the expression of Nrf2 as previously noted in other disease models [57,58]. Indeed, ATS-NPs reprogram splenic macrophage from a pro-inflammatory to a pro-resolving M2 profile [43].

In conclusion, we propose that Nrf2 is a key transcriptional factor in erythropoiesis protecting against oxidative damage during aging (Figure 7). In aging mice, the absence of Nrf2 results in severe oxidative damage and overactivation of the UPR and autophagy resulting in perturbation of proteostasis, leading to apoptosis of erythroblasts. The senostatic effect of ATS-NPs decreases anemia and ineffective erythropoiesis in Nrf2$^{-/-}$ mice, preventing the overwhelming of adaptative systems and ensuring erythroid survival and maturation.

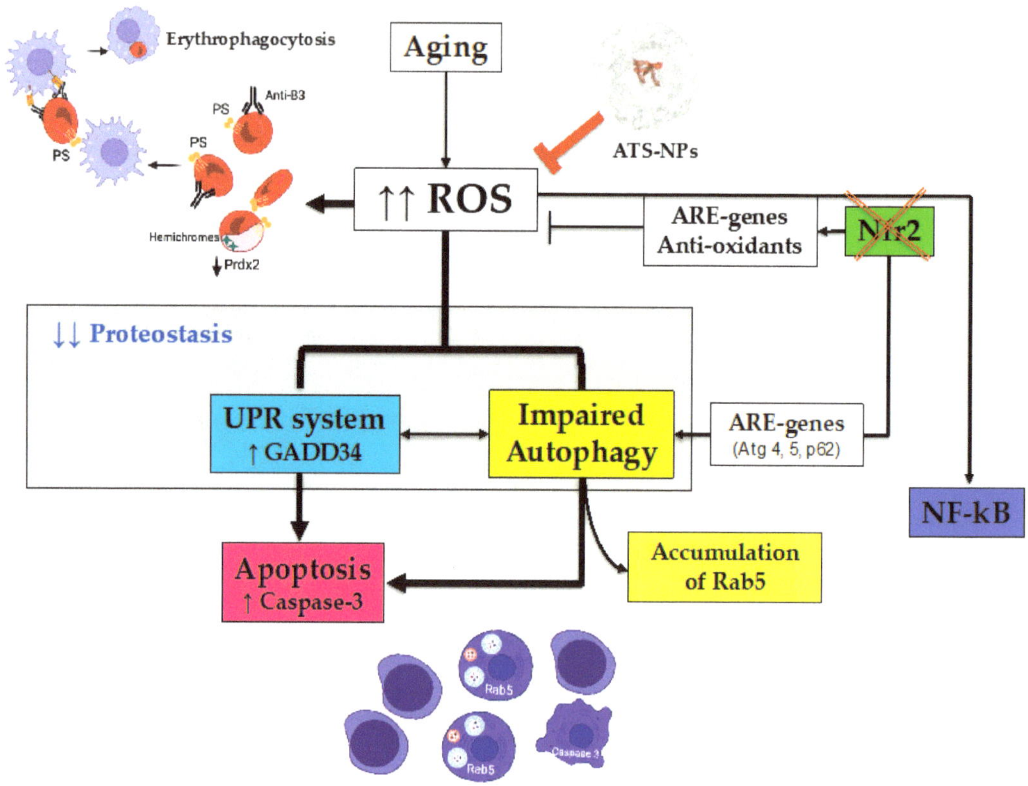

Figure 7. Schematic diagram of the role of Nrf2 in erythropoiesis during aging and the protective effects of astaxanthin PLGA nanoparticles. Aging is associated with increased ROS production, which is limited by the activation of Nrf2. This results in the upregulation of ARE genes encoding for antioxidants and cytoprotective systems as well as by the activation of adaptative mechanisms such as the UPR system to face ER stress and autophagy to clear damaged proteins. The absence of Nrf2 (Nrf2$^{-/-}$ mice) negatively affects the antioxidant cell machinery, resulting in severe and sustained oxidation. Nrf2$^{-/-}$ mouse red cells display severe membrane oxidation, exposition of phosphatidylserine, membrane binding of hemichromes and reduced expression of antioxidants and cytoprotective systems such as Prdx2. Red cell membrane protein oxidation favors band 3 protein clusterization, which is recognized by the naturally occurring anti-band 3 IgG antibodies. Both mechanisms drive Nrf2$^{-/-}$ mouse red cells towards erythrophagocytosis by splenic macrophages. In erythroblasts lacking Nrf2, the prolonged and severe oxidation due to the downregulation of antioxidants and cytoprotective systems promotes intense ER stress with overactivation of the UPR system and autophagy. Although the persistence of oxidative stress promotes compensatory activation of NF-kB, this is insufficient to prevent the overwhelming of proteostasis with impairment autophagy and accumulation of Rab5. This drives Nrf2$^{-/-}$ erythroblasts towards apoptosis via the caspase-3 pathway, resulting in ineffective erythropoiesis. ATS-NPs act as efficient antioxidants preventing the deleterious effects of the absence of Nrf2 on erythropoiesis and red cells during aging. PS: phosphatidylserine; ER: endoplasmic reticulum; UPR: unfolded protein response; ARE-: antioxidant-related element; ROS: reactive oxygen species; Prdx2: peroxiredoxin-2; Atg: autophagy-related protein; GADD34: growth arrest and DNA damage-inducible protein 34; PLGA: poly(lactic-co-glycolic acid).

Supplementary Materials: The following supporting information can be downloaded at https://www.mdpi.com/article/10.3390/antiox13040454/s1: Supplementary Methods; Figure S1. Expression of Nrf2 dependent antioxidant systems in red cell cytosol fraction and amounts of erythrocytes positive for naturally occurring antibody in young and old wild-type and $Nrf2^{-/-}$ mice; Figure S2. Profile or erythroblast subpopulations and expression of Nrf2-dependent antioxidant systems in sorted erythroblasts from wild-type and $Nrf2^{-/-}$ mice; Figure S3. Brightfield images, immunoblot analysis of peroxiredoxin-2 (Prdx2) nuclear translocation in sorted polychromatic and orthochromatic erythroblasts and *Erfe* gene expression in sorted erythroblasts from wild-type and $Nrf2^{-/-}$ mice; Figure S4. Effects of doxorubicin (Doxo) on hematologic parameters and erythropoiesis in wild-type and $Nrf2^{-/-}$ mice; Figure S5. Localization of HSP70, p62 gene expression and control brightfield in sorted erythroblasts from wild-type and $Nrf2^{-/-}$ mice; Figure S6. Characterization of ATS-NPs and calibration curve. Figure S7. Effects of ATS-NPs on $Nrf2^{-/-}$ mouse erythroblast subpopulations, red cell membrane oxidation, membrane association of HSP70 and 90 and macrophage profile; Table S1. Primers used in the present study. References [11,59–61] are cited in the supplementary materials.

Author Contributions: L.D.F., M.P., C.B., N.M. and A.I.; investigation, A.M., A.S., E.F., S.C.T.M., I.A., M.T.V., A.R., D.M. (Domenico Mattoscio), F.T. and V.R.; generation of ATS NPs, M.B. and A.P.; data curation, M.T.V., A.M. and E.F.; D.M. (Davide Melisiand): Chip analyses; mass spectrometric analysis, A.A.; writing—original draft preparation, L.D.F. and A.I.; writing—review and editing, C.B., N.M., M.P., A.R., F.T. and L.D.C.; funding acquisition: L.D.F. and A.I. All authors have read and agreed to the published version of the manuscript.

Funding: This research was funded by PRIN2020-MIUR (LDF and AI: 2020Z22PM7).

Institutional Review Board Statement: The Institutional Animal Experimental Committee of University of Verona (CIRSAL) and the Italian Ministry of Health approved the experimental protocols (prot. 56DC9.21), following the European directive 2010/63/EU and the FELASA guidelines and recommendations.

Informed Consent Statement: Not applicable.

Data Availability Statement: Detailed materials and methods are reported in Supplemental materials. All the data and protocols are stored in the Nas Synology DS216se Hard Disk, located at the University of Verona, 37134 Verona, Italy and will be available on request.

Acknowledgments: We are grateful to the "Centro Piattaforme Tecnologiche" of the University of Verona for giving access to DLS equipment.

Conflicts of Interest: The authors declare no conflicts of interest.

References

1. Hayes, J.D.; Dinkova-Kostova, A.T. The Nrf2 regulatory network provides an interface between redox and intermediary metabolism. *Trends Biochem. Sci.* **2014**, *39*, 199–218. [CrossRef] [PubMed]
2. Lee, J.M.; Chan, K.; Kan, Y.W.; Johnson, J.A. Targeted disruption of Nrf2 causes regenerative immune-mediated hemolytic anemia. *Proc. Natl. Acad. Sci. USA* **2004**, *101*, 9751–9756. [CrossRef] [PubMed]
3. Ishii, T.; Warabi, E.; Yanagawa, T. Novel roles of peroxiredoxins in inflammation, cancer and innate immunity. *J. Clin. Biochem. Nutr.* **2012**, *50*, 91–105. [CrossRef] [PubMed]
4. Leiser, S.F.; Miller, R.A. Nrf2 signaling, a mechanism for cellular stress resistance in long-lived mice. *Mol. Cell Biol.* **2010**, *30*, 871–884. [CrossRef] [PubMed]
5. Osburn, W.O.; Kensler, T.W. Nrf2 signaling: An adaptive response pathway for protection against environmental toxic insults. *Mutat. Res.* **2008**, *659*, 31–39. [CrossRef] [PubMed]
6. Baird, L.; Dinkova-Kostova, A.T. The cytoprotective role of the Keap1-Nrf2 pathway. *Arch. Toxicol.* **2011**, *85*, 241–272. [CrossRef]
7. Zhu, H.; Itoh, K.; Yamamoto, M.; Zweier, J.L.; Li, Y. Role of Nrf2 signaling in regulation of antioxidants and phase 2 enzymes in cardiac fibroblasts: Protection against reactive oxygen and nitrogen species-induced cell injury. *FEBS Lett.* **2005**, *579*, 3029–3036. [CrossRef] [PubMed]
8. Shin, B.Y.; Jin, S.H.; Cho, I.J.; Ki, S.H. Nrf2-ARE pathway regulates induction of Sestrin-2 expression. *Free Radic. Biol. Med.* **2012**, *53*, 834–841. [CrossRef]
9. Chan, K.; Lu, R.; Chang, J.C.; Kan, Y.W. NRF2, a member of the NFE2 family of transcription factors, is not essential for murine erythropoiesis, growth, and development. *Proc. Natl. Acad. Sci. USA* **1996**, *93*, 13943–13948. [CrossRef]

10. Matte, A.; De Falco, L.; Federti, E.; Cozzi, A.; Iolascon, A.; Levi, S.; Mohandas, N.; Zamo, A.; Bruno, M.; Lebouef, C.; et al. Peroxiredoxin-2: A Novel Regulator of Iron Homeostasis in Ineffective Erythropoiesis. *Antioxid. Redox Signal.* **2018**, *28*, 1–14. [CrossRef]
11. Matte, A.; De Falco, L.; Iolascon, A.; Mohandas, N.; An, X.; Siciliano, A.; Leboeuf, C.; Janin, A.; Bruno, M.; Choi, S.Y.; et al. The Interplay Between Peroxiredoxin-2 and Nuclear Factor-Erythroid 2 Is Important in Limiting Oxidative Mediated Dysfunction in beta-Thalassemic Erythropoiesis. *Antioxid. Redox Signal.* **2015**, *23*, 1284–1297. [CrossRef] [PubMed]
12. Barbosa, M.C.; Grosso, R.A.; Fader, C.M. Hallmarks of Aging: An Autophagic Perspective. *Front. Endocrinol.* **2018**, *9*, 790. [CrossRef]
13. Lopez-Otin, C.; Blasco, M.A.; Partridge, L.; Serrano, M.; Kroemer, G. The hallmarks of aging. *Cell* **2013**, *153*, 1194–1217. [CrossRef]
14. Minakshi, R.; Rahman, S.; Jan, A.T.; Archana, A.; Kim, J. Implications of aging and the endoplasmic reticulum unfolded protein response on the molecular modality of breast cancer. *Exp. Mol. Med.* **2017**, *49*, e389. [CrossRef] [PubMed]
15. Cominacini, L.; Mozzini, C.; Garbin, U.; Pasini, A.; Stranieri, C.; Solani, E.; Vallerio, P.; Tinelli, I.A.; Fratta Pasini, A. Endoplasmic reticulum stress and Nrf2 signaling in cardiovascular diseases. *Free Radic. Biol. Med.* **2015**, *88*, 233–242. [CrossRef]
16. Lupo, F.; Tibaldi, E.; Matte, A.; Sharma, A.K.; Brunati, A.M.; Alper, S.L.; Zancanaro, C.; Benati, D.; Siciliano, A.; Bertoldi, M.; et al. A new molecular link between defective autophagy and erythroid abnormalities in chorea-acanthocytosis. *Blood* **2016**, *128*, 2976–2987. [CrossRef] [PubMed]
17. Khungwanmaythawee, K.; Sornjai, W.; Paemanee, A.; Jaratsittisin, J.; Fucharoen, S.; Svasti, S.; Lithanatudom, P.; Roytrakul, S.; Smith, D.R. Mitochondrial changes in β^0-thalassemia/Hb E disease. *PLoS ONE* **2016**, *11*, e0153831. [CrossRef]
18. Zhang, S.; Macias-Garcia, A.; Velazquez, J.; Paltrinieri, E.; Kaufman, R.J.; Chen, J.J. HRI coordinates translation by eIF2alphaP and mTORC1 to mitigate ineffective erythropoiesis in mice during iron deficiency. *Blood* **2018**, *131*, 450–461. [CrossRef] [PubMed]
19. Naguib, Y.M. Antioxidant activities of astaxanthin and related carotenoids. *J. Agric. Food Chem.* **2000**, *48*, 1150–1154. [CrossRef]
20. Kamath, B.S.; Srikanta, B.M.; Dharmesh, S.M.; Sarada, R.; Ravishankar, G.A. Ulcer preventive and antioxidative properties of astaxanthin from Haematococcus pluvialis. *Eur. J. Pharmacol.* **2008**, *590*, 387–395. [CrossRef] [PubMed]
21. Zhang, Y.; Rogers, H.M.; Zhang, X.; Noguchi, C.T. Sex difference in mouse metabolic response to erythropoietin. *FASEB J.* **2017**, *31*, 2661–2673. [CrossRef] [PubMed]
22. Casarrubea, D.; Viatte, L.; Hallas, T.; Vasanthakumar, A.; Eisenstein, R.S.; Schumann, K.; Hentze, M.W.; Galy, B. Abnormal body iron distribution and erythropoiesis in a novel mouse model with inducible gain of iron regulatory protein (IRP)-1 function. *J. Mol. Med.* **2013**, *91*, 871–881. [CrossRef] [PubMed]
23. Beneduce, E.; Matte, A.; De Falco, L.; Mbiandjeu, S.; Chiabrando, D.; Tolosano, E.; Federti, E.; Petrillo, S.; Mohandas, N.; Siciliano, A.; et al. Fyn kinase is a novel modulator of erythropoietin signaling and stress erythropoiesis. *Am. J. Hematol.* **2019**, *94*, 10–20. [CrossRef] [PubMed]
24. Giarratana, M.C.; Kobari, L.; Lapillonne, H.; Chalmers, D.; Kiger, L.; Cynober, T.; Marden, M.C.; Wajcman, H.; Douay, L. Ex vivo generation of fully mature human red blood cells from hematopoietic stem cells. *Nat. Biotechnol.* **2005**, *23*, 69–74. [CrossRef] [PubMed]
25. Shi, J.; Kundrat, L.; Pishesha, N.; Bilate, A.; Theile, C.; Maruyama, T.; Dougan, S.K.; Ploegh, H.L.; Lodish, H.F. Engineered red blood cells as carriers for systemic delivery of a wide array of functional probes. *Proc. Natl. Acad. Sci. USA* **2014**, *111*, 10131–10136. [CrossRef] [PubMed]
26. Matte, A.; Low, P.S.; Turrini, F.; Bertoldi, M.; Campanella, M.E.; Spano, D.; Pantaleo, A.; Siciliano, A.; De Franceschi, L. Peroxiredoxin-2 expression is increased in β-thalassemic mouse red cells but is displaced from the membrane as a marker of oxidative stress. *Free Radic. Biol. Med.* **2010**, *49*, 457–466. [CrossRef] [PubMed]
27. De Franceschi, L.; Olivieri, O.; Miraglia del Giudice, E.; Perrotta, S.; Sabato, V.; Corrocher, R.; Iolascon, A. Membrane cation and anion transport activities in erythrocytes of hereditary spherocytosis: Effects of different membrane protein defects. *Am. J. Hematol.* **1997**, *55*, 121–128. [CrossRef]
28. Xu, L.; Gao, Z.; Yang, Z.; Qu, M.; Li, H.; Chen, L.; Lv, Y.; Fan, Z.; Yue, W.; Li, C.; et al. Evaluation of Reliable Reference Genes for In Vitro Erythrocyte Generation from Cord Blood $CD34^+$ Cells. *DNA Cell Biol.* **2021**, *40*, 1200–1210. [CrossRef]
29. Makadia, H.K.; Siegel, S.J. Poly Lactic-co-Glycolic Acid (PLGA) as Biodegradable Controlled Drug Delivery Carrier. *Polymers* **2011**, *3*, 1377–1397. [CrossRef] [PubMed]
30. Livak, K.J.; Schmittgen, T.D. Analysis of relative gene expression data using real-time quantitative PCR and the 2(-Delta Delta C(T)) Method. *Methods* **2001**, *25*, 402–408. [CrossRef]
31. Luo, W.; Chen, I.; Chen, Y.; Alkam, D.; Wang, Y.; Semenza, G.L. PRDX2 and PRDX4 are negative regulators of hypoxia-inducible factors under conditions of prolonged hypoxia. *Oncotarget* **2016**, *7*, 6379–6397. [CrossRef] [PubMed]
32. Leak, R.K.; Zhang, L.; Luo, Y.; Li, P.; Zhao, H.; Liu, X.; Ling, F.; Jia, J.; Chen, J.; Ji, X. Peroxiredoxin 2 battles poly(ADP-ribose) polymerase 1- and p53-dependent prodeath pathways after ischemic injury. *Stroke* **2013**, *44*, 1124–1134. [CrossRef]
33. Gan, Y.; Ji, X.; Hu, X.; Luo, Y.; Zhang, L.; Li, P.; Liu, X.; Yan, F.; Vosler, P.; Gao, Y.; et al. Transgenic overexpression of peroxiredoxin-2 attenuates ischemic neuronal injury via suppression of a redox-sensitive pro-death signaling pathway. *Antioxid. Redox Signal.* **2012**, *17*, 719–732. [CrossRef]
34. Rodier, F.; Munoz, D.P.; Teachenor, R.; Chu, V.; Le, O.; Bhaumik, D.; Coppe, J.P.; Campeau, E.; Beausejour, C.M.; Kim, S.H.; et al. DNA-SCARS: Distinct nuclear structures that sustain damage-induced senescence growth arrest and inflammatory cytokine secretion. *J. Cell Sci.* **2011**, *124*, 68–81. [CrossRef] [PubMed]

35. Fernandez, A.; Ordonez, R.; Reiter, R.J.; Gonzalez-Gallego, J.; Mauriz, J.L. Melatonin and endoplasmic reticulum stress: Relation to autophagy and apoptosis. *J. Pineal Res.* **2015**, *59*, 292–307. [CrossRef]
36. Inami, Y.; Waguri, S.; Sakamoto, A.; Kouno, T.; Nakada, K.; Hino, O.; Watanabe, S.; Ando, J.; Iwadate, M.; Yamamoto, M.; et al. Persistent activation of Nrf2 through p62 in hepatocellular carcinoma cells. *J. Cell Biol.* **2011**, *193*, 275–284. [CrossRef] [PubMed]
37. Hama, S.; Uenishi, S.; Yamada, A.; Ohgita, T.; Tsuchiya, H.; Yamashita, E.; Kogure, K. Scavenging of hydroxyl radicals in aqueous solution by astaxanthin encapsulated in liposomes. *Biol. Pharm. Bull.* **2012**, *35*, 2238–2242. [CrossRef]
38. Donoso, A.; Gonzalez-Duran, J.; Munoz, A.A.; Gonzalez, P.A.; Agurto-Munoz, C. Therapeutic uses of natural astaxanthin: An evidence-based review focused on human clinical trials. *Pharmacol. Res.* **2021**, *166*, 105479. [CrossRef]
39. Fakhri, S.; Abbaszadeh, F.; Dargahi, L.; Jorjani, M. Astaxanthin: A mechanistic review on its biological activities and health benefits. *Pharmacol. Res.* **2018**, *136*, 1–20. [CrossRef]
40. Augusti, P.R.; Quatrin, A.; Somacal, S.; Conterato, G.M.; Sobieski, R.; Ruviaro, A.R.; Maurer, L.H.; Duarte, M.M.; Roehrs, M.; Emanuelli, T. Astaxanthin prevents changes in the activities of thioredoxin reductase and paraoxonase in hypercholesterolemic rabbits. *J. Clin. Biochem. Nutr.* **2012**, *51*, 42–49. [CrossRef]
41. Genc, Y.; Bardakci, H.; Yucel, C.; Karatoprak, G.S.; Kupeli Akkol, E.; Hakan Barak, T.; Sobarzo-Sanchez, E. Oxidative Stress and Marine Carotenoids: Application by Using Nanoformulations. *Mar. Drugs* **2020**, *18*, 423. [CrossRef] [PubMed]
42. Nabar, G.M.; Mahajan, K.D.; Calhoun, M.A.; Duong, A.D.; Souva, M.S.; Xu, J.; Czeisler, C.; Puduvalli, V.K.; Otero, J.J.; Wyslouzil, B.E.; et al. Micelle-templated, poly(lactic-co-glycolic acid) nanoparticles for hydrophobic drug delivery. *Int. J. Nanomed.* **2018**, *13*, 351–366. [CrossRef] [PubMed]
43. Madhyastha, R.; Madhyastha, H.; Nurrahmah, Q.I.; Purbasari, B.; Maruyama, M.; Nakajima, Y. MicroRNA 21 Elicits a Pro-inflammatory Response in Macrophages, with Exosomes Functioning as Delivery Vehicles. *Inflammation* **2021**, *44*, 1274–1287. [CrossRef] [PubMed]
44. Aumann, K.; Frey, A.V.; May, A.M.; Hauschke, D.; Kreutz, C.; Marx, J.P.; Timmer, J.; Werner, M.; Pahl, H.L. Subcellular mislocalization of the transcription factor NF-E2 in erythroid cells discriminates prefibrotic primary myelofibrosis from essential thrombocythemia. *Blood* **2013**, *122*, 93–99. [CrossRef] [PubMed]
45. Jin, W.; Wang, H.; Yan, W.; Xu, L.; Wang, X.; Zhao, X.; Yang, X.; Chen, G.; Ji, Y. Disruption of Nrf2 enhances upregulation of nuclear factor-kappaB activity, proinflammatory cytokines, and intercellular adhesion molecule-1 in the brain after traumatic brain injury. *Mediat. Inflamm.* **2008**, *2008*, 725174. [CrossRef] [PubMed]
46. Gao, W.; Guo, L.; Yang, Y.; Wang, Y.; Xia, S.; Gong, H.; Zhang, B.K.; Yan, M. Dissecting the Crosstalk Between Nrf2 and NF-kappaB Response Pathways in Drug-Induced Toxicity. *Front. Cell Dev. Biol.* **2021**, *9*, 809952. [CrossRef]
47. Bolati, D.; Shimizu, H.; Yisireyili, M.; Nishijima, F.; Niwa, T. Indoxyl sulfate, a uremic toxin, downregulates renal expression of Nrf2 through activation of NF-kappaB. *BMC Nephrol.* **2013**, *14*, 56. [CrossRef]
48. Hamza, E.; Vallejo-Mudarra, M.; Ouled-Haddou, H.; Garcia-Caballero, C.; Guerrero-Hue, M.; Santier, L.; Rayego-Mateos, S.; Larabi, I.A.; Alvarez, J.C.; Garcon, L.; et al. Indoxyl sulfate impairs erythropoiesis at BFU-E stage in chronic kidney disease. *Cell Signal* **2023**, *104*, 110583. [CrossRef] [PubMed]
49. Kautz, L.; Jung, G.; Valore, E.V.; Rivella, S.; Nemeth, E.; Ganz, T. Identification of erythroferrone as an erythroid regulator of iron metabolism. *Nat. Genet.* **2014**, *46*, 678–684. [CrossRef] [PubMed]
50. Coffey, R.; Ganz, T. Erythroferrone: An Erythroid Regulator of Hepcidin and Iron Metabolism. *Hemasphere* **2018**, *2*, e35. [CrossRef]
51. Matte, A.; Federti, E.; Winter, M.; Koerner, A.; Harmeier, A.; Mazer, N.; Tomka, T.; Di Paolo, M.L.; Defalco, L.; Andolfo, I.; et al. Bitopertin, a selective oral GLYT1 inhibitor, improves anemia in a mouse model of β-thalassemia. *JCI Insight* **2019**, *4*, e130111. [CrossRef] [PubMed]
52. Zhang, K.; Kaufman, R.J. The unfolded protein response: A stress signaling pathway critical for health and disease. *Neurology* **2006**, *66*, S102–S109. [CrossRef]
53. Rhee, S.G.; Bae, S.H. The antioxidant function of sestrins is mediated by promotion of autophagic degradation of Keap1 and Nrf2 activation and by inhibition of mTORC1. *Free Radic. Biol. Med.* **2015**, *88*, 205–211. [CrossRef] [PubMed]
54. Pajares, M.; Jimenez-Moreno, N.; Garcia-Yague, A.J.; Escoll, M.; de Ceballos, M.L.; Van Leuven, F.; Rabano, A.; Yamamoto, M.; Rojo, A.I.; Cuadrado, A. Transcription factor NFE2L2/NRF2 is a regulator of macroautophagy genes. *Autophagy* **2016**, *12*, 1902–1916. [CrossRef] [PubMed]
55. Glick, D.; Barth, S.; Macleod, K.F. Autophagy: Cellular and molecular mechanisms. *J. Pathol.* **2010**, *221*, 3–12. [CrossRef] [PubMed]
56. Jain, A.; Lamark, T.; Sjottem, E.; Larsen, K.B.; Awuh, J.A.; Overvatn, A.; McMahon, M.; Hayes, J.D.; Johansen, T. p62/SQSTM1 is a target gene for transcription factor NRF2 and creates a positive feedback loop by inducing antioxidant response element-driven gene transcription. *J. Biol. Chem.* **2010**, *285*, 22576–22591. [CrossRef] [PubMed]
57. Sarad, K.; Stefanska, M.; Kraszewska, I.; Szade, K.; Sluimer, J.C.; Blyszczuk, P.; Dulak, J.; Jazwa-Kusior, A. Single-cell transcriptomics reveals subtype-specific molecular profiles in Nrf2-deficient macrophages from murine atherosclerotic aortas. *Front. Immunol.* **2023**, *14*, 1249379. [CrossRef]
58. Ruotsalainen, A.K.; Inkala, M.; Partanen, M.E.; Lappalainen, J.P.; Kansanen, E.; Makinen, P.I.; Heinonen, S.E.; Laitinen, H.M.; Heikkila, J.; Vatanen, T.; et al. The absence of macrophage Nrf2 promotes early atherogenesis. *Cardiovasc. Res.* **2013**, *98*, 107–115. [CrossRef] [PubMed]
59. Santoro, R.; Zanotto, M.; Carbone, C.; Piro, G.; Tortora, G.; Melisi, D. MEKK3 Sustains EMT and Stemness in Pancreatic Cancer by Regulating YAP and TAZ Transcriptional Activity. *Anticancer Res.* **2018**, *38*, 1937–1946. [CrossRef]

60. Colman-Martinez, M.; Martinez-Huelamo, M.; Miralles, E.; Estruch, R.; Lamuela-Raventos, R.M. A New Method to Simultaneously Quantify the Antioxidants: Carotenes, Xanthophylls, and Vitamin A in Human Plasma. *Oxid. Med. Cell. Longev.* **2015**, *2015*, 9268531. [CrossRef]
61. Dattilo, M.; Fontanarosa, C.; Spinelli, M.; Bini, V.; Amoresano, A. Modulation of Human Hydrogen Sulfide Metabolism by Micronutrients, Preliminary Data. *Nutr. Metab. Insights* **2022**, *15*, 11786388211065372. [CrossRef] [PubMed]

antioxidants

MDPI

Article

Ascorbic Acid Protects Bone Marrow from Oxidative Stress and Transient Elevation of Corticosterone Caused by X-ray Exposure in *Akr1a*-Knockout Mice

Tomoki Bo [1,*], Hidekazu Nohara [1], Ken-ichi Yamada [2], Satoshi Miyata [3] and Junichi Fujii [4,*]

1 Laboratory Animal Center, Institute for Promotion of Medical Science Research, Yamagata University Faculty of Medicine, 2-2-2 Iidanishi, Yamagata 990-9585, Japan
2 Physical Chemistry for Life Science Laboratory, Faculty of Pharmaceutical Sciences, Kyushu University, 3-1-1 Maidashi Higashi-ku, Fukuoka 812-8582, Japan; kenyamada@phar.kyushu-u.ac.jp
3 Miyata Diabetes and Metabolism Clinic, 5-17-21 Fukushima, Fukushima-ku, Osaka 553-0003, Japan
4 Department of Biochemistry and Molecular Biology, Graduate School of Medical Science, Yamagata University, 2-2-2 Iidanishi, Yamagata 990-9585, Japan
* Correspondence: bo.tomoki@med.id.yamagata-u.ac.jp (T.B.); jfujii@med.id.yamagata-u.ac.jp (J.F.)

Abstract: Bone marrow cells are the most sensitive to exposure to X-rays in the body and are selectively damaged even by doses that are generally considered permissive in other organs. Ascorbic acid (Asc) is a potent antioxidant that is reported to alleviate damages caused by X-ray exposure. However, rodents can synthesize Asc, which creates difficulties in rigorously assessing its effects in such laboratory animals. To address this issue, we employed mice with defects in their ability to synthesize Asc due to a genetic ablation of aldehyde reductase (*Akr1a*-KO). In this study, concentrations of white blood cells (WBCs) were decreased 3 days after exposure to X-rays at 2 Gy and then gradually recovered. At approximately one month, the recovery rate of WBCs was delayed in the *Akr1a*-KO mouse group, which was reversed via supplementation with Asc. Following exposure to X-rays, Asc levels decreased in plasma, bone marrow cells, and the liver during an early period, and then started to increase. X-ray exposure stimulated the pituitary gland to release adrenocorticotropic hormone (ACTH), which stimulated corticosterone secretion. Asc released from the liver, which was also stimulated by ACTH, appeared to be recruited to the bone marrow. Since corticosterone in high doses is injurious, these collective results imply that Asc protects bone marrow via its antioxidant capacity against ROS produced via exposure to X-rays and the cytotoxic action of transiently elevated corticosterone.

Keywords: reactive oxygen species; adrenal; adrenocorticotropic hormone

Citation: Bo, T.; Nohara, H.; Yamada, K.-i.; Miyata, S.; Fujii, J. Ascorbic Acid Protects Bone Marrow from Oxidative Stress and Transient Elevation of Corticosterone Caused by X-ray Exposure in *Akr1a*-Knockout Mice. *Antioxidants* **2024**, *13*, 152. https://doi.org/10.3390/antiox13020152

Academic Editors: Angelo D'Alessandro and Alkmini T. Anastasiadi

Received: 28 December 2023
Revised: 22 January 2024
Accepted: 24 January 2024
Published: 25 January 2024

1. Introduction

Individuals exposed to radiation from either a nuclear accident or during a medical treatment are at risk of developing acute radiation syndrome, which characteristically occurs in hematopoietic, gastrointestinal, neurovascular, and cutaneous systems [1,2]. Reactive oxygen species (ROS) produced upon exposure to X-rays are major components of tumoricidal action during radiation therapy and are known to oxidatively damage normal tissues [3]. Because bone marrow cells (BMCs) and white blood cells (WBCs), notably lymphocytes and granulocytes, are extremely sensitive to X-ray exposure [4,5], even otherwise permissive doses of X-rays may disrupt such immune systems, thereby creating a disproportionate susceptibility to infection. While oxygen concentrations largely affect the susceptibility of cells to radiation therapy [6], other humoral factors are also known to be involved in the sensitivity to and/or the protection against exposure to X-rays. Compounds that exhibit antioxidant activity are promising agents that tend to reduce oxidative damage to BMCs. Ascorbic acid (Asc) is condensed in some cellular organelles and is considered to be one of the strongest forms of antioxidants among nutritional compounds [7,8]. In fact, many studies have established the beneficial action of Asc in protecting cells from unfavorable reactions caused by exposure to X-rays [9–12]. The protective effects of Asc appear to be evident following low doses of irradiation, which is not the case following high doses [13].

Most animals are able to synthesize Asc, mainly in the liver, and this includes rodents [14,15]. The gene encoding L-gulono-γ-lactone oxidase (*Gulo*) that catalyzes the last step in the synthesis of Asc was, however, likely mutated approximately 63,000,000 years ago, rendering primates unable to synthesize it [16]. Two members of the aldo-keto reductase (AKR) superfamily, aldehyde reductase (AKR1A) and aldose reductase (AKR1B), catalyze the reductive conversion of D-glucuronate into L-gulonate in the Asc synthesis pathway and contribute 85–90% and 10–15% to total Asc synthesis, respectively [17,18]. *Akr1a*-knockout (*Akr1a*-KO) mice die within one year when growing while being fed an Asc-deficient diet, but supplementation with Asc (1.5 mg/mL) in drinking water extends their lifespan to an extent similar to that of wild-type (WT) mice [19]. In the meantime, *Akr1a*-KO mice show an increase in plasma levels of corticosterone, aggressive behavior and developmental retardation [20,21]. Asc is abundantly present in the adrenal glands [22] and is thought to stimulate steroidogenesis in the adrenal glomerular zone [23,24]. Nevertheless, Asc reportedly suppresses corticosterone secretion [25]. Cells take up Asc via sodium ascorbate co-transporters, SVCT1 and SVCT2, from nutrients in the intestine and in blood [26]. The ablation of *SVCT2* decreases catecholamine storage in the adrenal medulla and plasma levels of corticosterone [27]. Although developmental retardation is associated with high levels of corticosterone and low Asc concentrations [21], supplementation with Asc affects neither aggressive behavior nor high corticosterone levels in *Akr1a*-KO mice [20].

Exposure to X-rays stimulates pituitary glands to release adrenocorticotropic hormone (ACTH), which then stimulates adrenal glands to release corticosterone and also Asc [28–30]. In the meantime, X-ray exposure decreases Asc in the BMCs of rats [31]. For the clinical application of Asc, it is essential to understand the regulatory mechanism of Asc concentrations in BMCs in order to maintain the physiological relevance between Asc and the endocrine system during radiation therapy. To gain insights into these issues, we employed *Akr1a*-KO mice and WT mice that were exposed to non-lethal, permissive doses (2 Gy) of X-rays in this study.

2. Materials and Methods

2.1. Animals

Akr1a-KO mice with a C57BL/6 background were obtained using a gene-targeting technique [18], bred in our institution, and used throughout this study. The female WT and *Akr1a*-KO mice were weaned at 30 days of age and fed a standard diet without Asc (Picolab 5053, LabDiet, St. Louis, MO, USA) ad libitum with free access to either water or water containing 1.5 mg/mL of Asc (FUJIFILM Wako Pure Chemical, Osaka, Japan) until they were used. The animals used were at least 10 weeks old. Animal experiments were performed in accordance with the Declaration of Helsinki under protocols approved by the Animal Research Committee at Yamagata University.

2.2. X-ray Irradiation

During exposure to X-rays, the mice were anesthetized and immobilized. X-ray irradiation at 2 Gy was performed using a Shimadzu X-TITAN 225S X-ray generator (Shimazu, Kyoto, Japan) at 200 kV and 14 mA with a 1.0 mm aluminum filter.

2.3. Blood Cell Counts in Mice

Following exposure to X-rays, 100 μL of mouse blood was collected in EDTA tubes after 3 days and again every 7 days from the submandibular vein using 5 mm Goldenrod Animal Lancet (MEDIpoint, New York, NY, USA). Blood samples were analyzed with a hematology analyzer, VETSCAN HM5 (ABAXIS, Union City, CA, USA), to acquire lymphocytes and WBCs.

2.4. BMC Isolation

In order to isolate the BMCs, WT and *Akr1a*-KO mice were euthanized by cervical dislocation at indicated times following exposure to X-rays, and their femurs were removed. All bone marrow was flushed out with PBS and filtered using a 70 μm Cell Strainer (Corning, NY, USA). Following centrifugation, erythrocytes were lysed with 1 mL of RBC lysis buffer (155 mmol/L NH_4Cl, 10 mmol/L $KHCO_3$ and 0.1 mmol/L EDTA). Following centrifugation, the supernatant was aspirated. The BMCs were frozen with liquid nitrogen until use. Data using these cells were expressed per protein amount.

2.5. Measurement of the Reduced Form of Asc

To measure Asc levels, plasma, liver, adrenal glands and BMCs were collected and used for the experiments following exposure to X-rays. The tissue samples (10 mg) excluding plasma were homogenized with a glass–Teflon homogenizer in 100 μL of lysis buffer (50 mmol/L Tris-HCl, pH 7.5, 1% (*v*/*v*) Triton X-100, 5% (*v*/*v*) glycerol, 5 mmol/L EDTA, and 150 mmol/L NaCl). Samples were incubated with 50 μmol/L of Naph-DiPy provided by Dr. Ken-ichi Yamada [32] for 30 min at room temperature, and the fluorescence intensity was measured using a microplate reader, Varioskan Flash (Ex: 310 nm, Em: 410 nm; Thermo Fisher Scientific, Waltham, MA, USA). The protein concentration of the samples was measured using Bio-Rad Protein Assay Dye Reagent (Bio-Rad, Hercules, CA, USA). The results were corrected per unit of protein.

2.6. Measurement of Blood Biochemical Parameters

Blood (100 μL) was collected from the submandibular vein using 5 mm Goldenrod Animal Lancet on the first day following exposure to X-rays and again 3 days later. Following centrifugation at 800× *g* for 5 min at 4 °C, plasma was collected. Blood biochemical parameters were measured using DRI-CHEM NX500V (FUJIFILM Wako Pure Chemical) in accordance with the manufacturer's protocol. The following parameters were measured: blood urea nitrogen (BUN), creatinine (CRE), aspartate aminotransferase (AST), and alanine aminotransferase (ALT).

2.7. Measurements of ACTH and Corticosterone

Blood was collected from mice at the indicated time points following exposure to X-rays. Following centrifugation at 800× *g* for 5 min at 4 °C, plasma was collected; then, ACTH and corticosterone levels were measured using an ACTH ELISA kit (M046006; MDB, Zürich, Switzerland) and Corticosterone EIA Kit (YK240; Yanaihara Institute Inc., Shizuoka, Japan), respectively, in accordance with the manufacturer's protocol. To measure corticosterone levels in the adrenal grands, the glands were homogenized in lysis buffer. Following centrifugation at 18,000× *g* for 15 min at 4 °C, supernatants were collected and used for corticosterone measurement. The protein concentration of the samples was measured using Bio-Rad Protein Assay Dye Reagent. The results were corrected per unit of protein.

2.8. Western Blotting

Tissue samples were homogenized with a glass–Teflon homogenizer in lysis buffer. After centrifugation at 18,000× *g* for 15 min at 4 °C, supernatants were collected. A 3-fold-concentration of Laemmli's sample buffer (0.1875 mol/L Tris-HCl, pH 6.8, 15% (*v*/*v*) β-mercaptoethanol, 6% (*w*/*v*) SDS, 30% (*v*/*v*) glycerol, and 0.006% (*w*/*v*) bromophenol blue) was added to the supernatants, and the samples were boiled for 3 min. Proteins were separated via SDS–PAGE and transferred onto a nitrocellulose membrane (Advantec TOYO, Tokyo, Japan). The membrane was blocked with TBST (10 mmol/L Tris-HCl, pH 7.4, 0.1 mol/L NaCl, and 0.1% Tween-20) containing 5% (*w*/*v*) nonfat skim milk and probed with specific antibodies diluted with TBST containing 5% (*w*/*v*) nonfat skim milk overnight at 4 °C. After probing with HRP-conjugated secondary antibodies, the bound antibodies were detected using a Western Lightning Plus-ECL substrate (Merck Millipore,

Burlington, MA, USA). Image acquisition was performed using an image analyzer (ImageQuant LAS500, GE Healthcare, Chicago, IL, USA), and image analysis was performed using ImageJ software version 1.54f. The following antibodies were used in this study: StAR (Santa Cruz Biotechnology, Santa Cruz, CA, USA), Actin (Santa Cruz Biotechnology), $P450_{scc}$ (Proteintech, Rosemont, IL, USA), AKR1A [33], mouse anti-rabbit IgG-HRP (Santa Cruz Biotechnology), and m-IgGκ BP-HRP (Santa Cruz Biotechnology). The dilution ratios for the primary and secondary antibodies are 1:1000 and 1:2000, respectively.

2.9. Quantitative Reverse-Transcription Polymerase Chain Reaction (RT-PCR)

RNA from mouse livers (20 mg) and BMCs (10 mg) were purified using ISOGEN II (Nippongene, Tokyo, Japan). Total RNA concentration was measured using BMe-spect2 (Malcom, Tokyo, Japan), and cDNA was prepared from 500 ng of RNA using a primescript cDNA synthesis kit (TaKaRa, Shiga, Japan). Amplification of the cDNAs using the corresponding primers (Supplementary Table S1) followed by separation on agarose gels was performed. Quantitative RT-PCR analyses were performed using the CFX96 real-time PCR system (Bio-Rad) and a Thunderbird SYBR qPCR mix (TOYOBO, Osaka, Japan) in accordance with the manufacturer's recommendations. The relative mRNA levels of *SVCT1* and *SVCT2* were calculated using the ΔΔCt method with normalization to the level of HPRT1, which was used as the internal control.

2.10. Adrenalectomy (ADX)

Mice were anesthetized using a mixture of 0.3 mg/kg of medetomidine, 4 mg/kg of midazolam, and 5 mg/kg of butorphanol. The left sides of the dorsal skin and peritoneum were sequentially incised. Then, the adrenal gland was gripped with forceps and cut off using scissors. The peritoneum and skin were sequentially stitched. The right-side operation followed the same procedure as that on the left side. The animals were used in the experiment at least 2 weeks following surgery.

2.11. Cultivation of BMCs and Lactate Dehydrogenase (LDH) Release Assay

BMCs were isolated from WT mice and were incubated in RPMI1640 culture media (Thermo Fisher Scientific) containing 10% FBS (Sigma Aldrich, St. Louis, MO, USA) and 50 U/mL of penicillin and streptomycin (FUJIFILM Wako Pure Chemical) overnight. When necessary, the BMCs were treated with 50 μmol/L Asc or 200 nmol/L dexamethasone (FUJIFILM Wako Pure Chemical) 2 h before X-irradiation. One day, following the exposure to X-rays, the medium containing BMCs was collected and centrifuged at 15,000× *g* for 5 min at 4 °C. The supernatant was subjected to Cytotoxicity LDH Assay Kit-WST (Dojindo, Kumamoto, Japan) in accordance with the manufacturer's instructions.

2.12. Statistical Analysis

All results are expressed as the mean ± the standard error (SE) of experiments performed in triplicate at least. Comparisons between the two groups were performed using Student's *t*-test. For multiple comparisons, the Tukey–Kramer test was used. The minimum level of significance was set at $p < 0.05$.

3. Results

3.1. Akr1a-KO Delayed the Recovery of WBC Counts following Exposure to X-rays

To examine the effect of Asc on the WBC population following exposure to X-rays, we initially measured the blood cells of WT and *Akr1a*-KO mice post-exposure. Following exposure at 2 Gy, WBC counts immediately decreased and then gradually recovered (Figure 1A). Since the majority of WBCs comprise lymphocytes, the radiation-induced decrease in WBCs was attributed to changes in lymphocytes (Figure 1B, Supplementary Figure S1). While there was no difference between WT and *Akr1a*-KO mice 21 days following the initial exposure to X-rays, delayed recovery was observed in the *Akr1a*-KO mice at 28 and 35 days post-exposure. Supplementing Asc significantly alleviated the radiation-induced decrease in WBC counts, and the improving effect of Asc was comparable between both WT and *Akr1a*-KO mice.

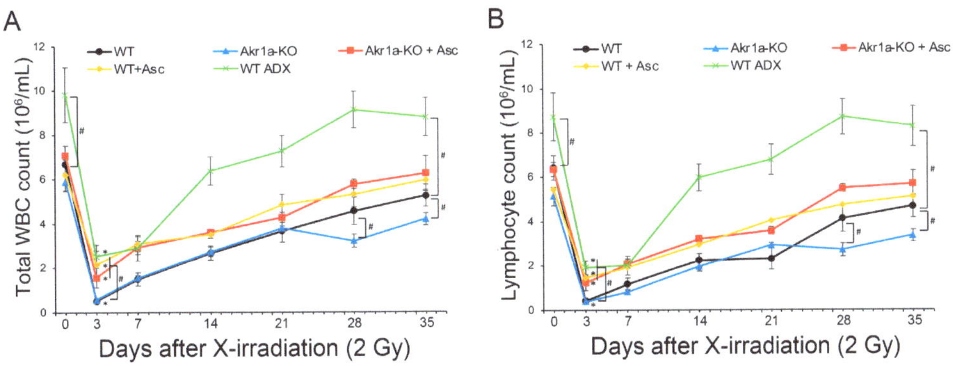

Figure 1. *Akr1a*-KO delayed the recovery of WBCs following exposure to X-rays. Blood samples were analyzed using a hematology analyzer following exposure to X-rays at 2 Gy in WT and *Akr1a*-KO mice with or without Asc. In WT ADX mice, the measurement was begun 2 weeks after adrenalectomy. (**A**) Total WBC counts were measured following exposure. (**B**) Lymphocyte counts were measured after exposure. Data are expressed as the mean ± SE for 5–7 animals per group. * $p < 0.05$ vs. 0 Gy and # $p < 0.05$ vs. WT mice. WBC, white blood cells; WT ADX, mouse with adrenalectomy.

Previous studies have reported that corticosterone impairs WBC recovery [34], and we assumed that corticosterone might affect the results of the current study. To investigate the effect of corticosterone on WBC recovery following exposure to X-rays, WT ADX mice were used. The recovery rate of the WBC count post-exposure in WT ADX mice was significantly increased in ADX mice compared with that measured in WT mice. We also measured the blood biochemical parameters associated with liver and kidney function, but exposure to X-rays at 2 Gy induced neither liver nor kidney damage in either WT or *Akr1a*-KO mice (Supplementary Figure S2). These results suggest that the recovery of WBCs following exposure to a permissive dose of X-rays was delayed in *Akr1a*-KO mice, and that the secretion of corticosterone following exposure could have adversely affected the recovery of WBC counts.

3.2. X-ray Irradiation Decreased Asc Levels in the Plasma, Liver and BMCs

Asc is synthesized mainly in the liver [14,15] and is a potent antioxidant that reportedly alleviates damages caused by exposure to X-rays [12]. Although it is known that such exposure decreases Asc levels in the adrenal glands of rats [31], this phenomenon has not yet been confirmed in mice. We then measured Asc levels in the plasma, BMCs, liver, and adrenal glands following exposure to X-rays (Figure 2). We found that Asc levels were decreased in the plasma, BMCs, and livers when measured one day following exposure to X-rays, after which the levels began to increase. On the other hand, Asc levels in the adrenal grands were unchanged following exposure to X-rays. The mRNAs for *SVCT1* and *SVCT2*, which act to uptake Asc, were increased in the livers following exposure to X-rays in WT mice (Figure 3). Surprisingly, the *SVCT1* and *SVCT2* mRNAs were undetectable in BMCs both before and after X-ray exposure. In the BMCs of *Akr1a*-KO mice, Asc levels remained low before and after exposure to X-rays. These results suggest that released Asc from the liver is supplied to BMCs to support their protection and regeneration.

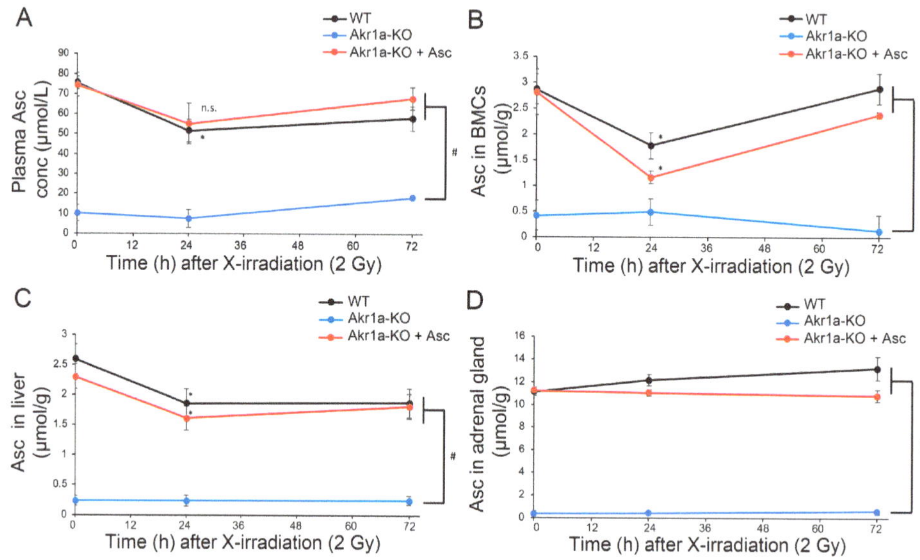

Figure 2. Exposure to X-rays decreasing the Asc levels in plasma, BMCs and liver. Asc levels in plasma (**A**), BMCs (**B**), liver (**C**), and adrenal glands (**D**) were measured using Naph-DiPy in WT and *Akr1a*-KO mice with or without Asc following exposure to X-rays at 2 Gy. The fluorescence intensity was measured using a microplate reader (Ex: 310 nm, Em: 410 nm). Data are expressed as the mean ± SE for 3–5 animals per group. * $p < 0.05$ vs. 0 Gy and # $p < 0.05$ vs. WT mice and *Akr1a*-KO mice supplemented with Asc. n.s., not significant.

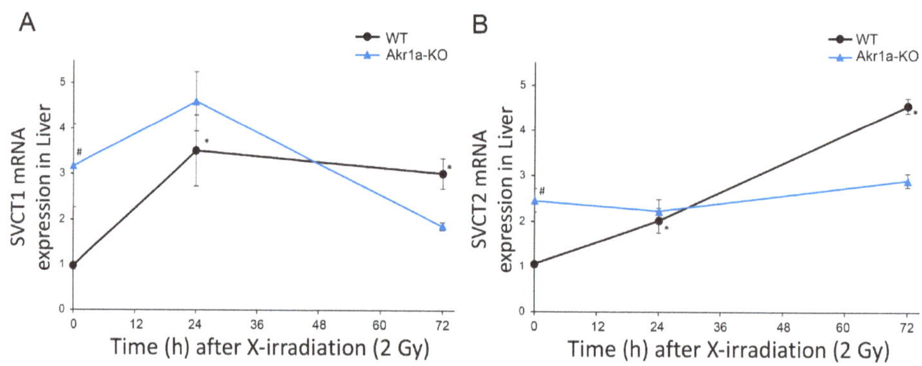

Figure 3. Increased *SVCT1* and *SVCT2* mRNA levels in the liver following exposure to X-rays. The expression of *SVCT1* (**A**) and *SVCT2* (**B**) mRNA in the liver was measured via quantitative RT-PCR following exposure to X-rays in WT and *Akr1a*-KO mice. Data are expressed as the mean ± SE for 5 animals per group. * $p < 0.05$ vs. 0 Gy and # $p < 0.05$ vs. WT mice.

3.3. *AKr1a-KO Increased Corticosterone Secretion Post-Exposure to X-rays*

Stress caused by X-rays reportedly promotes ACTH release from pituitary glands, which results in the secretion of corticosterone from adrenal glands [28]. Because corticosterone secretion could impair the recovery of WBCs following exposure to X-rays (Figure 1), we measured the ACTH and corticosterone levels post-exposure in WT and *Akr1a*-KO mice. As reported in a previous study [28], ACTH levels were increased immediately following X-ray exposure in both WT and *Akr1a*-KO mice, but there was no difference in the concentrations between the mouse groups (Figure 4A). Consistent with the increase in ACTH levels, corticosterone levels showed a dramatic increase immediately following exposure to X-rays, followed by a decrease 24 h post-exposure (Figure 4B). In *Akr1a*-KO mice, corticosterone levels were higher compared with those in WT and *Akr1a*-KO mice that were supplemented with Asc before and 3 h following exposure to X-rays, but the levels were decreased to a similar extent one day later. These results suggest that corticosterone released from adrenal glands is associated with Asc levels, and that Asc deficiency adversely affects BMC proliferation.

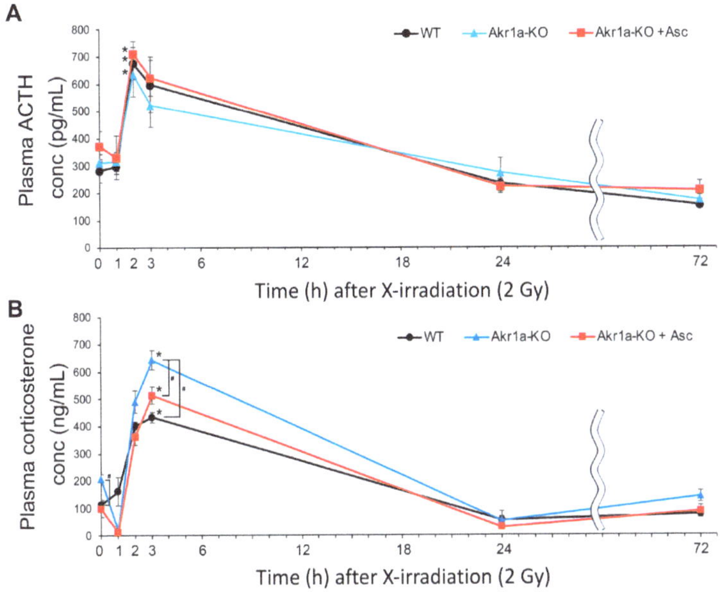

Figure 4. *Akr1a*-KO enhanced the radiation-induced increase in plasma corticosterone levels following exposure to X-rays. ACTH levels (**A**) and corticosterone levels (**B**) in plasma were measured using an ELISA kit in WT and *Akr1a*-KO mice with or without Asc following X-ray exposure at 2 Gy. Data are expressed as the mean ± SE for 5 animals per group. * $p < 0.05$ vs. 0 Gy and # $p < 0.05$ vs. WT mice and *Akr1a*-KO mice supplemented with Asc.

To investigate corticosterone synthesis capacity following exposure to X-rays, we examined the levels of two key proteins in the adrenal glands that are associated with steroid synthesis in WT and *Akr1a*-KO mice before and after exposure. StAR is a steroidogenic acute regulatory protein that is involved in cholesterol transport from the outer to inner mitochondrial membranes [35], and $P450_{scc}$ (CYP11A1) is a cholesterol side-chain cleavage enzyme that regulates the first and rate-limiting steps in the synthesis of steroid hormones [36]. To examine the proteins involved in steroid hormone synthesis following exposure to X-rays, we performed Western blotting on the adrenal glands (Figure 5A). These protein expression levels were unchanged in both WT and *Akr1a*-KO mice before and after X-ray exposure. This result suggests that exposure to X-rays did not affect the protein expression associated with corticosterone synthesis. As shown in Figure 4B, corticosterone levels immediately increased in a few hours and were then decreased one day following exposure to X-rays, which suggests that corticosterone secretion from the adrenal glands had been stimulated by this exposure, but this secretion subsequently declined after 24 h due to excretion. We then evaluated corticosterone levels in the adrenal glands after exposure to X-rays (Figure 5B). Adrenal corticosterone levels had significantly decreased one day following the exposure to X-rays in both WT and *Akr1a*-KO mice. However, the amount of corticosterone in the adrenal glands was lower compared with that in plasma, which indicated that the corticosterone stored in the adrenal glands was low. These results suggest that, following exposure to X-rays, adrenal corticosterone synthesis was initially activated in response to ACTH and then was subsequently reduced, probably through a feedback mechanism, resulting in a decrease in plasma corticosterone levels.

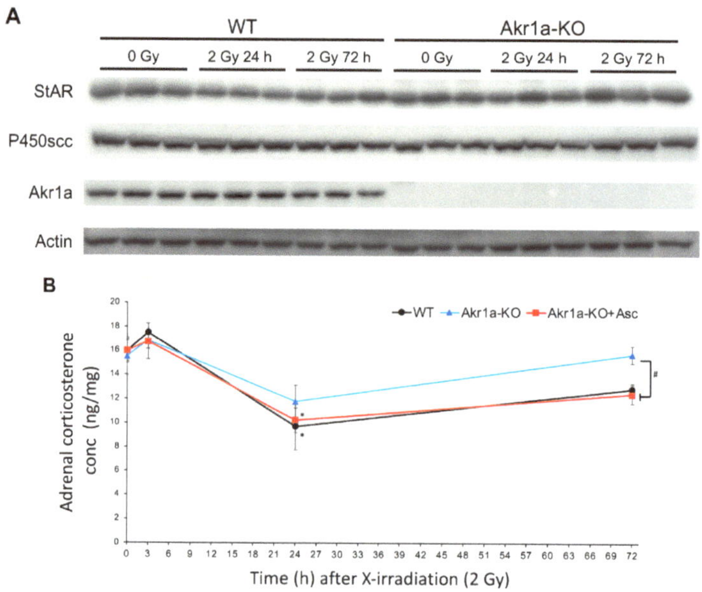

Figure 5. Exposure to X-rays, which did not affect the levels of the primary steroidogenic proteins but transiently decreased corticosterone concentrations. (**A**) The proteins associated with steroid hormone synthesis evaluated via Western blotting in representative blots of StAR, $P450_{scc}$, AKR1A, and Actin. (**B**) Corticosterone levels in adrenal glands measured using an ELISA kit in WT and *Akr1a*-KO mice with or without Asc following X-ray exposure at 2 Gy. Data are expressed as the mean ± SE for 3–5 animals per group. * $p < 0.05$ vs. 0 Gy and # $p < 0.05$ vs. WT mice and *Akr1a*-KO mice supplemented with Asc.

3.4. Asc Treatment Mitigated Radiation-Induced Damage, While Dexamethasone Treatment Enhanced It in Cultured BMCs

To confirm the effect of Asc and corticosterone on the viability of BMCs following exposure to X-rays, we performed an LDH assay using cultured BMCs from WT mice. Exposure to X-rays at 2 Gy had not damaged the cultured BMCs, and, therefore, we separately administered Asc and dexamethasone 2 h before exposure to 5 Gy. After 24 h of incubation, the medium containing the BMCs was collected and used for the LDH assay (Figure 6). X-ray exposure significantly increased LDH release. Asc treatment reduced the radiation-induced increase in LDH release, while dexamethasone treatment increased it. This result indicated that Asc was protective against radiation-induced damage and that corticosterone had adversely affected the viability of BMCs.

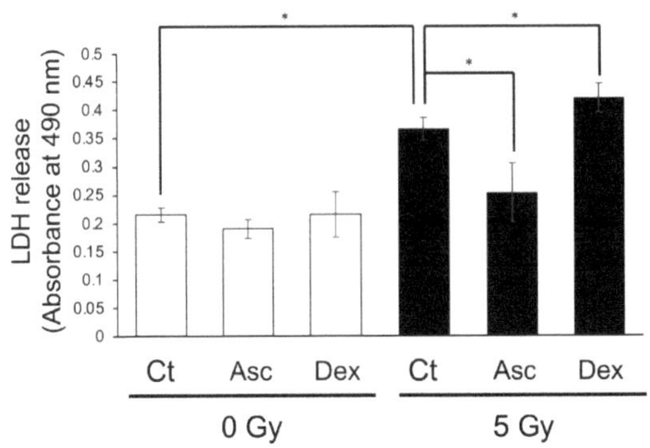

Figure 6. Radiation-induced LDH release from cultured BMCs reduced via Asc treatment but increased via dexamethasone treatment. BMCs were isolated and cultured overnight. The BMCs were treated either with 50 μmol/L Asc or 200 nmol/L dexamethasone (DEX) 2 h before exposure to X-rays at 5 Gy. LDH release from cultured BMCs was measured one day following exposure. Data are expressed as the mean ± SE for 3 samples per group. * $p < 0.05$.

4. Discussion

In the current study, we used *Akr1a*-KO mice, which are defective in synthesizing Asc and show elevations in plasma corticosterone concentrations under ordinary conditions [20]. After exposure to X-rays at 2 Gy, plasma Asc levels were decreased temporarily on the first day, and then tended to recover gradually in all groups of mice (Figure 2). Since supplementation with Asc accelerated the recovery of WBC concentrations, mostly lymphocytes, following the exposure to X-rays in both *Akr1a*-KO and WT mice, low Asc concentrations appeared to be responsible for the delayed recovery in *Akr1a*-KO mice (Figure 1). Following exposure to X-rays, ACTH and corticosterone levels were soon increased in the blood (Figure 4), which appeared to be somewhat lessened in *Akr1a*-KO mice via their Asc status. Asc appeared to be protective against exposure to X-rays also in the primary culture of BMCs (Figure 6), which supports the beneficial action of Asc.

Asc concentrations in both BMCs and blood plasma were transiently decreased on day 1 and then gradually recovered in all groups of mice (Figure 2). The decline in Asc concentrations in BMCs has also been reported in rats following exposure to X-rays, but no change has been observed in the blood levels of Asc [31]. In our study, WBCs were decreased similarly in all experimental groups, but the recovery was delayed in *Akr1a*-KO mice at one month following their exposure to X-rays (Figure 1). Cells either with a low degree of differentiation or those in proliferation seem to be more susceptible to radiation damage. Therefore, a delay in the recovery of WBC concentrations appears to reflect a degree of selective damage to the less-differentiated progenitor cells such as hematopoietic stem cells (HSCs) in BMCs. Since supplementation with Asc rescued the abnormality of the *Akr1a*-KO mice to an extent similar to that in the WT mice (Figure 1), we assumed that Asc either protected the BMCs from oxidative damage or stimulated the hematopoietic process. In fact, a recent clinical study has reported that intravenous administration of Asc (50 mg/kg/day, divided into three doses given on days 1–14) to patients undergoing allogeneic HSC transplant is safe and reduces non-relapse mortality, resulting in the improvement of overall survival [37]. The intravenous administration of mesenchymal stem cells overexpressing extracellular superoxide dismutase (*SOD3*) to the mice after 9-Gy γ-ray irradiation reportedly improved their survival [38]. The results imply that SOD3 was produced by the homed cells and protected the irradiated tissues including bone marrow through the scavenging of superoxide by SOD3. Since Asc also effectively scavenges radicals including superoxide [7,8], it is thought that HSCs were protected from the cytotoxic effects of oxygen radicals by Asc in our study.

The adrenal gland is the first organ that was discovered to be enriched in Asc [22], but the roles of Asc have long been debated. Asc reportedly stimulates steroidogenesis in the adrenal glomerular zone of rats [23]. Genes responsible for steroidogenesis, including those encoding $P450_{SCC}$, 3-hydroxysteroid dehydrogenase type 1, and aromatase, are induced by Asc in human choriocarcinoma cells [24]. Consistent with these reports, the knockout of *SVCT2* impairs adrenal chromaffin cells and decreases plasma levels of corticosterone as well as tissue levels of adrenalin/noradrenalin [39]. In the present study, however, the levels of corticosterone in the plasma of *Akr1a*-KO mice were originally high and were transiently increased following exposure to X-rays, despite levels of Asc that were quite low in the plasma and adrenal glands regardless of exposure (Figure 2). Similar phenomena have been found in other animal models with Asc deficiency as follows. Rats with an osteogenic disorder (ODS), which presents a hereditary defect in Asc-synthesizing ability due to a mutation of the *Gulo* gene, show higher plasma and adrenal levels of corticosterone [40]. Guinea pigs, which also carry a defect in the *Gulo* gene, show higher levels of glucocorticoid in the blood due to Asc deficiency [41]. Thus, high Asc levels seem to rather suppress glucocorticoid levels in the blood of these animal species, and it is unlikely that Asc supports steroidogenesis in the adrenal glands.

Whole-body exposure to X-rays causes stress that stimulates the pituitary–adrenal axis. As a result, exposure to X-rays promotes the secretion of ACTH, which leads to a release of adrenocortical hormones, mainly corticosterone, in mice [28]. Given the rapid secretion of ACTH from pituitary glands [29] and the short half-life of corticosterone in the blood of only a few hours in humans [42], lower corticosterone concentrations after one day would reflect fatigue in the steroidogenic ability of the cells in adrenal glands. Protection of the lower portion of the body by a lead shield causes a greater release of corticosterone and adrenal Asc in comparison with that under total-body irradiation in rats [2]. These observations demonstrate that exposure to X-rays first stimulates the release of ACTH from pituitary glands, which leads to the stimulation of corticosterone production [29]. ACTH also stimulates the release of Asc from the liver and adrenal glands in rats [43]. Therefore, Asc recovery in plasma following X-ray exposure could be attributable to recruitment from the liver, which is the main organ that produces Asc in rodents [14]. A following study has shown that ACTH also stimulates the release of Asc from the adrenal glands in humans [30]. Based on these findings, we know that upon exposure to X-rays, ACTH release from pituitary glands then stimulates not only corticosterone secretion but also Asc release from both the adrenal glands and the liver.

Anti-inflammation is the well-established function of corticosterone, and the inhibition of a transcriptional regulatory factor NF-κB, which dominantly induces expression of pro-inflammatory genes, is one of its mechanisms [44]. On the other hand, BMCs are prone to apoptosis under exposure to corticosterone as well as under exposure to X-rays [5,45]. Glucocorticoid stimulates energy metabolism to cope with stress conditions, but, at the same time, it adversely affects BMCs, notably in lymphocytes [46,47]. Glucocorticoid is known to increase ROS in part by suppressing the function of Nrf2, which is a key transcriptional activator for the expression of some of the antioxidant genes [48,49]. Since the replenishment of Asc leads to a decreased glucocorticoid response in humans and in animal models after exposure to a psychological or physical stressors [50], Asc likely acts to suppress the injurious effects of the glucocorticoid that is released under stress conditions. Because Asc is a potent antioxidant, scavenging oxygen radicals with electron donation [7,8], the primary action of Asc appears to be that of protecting BMCs against ROS that are produced by ionizing radiation and by corticosterone response. Since Asc exerts pleiotropic functions, it is also possible that Asc stimulates the hematopoietic process by supporting enzymatic reactions, such as prolyl hydroxylase, or by epigenetically regulating gene expression [51]. Considering the cytotoxic action of corticosterone, the early recovery of WBCs in ADX mice following exposure to X-rays might be associated with an absence of cytotoxic effects of plasma corticosterone.

5. Conclusions

In this study, Asc deficiency delayed the recovery of WBCs in *Akr1a*-KO mice following exposure to a permissive dose of X-rays, which are known to increase ROS production via the interaction of oxygen molecules, but also trigger the secretion of corticosterone through the stimulation of ACTH released from pituitary glands. Asc deficiency in the *Akr1a*-KO mice elevated corticosterone levels, and oxidative damage was increased via such suppression of antioxidant enzyme expression, which appears to also be partially responsible for aggravating the damage to BMCs. It is conceivable that physiological levels of Asc alleviate the damage to BMCs through their antioxidant capacity against ROS that are produced by exposure to X-rays and via the attendant cytotoxic action of corticosterone (Figure 7).

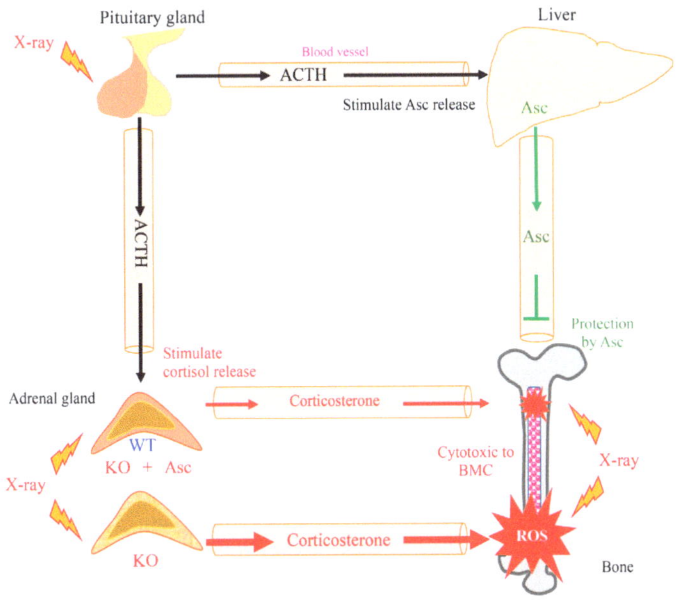

Figure 7. Schematic diagram outlining the results of this study.

Supplementary Materials: The following supporting information can be downloaded at https://www.mdpi.com/article/10.3390/antiox13020152/s1. Figure S1: Changes in count of granulocytes following X-ray irradiation; Figure S2: Liver and kidney damages not observed after Xray irradiation in WT and *Akr1a*-KO mice; Table S1: Primer sequence for quantitative-RT PCR.

Author Contributions: Conceptualization and writing, T.B. and J.F.; investigation, T.B. with the help of H.N.; Providing the probe for detecting ascorbic acid, K.-i.Y.; provision of *Akr1a*-knockout mice, S.M. All authors have read and agreed to the published version of the manuscript.

Funding: This research was funded by a JSPS KAKENHI Grant-in-Aid for Scientific Research from the Japan Society for the Promotion of Science (21K14971) for T.B. and by the YU-COE program (S6), Yamagata University, for J.F.

Institutional Review Board Statement: The animal study protocol was approved by the Animal Research Committee at Yamagata University. All procedures on mice were conducted in accordance with the Declaration of Helsinki under protocols approved by the Animal Research Committee at Yamagata University.

Informed Consent Statement: Not applicable.

Data Availability Statement: The data underlying this article are available in the article and Supplementary Materials.

Conflicts of Interest: The authors declare no conflicts of interest.

Abbreviations

ACTH	adrenocorticotropic hormone
ADX	adrenalectomy
AKR	aldo-keto reductase
AKR1A	aldehyde reductase
Akr1a	gene encoding adehyde reductase
AKR1B	aldose reductase
ALT	alanine aminotransferase
Asc	ascorbic acid
AST	aspartate aminotransferase
BMCs	bone marrow cells
BUN	blood urea nitrogen
CRE	creatinine
Gulo	gene encoding L-gulono-γ-lactone oxidase
KO	knockout
LDH	lactate dehydrogenase
ROS	reactive oxygen species
SE	standard error
WBCs	white blood cells
WT	wild-type

References

1. Dainiak, N.; Albanese, J. Medical management of acute radiation syndrome. *J. Radiol. Prot.* **2022**, *42*, 031002. [CrossRef]
2. Orschell, C.M.; Wu, T.; Patterson, A.M. Impact of Age, Sex, and Genetic Diversity in Murine Models of the Hematopoietic Acute Radiation Syndrome (H-ARS) and the Delayed Effects of Acute Radiation Exposure (DEARE). *Curr. Stem Cell Rep.* **2022**, *8*, 139–149. [CrossRef]
3. McBride, W.H.; Schaue, D. Radiation-induced tissue damage and response. *J. Pathol.* **2020**, *250*, 647–655. [CrossRef] [PubMed]
4. Till, J.E.; McCulloch, E.A. A direct measurement of the radiation sensitivity of normal mouse bone marrow cells. *Radiat. Res.* **2012**, *178*, Av3–Av7. [CrossRef]
5. Garvy, B.A.; Telford, W.G.; King, L.E.; Fraker, P.J. Glucocorticoids and irradiation-induced apoptosis in normal murine bone marrow B-lineage lymphocytes as determined by flow cytometry. *Immunology* **1993**, *79*, 270–277.
6. Ogawa, Y. Paradigm Shift in Radiation Biology/Radiation Oncology-Exploitation of the "H_2O_2 Effect" for Radiotherapy Using Low-LET (Linear Energy Transfer) Radiation such as X-rays and High-Energy Electrons. *Cancers* **2016**, *8*, 28. [CrossRef] [PubMed]
7. Njus, D.; Kelley, P.M.; Tu, Y.J.; Schlegel, H.B. Ascorbic acid: The chemistry underlying its antioxidant properties. *Free Radic. Biol. Med.* **2020**, *159*, 37–43. [CrossRef] [PubMed]
8. Fujii, J.; Osaki, T.; Bo, T. Ascorbate Is a Primary Antioxidant in Mammals. *Molecules* **2022**, *27*, 6187. [CrossRef] [PubMed]
9. el-Nahas, S.M.; Mattar, F.E.; Mohamed, A.A. Radioprotective effect of vitamins C and E. *Mutat. Res.* **1993**, *301*, 143–147. [CrossRef] [PubMed]
10. Koyama, S.; Kodama, S.; Suzuki, K.; Matsumoto, T.; Miyazaki, T.; Watanabe, M. Radiation-induced long-lived radicals which cause mutation and transformation. *Mutat. Res.* **1998**, *421*, 45–54. [CrossRef]
11. Ito, Y.; Kinoshita, M.; Yamamoto, T.; Sato, T.; Obara, T.; Saitoh, D.; Seki, S.; Takahashi, Y. A combination of pre- and post-exposure ascorbic acid rescues mice from radiation-induced lethal gastrointestinal damage. *Int. J. Mol. Sci.* **2013**, *14*, 19618–19635. [CrossRef]
12. Sato, T.; Kinoshita, M.; Yamamoto, T.; Ito, M.; Nishida, T.; Takeuchi, M.; Saitoh, D.; Seki, S.; Mukai, Y. Treatment of irradiated mice with high-dose ascorbic acid reduced lethality. *PLoS ONE* **2015**, *10*, e0117020. [CrossRef]
13. Harapanhalli, R.S.; Yaghmai, V.; Giuliani, D.; Howell, R.W.; Rao, D.V. Antioxidant effects of vitamin C in mice following X-irradiation. *Res. Commun. Mol. Pathol. Pharmacol.* **1996**, *94*, 271–287.
14. Linster, C.L.; Van Schaftingen, E. Vitamin C. Biosynthesis, recycling and degradation in mammals. *FEBS J.* **2007**, *274*, 1–22. [CrossRef]
15. Mandl, J.; Szarka, A.; Bánhegyi, G. Vitamin C: Update on physiology and pharmacology. *Br. J. Pharmacol.* **2009**, *157*, 1097–1110. [CrossRef] [PubMed]
16. Nishikimi, M.; Fukuyama, R.; Minoshima, S.; Shimizu, N.; Yagi, K. Cloning and chromosomal mapping of the human nonfunctional gene for L-gulono-gamma-lactone oxidase, the enzyme for L-ascorbic acid biosynthesis missing in man. *J. Biol. Chem.* **1994**, *269*, 13685–13688. [CrossRef] [PubMed]
17. Gabbay, K.H.; Bohren, K.M.; Morello, R.; Bertin, T.; Liu, J.; Vogel, P. Ascorbate synthesis pathway: Dual role of ascorbate in bone homeostasis. *J. Biol. Chem.* **2010**, *285*, 19510–19520. [CrossRef] [PubMed]

18. Takahashi, M.; Miyata, S.; Fujii, J.; Inai, Y.; Ueyama, S.; Araki, M.; Soga, T.; Fujinawa, R.; Nishitani, C.; Ariki, S.; et al. In vivo role of aldehyde reductase. *Biochim. Biophys. Acta* **2012**, *1820*, 1787–1796. [CrossRef] [PubMed]
19. Fujii, J.; Homma, T.; Miyata, S.; Takahashi, M. Pleiotropic Actions of Aldehyde Reductase (AKR1A). *Metabolites* **2021**, *11*, 343. [CrossRef] [PubMed]
20. Homma, T.; Akihara, R.; Okano, S.; Shichiri, M.; Yoshida, Y.; Yamada, K.I.; Miyata, S.; Nakajima, O.; Fujii, J. Heightened aggressive behavior in mice deficient in aldo-keto reductase 1a (Akr1a). *Behav. Brain Res.* **2017**, *319*, 219–224. [CrossRef] [PubMed]
21. Ishii, N.; Homma, T.; Takeda, Y.; Aung, N.Y.; Yamada, K.I.; Miyata, S.; Asao, H.; Yamakawa, M.; Fujii, J. Developmental retardation in neonates of aldehyde reductase (AKR1A)-deficient mice is associated with low ascorbic acid and high corticosterone levels. *J. Nutr. Biochem.* **2021**, *91*, 108604. [CrossRef]
22. Szent-Györgyi, A. Observations on the function of peroxidase systems and the chemistry of the adrenal cortex: Description of a new carbohydrate derivative. *Biochem. J.* **1928**, *22*, 1387–1409. [CrossRef] [PubMed]
23. Mitani, F.; Ogishima, T.; Mukai, K.; Suematsu, M. Ascorbate stimulates monooxygenase-dependent steroidogenesis in adrenal zona glomerulosa. *Biochem. Biophys. Res. Commun.* **2005**, *338*, 483–490. [CrossRef] [PubMed]
24. Wu, X.; Iguchi, T.; Itoh, N.; Okamoto, K.; Takagi, T.; Tanaka, K.; Nakanishi, T. Ascorbic acid transported by sodium-dependent vitamin C transporter 2 stimulates steroidogenesis in human choriocarcinoma cells. *Endocrinology* **2008**, *149*, 73–83. [CrossRef] [PubMed]
25. Wexler, B.C.; Pencharz, R.; Thomas, S.F. Adrenal ascorbic acid and histological changes in male and female rats after half-body x-ray irradiation. *Am. J. Physiol.* **1955**, *183*, 71–74. [CrossRef] [PubMed]
26. Savini, I.; Rossi, A.; Pierro, C.; Avigliano, L.; Catani, M.V. SVCT1 and SVCT2: Key proteins for vitamin C uptake. *Amino Acids* **2008**, *34*, 347–355. [CrossRef] [PubMed]
27. Patak, P.; Willenberg, H.S.; Bornstein, S.R. Vitamin C is an important cofactor for both adrenal cortex and adrenal medulla. *Endocr. Res.* **2004**, *30*, 871–875. [CrossRef] [PubMed]
28. Mateyko, G.M.; Edelmann, A. The effects of localized cathode-ray particle irradiation of the hypophysis and whole-body x-irradiation on gonadotrophin, thyrotrophin, and adrenocorticotrophin of the rat pituitary. *Radiat. Res.* **1954**, *1*, 470–486. [CrossRef]
29. Hameed, J.M.; Haley, T.J. Plasma and adrenal gland corticosterone levels after X-ray exposure in rats. *Radiat. Res.* **1964**, *23*, 620–629. [CrossRef]
30. Padayatty, S.J.; Doppman, J.L.; Chang, R.; Wang, Y.; Gill, J.; Papanicolaou, D.A.; Levine, M. Human adrenal glands secrete vitamin C in response to adrenocorticotrophic hormone. *Am. J. Clin. Nutr.* **2007**, *86*, 145–149. [CrossRef]
31. Umegaki, K.; Aoki, S.; Esashi, T. Whole body X-ray irradiation to mice decreases ascorbic acid concentration in bone marrow: Comparison between ascorbic acid and vitamin E. *Free Radic. Biol. Med.* **1995**, *19*, 493–497. [CrossRef] [PubMed]
32. Matsuoka, Y.; Yamato, M.; Yamasaki, T.; Mito, F.; Yamada, K. Rapid and convenient detection of ascorbic acid using a fluorescent nitroxide switch. *Free Radic. Biol. Med.* **2012**, *53*, 2112–2118. [CrossRef] [PubMed]
33. Takahashi, M.; Fujii, J.; Teshima, T.; Suzuki, K.; Shiba, T.; Taniguchi, N. Identity of a major 3-deoxyglucosone-reducing enzyme with aldehyde reductase in rat liver established by amino acid sequencing and cDNA expression. *Gene* **1993**, *127*, 249–253.
34. Dhabhar, F.S.; Miller, A.H.; McEwen, B.S.; Spencer, R.L. Stress-induced changes in blood leukocyte distribution. Role of adrenal steroid hormones. *J. Immunol.* **1996**, *157*, 1638–1644. [CrossRef]
35. Galano, M.; Venugopal, S.; Papadopoulos, V. Role of STAR and SCP2/SCPx in the Transport of Cholesterol and Other Lipids. *Int. J. Mol. Sci.* **2022**, *23*, 12115. [CrossRef]
36. Arukwe, A. Steroidogenic acute regulatory (StAR) protein and cholesterol side-chain cleavage (P450scc)-regulated steroidogenesis as an organ-specific molecular and cellular target for endocrine disrupting chemicals in fish. *Cell Biol. Toxicol.* **2008**, *24*, 527–540. [CrossRef]
37. Toor, A.; Simmons, G.; Sabo, R.; Aziz, M.; Martin, E.; Bernard, R.; Sriparna, M.; McIntire, C.; Kreiger, E.; Brophy, D.; et al. Intravenous Vitamin C Supplementation in Allogeneic Hematopoietic Cell Transplant Recipients: Salutary Impact on Clinical Outcomes. *Res. Sq.* **2023**, preprint. [CrossRef]
38. Abdel-Mageed, A.S.; Senagore, A.J.; Pietryga, D.W.; Connors, R.H.; Giambernardi, T.A.; Hay, R.V.; Deng, W. Intravenous administration of mesenchymal stem cells genetically modified with extracellular superoxide dismutase improves survival in irradiated mice. *Blood* **2009**, *113*, 1201–1203. [CrossRef]
39. Bornstein, S.R.; Yoshida-Hiroi, M.; Sotiriou, S.; Levine, M.; Hartwig, H.G.; Nussbaum, R.L.; Eisenhofer, G. Impaired adrenal catecholamine system function in mice with deficiency of the ascorbic acid transporter (SVCT2). *FASEB J.* **2003**, *17*, 1928–1930. [CrossRef]
40. Horio, F.; Ozaki, K.; Yoshida, A.; Makino, S.; Hayashi, Y. Requirement for ascorbic acid in a rat mutant unable to synthesize ascorbic acid. *J. Nutr.* **1985**, *115*, 1630–1640. [CrossRef]
41. Enwonwu, C.O.; Sawiris, P.; Chanaud, N. Effect of marginal ascorbic acid deficiency on saliva level of cortisol in the guinea pig. *Arch. Oral Biol.* **1995**, *40*, 737–742. [CrossRef]
42. Michalowski, A.S. On radiation damage to normal tissues and its treatment. II. Anti-inflammatory drugs. *Acta Oncol.* **1994**, *33*, 139–157. [CrossRef]
43. Overbeek, G.A. Hormonal regulation of ascorbic acid in the adrenal of the rat. *Acta Endocrinol.* **1985**, *109*, 393–402. [CrossRef]

44. Gerber, A.N.; Newton, R.; Sasse, S.K. Repression of transcription by the glucocorticoid receptor: A parsimonious model for the genomics era. *J. Biol. Chem.* **2021**, *296*, 100687. [CrossRef]
45. Sainz, R.M.; Mayo, J.C.; Reiter, R.J.; Tan, D.X.; Rodriguez, C. Apoptosis in primary lymphoid organs with aging. *Microsc. Res. Tech.* **2003**, *62*, 524–539. [CrossRef]
46. Thiersch, J.B.; Conroy, L.; Stevens, A.R., Jr.; Finch, C.A. Adverse effect of cortisone on marrow regeneration following irradiation. *J. Lab. Clin. Med.* **1952**, *40*, 174–181.
47. Ashwell, J.D.; Lu, F.W.; Vacchio, M.S. Glucocorticoids in T cell development and function*. *Annu. Rev. Immunol.* **2000**, *18*, 309–345. [CrossRef] [PubMed]
48. Kratschmar, D.V.; Calabrese, D.; Walsh, J.; Lister, A.; Birk, J.; Appenzeller-Herzog, C.; Moulin, P.; Goldring, C.E.; Odermatt, A. Suppression of the Nrf2-dependent antioxidant response by glucocorticoids and 11β-HSD1-mediated glucocorticoid activation in hepatic cells. *PLoS ONE* **2012**, *7*, e36774. [CrossRef] [PubMed]
49. Alam, M.M.; Okazaki, K.; Nguyen, L.T.T.; Ota, N.; Kitamura, H.; Murakami, S.; Shima, H.; Igarashi, K.; Sekine, H.; Motohashi, H. Glucocorticoid receptor signaling represses the antioxidant response by inhibiting histone acetylation mediated by the transcriptional activator NRF2. *J. Biol. Chem.* **2017**, *292*, 7519–7530. [CrossRef] [PubMed]
50. Brody, S.; Preut, R.; Schommer, K.; Schürmeyer, T.H. A randomized controlled trial of high dose ascorbic acid for reduction of blood pressure, cortisol, and subjective responses to psychological stress. *Psychopharmacology* **2002**, *159*, 319–324. [CrossRef] [PubMed]
51. Mikkelsen, S.U.; Gillberg, L.; Lykkesfeldt, J.; Grønbæk, K. The role of vitamin C in epigenetic cancer therapy. *Free Radic. Biol. Med.* **2021**, *170*, 179–193. [CrossRef] [PubMed]

antioxidants

MDPI

Review

Vitamin C-Dependent Uptake of Non-Heme Iron by Enterocytes, Its Impact on Erythropoiesis and Redox Capacity of Human Erythrocytes

Xia Pan [1], Martin Köberle [2] and Mehrdad Ghashghaeinia [1,*]

1 Physiological Institute, Department of Vegetative and Clinical Physiology, Eberhard Karls University of Tübingen, 72074 Tübingen, Germany
2 Department of Dermatology and Allergology, School of Medicine and Health, Technical University of Munich, Biedersteinerstr. 29, 80802 München, Germany
* Correspondence: m.ghashghaeinia@uni-tuebingen.de

Abstract: In the small intestine, nutrients from ingested food are absorbed and broken down by enterocytes, which constitute over 95% of the intestinal epithelium. Enterocytes demonstrate diet- and segment-dependent metabolic flexibility, enabling them to take up large amounts of glutamine and glucose to meet their energy needs and transfer these nutrients into the bloodstream. During glycolysis, ATP, lactate, and H^+ ions are produced within the enterocytes. Based on extensive but incomplete glutamine oxidation large amounts of alanine or lactate are produced. Lactate, in turn, promotes hypoxia-inducible factor-1α (Hif-1α) activation and Hif-1α-dependent transcription of various proton channels and exchangers, which extrude cytoplasmic H^+-ions into the intestinal lumen. In parallel, the vitamin C-dependent and duodenal cytochrome b-mediated conversion of ferric iron into ferrous iron progresses. Finally, the generated electrochemical gradient is utilized by the divalent metal transporter 1 for H^+-coupled uptake of non-heme Fe^{2+}-ions. Iron efflux from enterocytes, subsequent binding to the plasma protein transferrin, and systemic distribution supply a wide range of cells with iron, including erythroid precursors essential for erythropoiesis. In this review, we discuss the impact of vitamin C on the redox capacity of human erythrocytes and connect enterocyte function with iron metabolism, highlighting its effects on erythropoiesis.

Keywords: cationic amphiphilic drugs; desipramine; pregnancy; enterocytes; erythrocytes; testosterone; hepcidin; folates; erythropoietin

Citation: Pan, X.; Köberle, M.; Ghashghaeinia, M. Vitamin C-Dependent Uptake of Non-Heme Iron by Enterocytes, Its Impact on Erythropoiesis and Redox Capacity of Human Erythrocytes. *Antioxidants* **2024**, *13*, 968. https://doi.org/10.3390/antiox13080968

Academic Editor: Angelo D'Alessandro

Received: 25 July 2024
Revised: 6 August 2024
Accepted: 7 August 2024
Published: 9 August 2024

1. Introduction

A considerable proportion of the cell volume of organelle-free mature human erythrocytes (hRBC) consists of hemoglobin, the molecule responsible for respiratory gas exchange of CO_2, O_2, and CO. Heme is an iron-containing heterocyclic molecule with an iron ion (Fe^{2+}) at its center. Iron has the ability to reversibly switch between its two most common oxidation states: ferrous (Fe^{2+}) and ferric (Fe^{3+}) forms. Both CO and O_2 solely bind to Fe^{2+}, whereas NO binds to both Fe^{2+} and Fe^{3+}. Methemoglobin (Hb-Fe^{3+}), produced during the auto-oxidation of oxyhemoglobin (HbFe^{2+}-O_2), has an essential physiological function. hRBCs as well as diverse cell types in our body (e.g., hepatocytes and astrocytes), produce the anti-inflammatory signaling molecule hydrogen sulfide (H_2S) [1–3]. hRBCs with their methemoglobin, are significantly involved in the degradation of H_2S and thus regulate H_2S levels in blood and tissues [4]. A remarkable property of hydrophobic and cell membrane permeable H_2S is its carrier- and facilitator-independent transmembrane diffusion [5]. hRBCs are versatile, polyfunctional, and highly complex. They interact with endogenous cells (platelets and lymphocytes) as well as with pathogens (bacteria and viruses). Both complement receptor 1 (CR-1) and glycophorin A (GPA) mediate the attachment of hRBCs to bacteria and viruses [6], respectively, leading to phagocytosis and final elimination of RBCs-bound bacteria and viruses in the liver and spleen [7]. hRBCs also promote the proliferation of activated CD8+ T cells [8] and actively absorb infectious HIV-1 virions [9]. Thus, RBCs considerably relieve our immune system. hRBCs can rapidly and reversibly bind an array of chemokines, including IL-8, a leukocyte chemotaxin [10–12]. As a result, excessive stimulation of leukocytes and uncontrolled inflammatory responses are avoided. Hemoglobin α and β chains bind LPS (lipopolysaccharide, an endotoxin of Gram-negative bacteria) and neutralize its activity [13,14]. Functioning hRBCs and their production (erythropoiesis), however, require a functioning gastrointestinal tract that makes the components of the ingested food, such as ions and vitamins, available systemically. In this review, we connect vitamin C-dependent uptake of non-heme iron in enterocytes, highlighting their systemic effects on erythropoiesis (Figure 1).

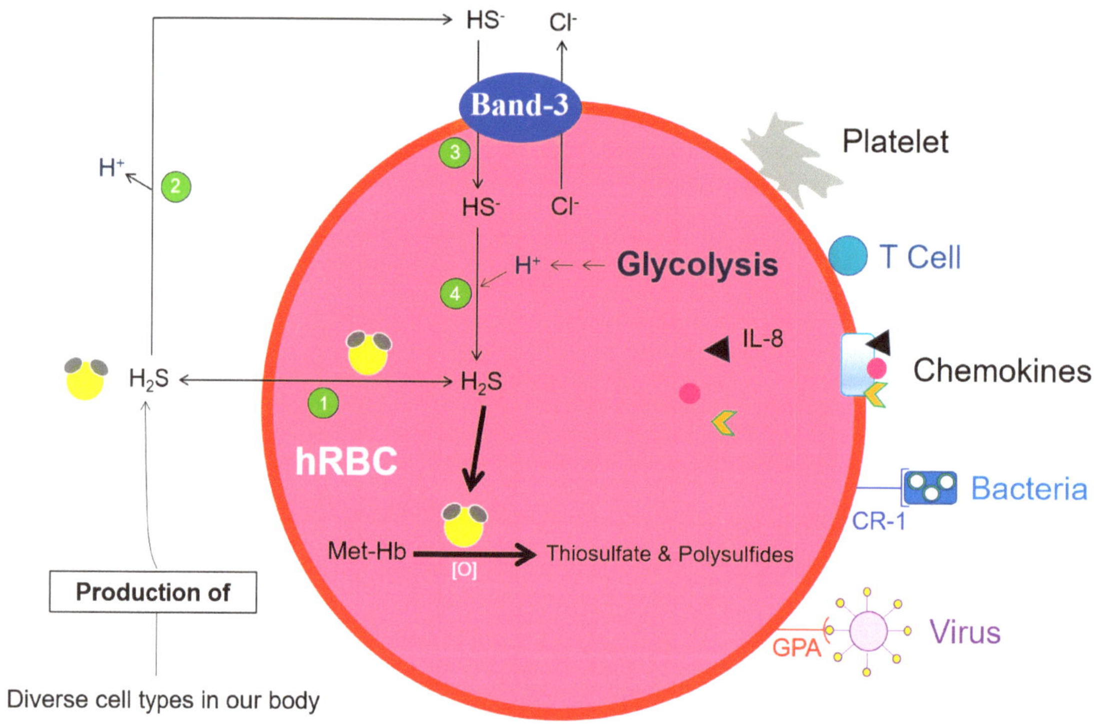

Figure 1. Transmembrane H_2S diffusion and Band-3 mediated Cl^-/HS^- exchange in hRBCs. Methemoglobin (Hb-Fe^{3+})-mediated H_2S degradation ensures the maintenance of physiological plasma and tissue concentration of free H_2S. The $Cl^-/HS^-/H_2S$ cycle is also efficiently involved in net acid (H^+-ions) efflux; for more details, see the following review [15]. For interactions of hRBCs with endogenous cells and pathogens, see the main text.

2. pH-Dependent Solubility and Uptake of Vitamin C and Dietary Non-Heme Iron

The proper uptake, storage, systemic distribution, and utilization of iron are prerequisites for general health and well-being. For hemoproteins formation (e.g., hemoglobin [16] and ascorbic acid-dependent transmembrane ferrireductases of the cytochrome b_{561} class [17]), various body tissues store iron in the cytosolic protein complex ferritin. Macrophages of the spleen, liver, bone marrow, and skeletal muscle are further prominent storage sites for iron. Regarding one electron transport capacity, iron shows mixed valence states, designated as ferrous (Fe^{2+}) and ferric (Fe^{3+}) forms. Freely available ferrous iron becomes cytotoxic in the presence of the respiratory by-product hydrogen peroxide (H_2O_2) through highly reactive hydroxyl radical formation via Fenton's reaction [18]. The environment of the digestive tract, especially of the duodenum is complex and adaptable. Following ingestion and upon entry of food into the duodenum, a dual and coincident stimulation of the external secretory functions of the liver and pancreas occurs, i.e., the flow of bile and pancreatic juice into the duodenum. Enterocytes covering the lumen of the intestinal mucosa can absorb ions, nutrients, vitamins, hormones, and water and transfer them to the blood. These cells non-competitively absorb both forms of dietary iron: heme and non-heme iron. Heme, released by hydrolysis of hemoproteins through intraluminal proteases and maintained in a soluble form by globin breakdown products, is absorbed intact by the enterocytes [19,20]. Afterwards, microsomal heme oxygenase-catalyzed heme degradation releases inorganic iron [21]. The latter is then either stored in ferritin molecules or transported to the basolateral membrane of the enterocytes for subsequent release into the blood.

The following equation shows the mucosal heme-splitting activity:

$$\text{Heme} + 3\,O_2 + 3^{1/2}\ \text{NADPH} + 3^{1/2}\ H^+ + 7\,e^- \rightarrow \text{biliverdin} + CO + 3^{1/2}\ \text{NADP}^+ + 3H_2O + Fe^{2+}.$$

However, iron intake does not directly reflect iron bioavailability. Uptake by enterocytes of dietary non-heme iron, with ferric iron (Fe^{3+}) being the most prevalent, is vitamin C-dependent. Ascorbic acid (AA) pH-dependently exerts both a reducing and a chelating effect on iron salts [22]. Both AA and ferric chloride are totally soluble in the acidic milieu of the stomach. This acidic pH causes the displacement of hydrogen ions from AA to ferric iron, leading to AA-iron chelate formation, which remains in solution over a pH range of 2 to 11. This iron chelate can thus mainly be absorbed at the slightly acidic pH of the duodenum [22,23]. In contrast to iron in heme complex, the uptake of non-heme iron is strongly regulated by dietary constituents.

3. Roles of Copper Ion (Cu^+), Regulatory Proteins, and Vitamin C in Non-Heme Iron Absorption across Human Enterocytes

Two carrier systems accomplish heme-bound iron absorption: (1) heme-carrier protein 1 (HCP-1), a primarily H^+-coupled folate transporter, and (2) receptor-mediated endocytosis. The apical influx of non-heme iron, especially ferrous iron (Fe^{2+}) into the human enterocytes, its basolateral efflux, and re-oxidation to ferric iron (Fe^{3+}), engages several regulatory and transporter proteins including their coordinated interactions. A sequence of steps is required prior to loading of monomeric plasma protein transferrin with two ferric iron ions. Both transmembrane proteins duodenal cytochrome B (Dcytb) and divalent metal transporter 1A-I (DMT1), with the former being ascorbic acid (AA)-reducible [17,24–26], are highly abundant in the brush-border membrane of duodenal enterocytes. The mammalian di-heme-containing [25,27] and iron-regulated ferric reductase Dcytb reduces ferric iron (Fe^{3+}) to ferrous iron (Fe^{2+}) prior to its transport by DMT1 [28]. The secondary active transporter DMT1, displays a pH dependence and, in an acidic environment operates as an H^+/Fe^{2+} cotransporter [29–31]. This leads, in the case of enterocytes located at the proximal duodenum (pH 6.0) to rapid intracellular acidification (Figure 2).

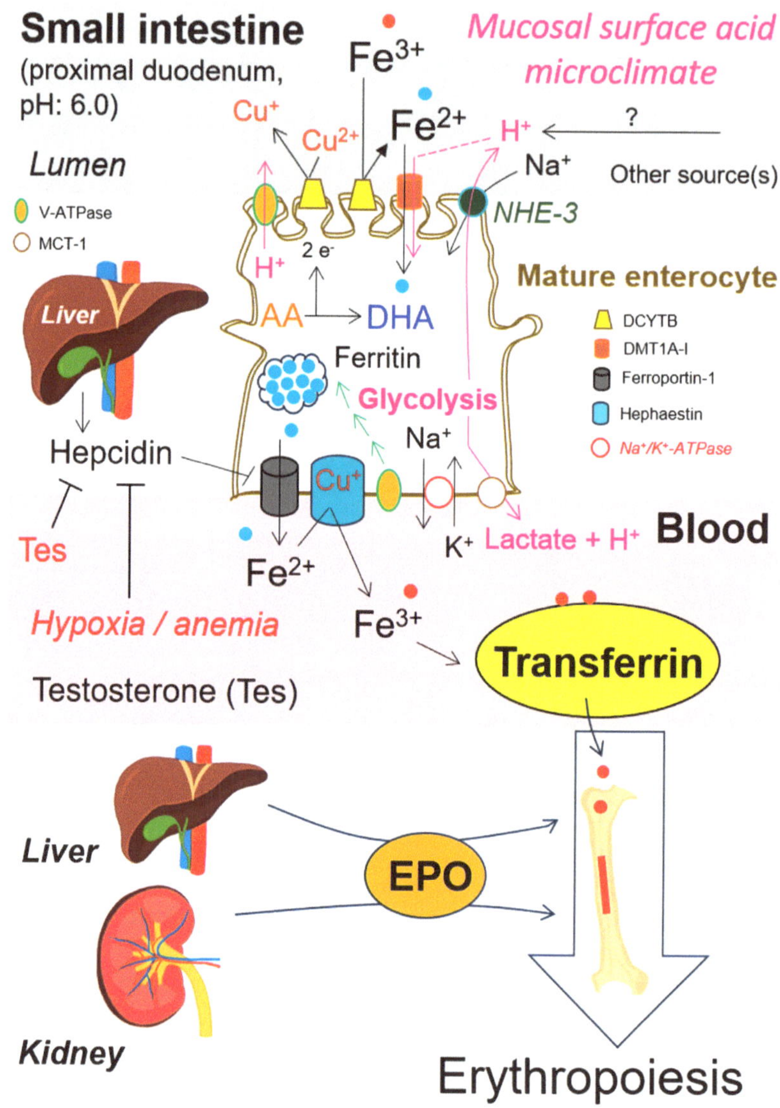

Figure 2. Uptake and systemic circulation of non-heme iron require several carriers located in the cell membrane of human enterocytes, plasma protein transferrin, and transferrin receptor. Erythropoiesis requires liver- and kidney-dependent production of erythropoietin.

4. Intracellular Enterocytes' Lactate Production by Glutamine and Glucose Metabolism

Enterocytes absorb large amounts of glutamine, glucose, and ketone bodies to cover their energy needs. However, they show substrate preference for oxidative metabolism which can be altered by the availability of other substrates. Both villus and crypt cells have mitochondrial glutaminase activity. Glutamate generated by glutaminase can be transaminated to produce alanine, aspartate, and α-ketoglutarate. The latter an intermediate in the Krebs cycle, contributes to ATP production. The intestinal mucosa consisting of 75% non-lymphoid and 25% lymphoid tissues by mass [32], plays an important role in mucosal immunity [33–36]. Both cell types—enterocytes and intraepithelial immune cells—residing in this area, utilize glutamine at high and comparable intensity [32,37]. Based on extensive but incomplete glutamine oxidation in the intestinal mucosa, large amounts of glutamine undergo two steps of decarboxylation with the final product being either alanine or lactate depending on the pyruvate pool [38,39]. It is important to mention the gastrointestinal pH profile of healthy subjects. The stomach has a pH of 1.3 to 2. In the small intestine, consisting of three successive sections (duodenum, jejunum, and ileum), the pH values increase. These are about 6.0 in the proximal part of the duodenum, 7.0 at the duodenojejunal junction, and 7.4 in the terminal ileum [40,41]. Figure 3A shows the inverse correlation between DMT1 abundance and pH along the small intestine. Figure 3B illustrates the uptake of glucose in human enterocytes and its utilization by the glycolysis pathway which culminates in the production of two ATP, lactate and H^+-ions each. Glycolysis is more intense in enterocytes of the proximal than in the distal intestine. Thus, glutamine and glucose metabolism are mainly involved in intracellular enterocyte lactate production.

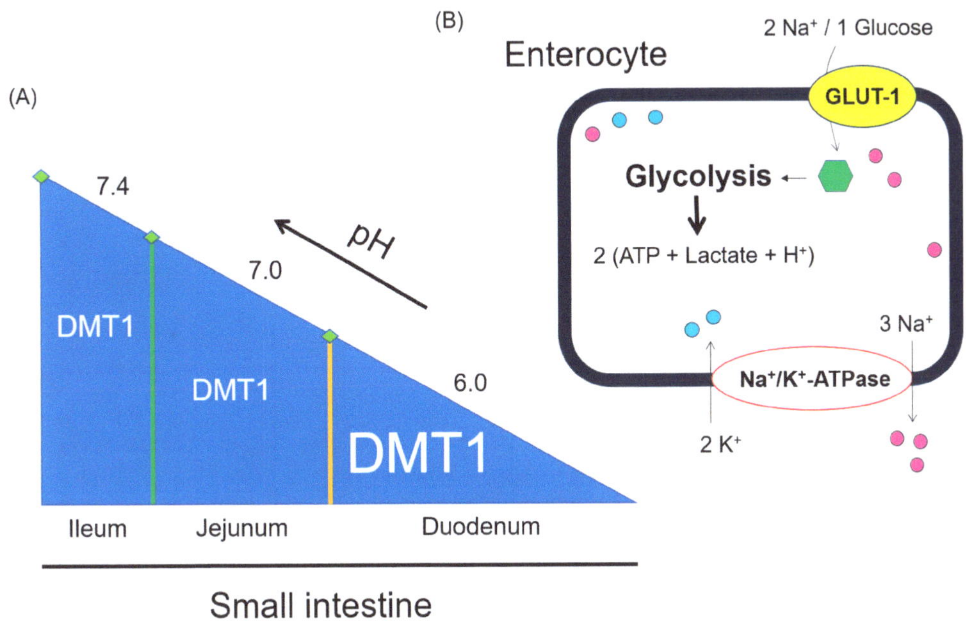

Figure 3. (A) Inverse correlation between DMT1 abundance and pH along the small intestinal. (B) GLUT-1-dependent influx of glucose into human enterocytes. DMT1: divalent metal transporter 1; GLUT-1: glucose transporter-1.

5. Lactate-Induced Activation of Hypoxia-Inducible Factor 1-Alpha (Hif-1α) in Enterocytes and Other Cell Types

Glycolysis, which is independent of intact mitochondrial function, represents a positive selective pressure of evolution for obtaining energy (ATP) within the shortest time and is not restricted to enterocytes. Several healthy cells, e.g., human cytotoxic $CD8^+$ T lymphocytes [42] and murine embryonic stem cells [43] temporarily exhibit a significant increase in glycolysis rate during activation and proliferation. This also applies to growing and proliferating tumor cells. However, for the continuation of glycolysis, the extrusion of lactate and H^+-ions into the extracellular environment is mandatory [44]. This is performed by H^+-linked monocarboxylate transporters (MCTs) [45]. The generated lactate also activates and stabilizes heme-containing hypoxia-inducible factor 1-α (Hif-1α) [46–48], a master regulator of glycolysis and oxygen homeostasis [49]. Hif-1α, discovered by Goldberg et al. [50], in turn, activates the transcription of numerous genes encoding MCT1, MCT4, Na^+/H^+-antiporter 1 (NHE-1) [51], vacuolar-type proton pump ATPase (V-ATPase), inducible nitric oxide synthase (i-NOS), heme oxygenase 1 (HO-1), transferrin (Tf), erythropoietin (EPO), ecto-enzyme carbonic anhydrase IX and glucose transporters 1 and 3 (GLUT-1 & -3), of which some are important contributors to erythropoiesis. For lactate-dependent and Hif-1α-mediated control of pH regulating pathways see the following review: [52].

6. Hif-1α Mediated Control of pH Regulating Pathways and Their Interplay with Divalent Metal Transporter 1 (DMT1) for Non-Heme Iron Transport

Intracellular pH $[pH]_i$ homeostasis is vital to the functioning of cells. Several ion transport mechanisms are involved in this process, e.g., exchangers (NHEs), proton (H^+) pumps (V-ATPases), and H^+-MCTs co-transporters, resulting in an alkaline shift in $[pH]_i$. As described above in Section 3, DMT1 displays pH dependence and, in an acidic environment (for instance, in the duodenum), operates as an H^+/Fe^{2+} cotransporter, leading to rapid intracellular acidification of enterocytes. To counteract this alteration of $[pH]_i$, reciprocal NHE-3/Na^+-K^+-ATPase interplay (concerning Na^+-ions translocation) as well as the contribution of V-ATPase are needed to apically efflux cytoplasmic H^+-ions into the intestine lumen. This leads to the generation of an acidic microclimate at the brush border membrane of duodenal enterocytes and the formation of an H^+-electrochemical gradient. The latter is subsequently used by enterocytes for a DMT1-dependent and H^+-coupled uptake of Fe^{2+}-ions. The most important physiological role of Na^+-K^+-ATPase is to channel the free energy of ATP-hydrolysis to intracellularly keep K^+-ions at high and Na^+-ions at low concentrations. On the basolateral site of the enterocyte membrane, MCT-1 regulates the equimolar and electroneutral co-extrusion of lactate and H^+-ions into the circulation. For more details see Figure 2.

7. Iron Flux across Enterocytes Membrane, Its Release into the Blood and Distribution by Plasma Transferrin

For iron flux across enterocytes, they require not only apically located influx carriers Dcytb and DMT1 but also basolaterally located transmembrane efflux proteins ferroportin-1 (FPN1) [53] and Cu^{1+}-dependent ferroxidase hephaestin [54–62]. Fe^{2+} exported by FPN1 is rapidly oxidized back to Fe^{3+} by hephaestin. Subsequently, Fe^{3+}-binding plasma protein transferrin (Tf), responsible for systemic iron circulation—Tf-$(Fe^{3+})_2$—supplies a wide range of cells with iron, including erythroid precursors essential for erythropoiesis [63,64] and Figure 2.

8. Roles of Folates, Vitamin B12, Ferrous Iron (Fe^{2+}), Erythropoietin, Testosterone and Hepcidin in Erythropoiesis

The progressive differentiation of short-term hematopoietic stem cells in the bone marrow leads, among other things, to the formation of an erythroid lineage from which terminally differentiated hemoglobin-containing hRBCs arise. This dynamic production process of erythrocytes, referred to as erythropoiesis [65–68], requires an adequate supply of folates, vitamin B12 (cobalamin), and ferrous iron (Fe^{2+}). Deficiency in one or more of these substances results in nutrition-related anemia. Folates are primarily absorbed in the duodenum and jejunum [69,70], whereas intrinsic factor-bound vitamin B12 (vit B12) is mainly absorbed in the terminal ileum [71–73]. Stomach acid (which is decreased in subjects with atrophic gastritis), its digestive enzymes (e.g., pepsin), and vit B12-binding glycoproteins haptocorrin and intrinsic factor positively regulate vit B12 absorption [71]. In adults, the kidney serves as the major site (peritubular fibroblasts in the renal cortex) [74–77], and the liver to a much lesser extent (hepatocytes and perisinusoidal Ito cells) [78–80] produces the circulating plasma protein hormone erythropoietin (EPO), the principal regulator of erythropoiesis. For reviews see [81–83]. Testosterone suppresses hepcidin [84,85]. The treatment of hypogonadal and especially middle-aged and elderly men with testosterone increases hematocrit levels [86]. This might explain gender-based differences in hematocrit content. The hepatic peptide hormone hepcidin, first discovered by Park et al. 2001 and Pigeon et al. 2001 [87,88], is directly involved in the maintenance of iron homeostasis, and its regulation is tightly controlled at the transcriptional level [89]. Hepcidin synthesis and release from the liver are positively correlated with inflammation and increased plasma and tissue iron levels. Under iron overload, hepcidin binds to its receptor FPN-1, leading to FPN-1 internalization [90,91], ubiquitination, and subsequent lysosomal degradation [91]. Thus, hepcidin-mediated FPN1-downregulation leads to diminished iron efflux, resulting in intracellular iron retention in iron-releasing target cells, e.g., hepatocytes, tissue macrophages, duodenal enterocytes, and placental cells; see also Figure 2. If persistent, this condition impairs iron-dependent erythropoiesis, as systemic iron levels decrease. The following is also of physiological importance: anemia and hypoxia significantly inhibit hepatocellular hepcidin gene expression [92]. Thus, the sophisticated interplay between plasma and tissue iron, testosterone, EPO, hepcidin, anemia, and hypoxia might be understood as a homeostatic loop to maintain the dynamic balance between iron deficiency and overload.

9. Glutathione and NADH-Dependent Vitamin C Reduction, Essential Contributors to Maintaining the Redox Capacity of hRBCs

In mammals, the plasma concentration of the antioxidant glutathione (GSH), a tripeptide with the structure γ-L-glutamyl-L-cysteinylglycine, is about 25 μM, with typical intracellular concentrations between 1 and 5 mM. The low micromolar plasma concentration of ascorbic acid (AA) is slightly higher than that of GSH, i.e., 40–60 μM. Intraerythrocytic concentrations of GSH and AA are considerably high and amount to 1–2 mM each. AA and DHA uptake are carrier-mediated: the former Na^+-dependent and the latter Na^+-independent, carried out by sodium-dependent vitamin C transporters (SVCTs) [93–96] and members of the glucose-transporter family (GLUT-1, -3 and -4) [94,97–100], respectively. The driving force for such substrate movements is the protein carrier-mediated secondary active transport with net accumulation of substrate on the other side of the membrane (here: outside → inside direction). Primarily, the mammalian Na^+/K^+-ATPase, discovered in 1965 [101], utilizes the glycolytically produced ATP to generate Na^+/K^+ asymmetry between cells and their surroundings, i.e., low Na^+/high K^+ content within the cell. This gradient is then used to drive diverse secondary active transports [102]. In contrast to nucleated cells, AA is a poor substrate for hRBCs. GLUT-1 transports DHA into hRBCs [103–105]. Once within the cells, GSH-dependent two-electron regeneration of AA occurs (DHA + 2 GSH → AA + GSSG), i.e., without involving the monoascorbyl free radical (AFR) intermediate [106–108]. This GSH-dependent reduction in DHA is not solely restricted to hRBCs [109]. GSH and DHA are interconnected and form a functional unit. The rapid entry of DHA inflicts on cells a high need for GSH for its reduction to AA. In this context, DHA stimulates the NADPH-generating pentose phosphate pathway (PPP) [110,111]. Subsequently, glutathione reductase (GR) catalyzes NADPH-dependent glutathione disulfide reduction (GSSG + NADPH + H^+ → 2 GSH + $NADP^+$). This self-supporting machinery, as a positive feedback loop, ensures permanent AA regeneration and accumulation within the hRBCs. The subsequent intracellular consumption and extracellular transport of AA (inside → outside direction) and the re-entry of its two-electron oxidized form DHA back into the cells (outside → inside direction) lead to the maintenance of a large intracellular electron pool. This culminates in the vitamin C-dependent high redox capacity of hRBCs; see also Figure 4. NADH generated during glycolysis represents another endogenous source for DHA reduction [112]. The groundbreaking discovery of this group was that a mixture of each mole of AA with 2 moles of ferricyanide instantly resulted in the generation of one mole DHA and two moles of forrocyanide, i.e., without involving the AFR intermediate (AA + 2 ferricyanide → DHA + 2 ferrocyanide) (see Figure 4).

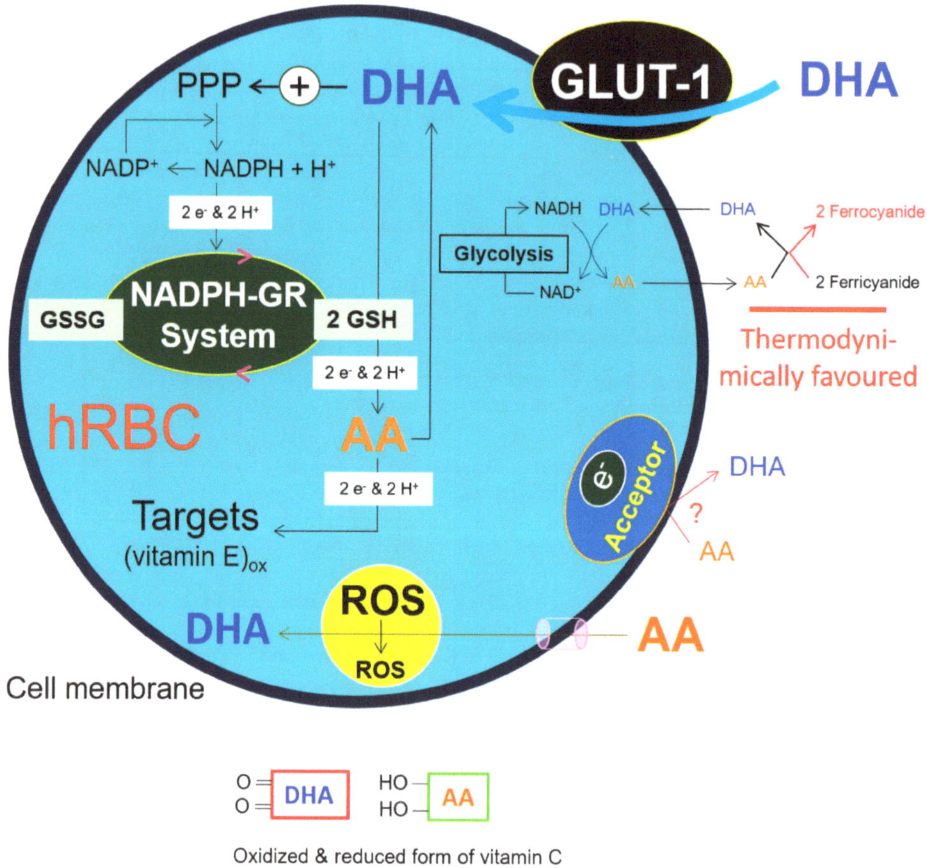

Figure 4. GLUT-1-mediated influx of oxidized form of vitamin C (DHA) into the mature hRBCs. The interplay between DHA, PPP, GSH, AA, and the subsequent reduction of vitamin E prevents lipid peroxidation. As a result, cell membrane integrity is maintained and in vivo hemolysis of erythrocytes is minimized. Recently, a link between iron metabolism, lipid peroxidation, and hemolysis was found in stored human and mice erythrocytes [113,114]. Vitamin E is located inside the cell membrane.

10. Impact of V-ATPase on Endosomal pH and DMT1B-II Mediated Iron (Fe^{2+}) Release into the Cytosol, and Its Relevance for Erythropoiesis

In previous sections, we summarized the apical uptake and basolateral efflux of iron across the cell membrane of enterocytes. Now, we will address its subsequent distribution into the blood and utilization in different target cells using the example of erythroid precursors. With almost 30 trillion eukaryotic cells in our body, RBCs, with ~25 trillion, represent nearly 84% of the total cells [115]. Circulating hRBCs have a lifespan of approximately 120 days [116]. Daily, about 1% (~250 billion) of senescent hRBCs are engulfed and degraded by macrophages and replaced to the same extent through erythropoiesis, i.e., two million RBCs are produced per second. A single intact hRBC contains over 270 million hemoglobin molecules. Hemoglobin (Hb) is an iron-porphyrin protein complex consisting of four polypeptide chains, each having a heme prosthetic group with a ferrous iron (Fe^{2+}) at its center. Thus, a single hRBC possesses 1.1 billion heme groups or 1.1 billion Fe^{2+} ions. In other words, humans acquire the major part of body iron by catabolizing Hb obtained from senescent RBCs. Dietary iron absorbed in the small intestine and excess iron stored within ferritin in the liver hepatocytes are also available to most body cells. The diferric transferrin—Tf-(Fe^{3+})$_2$—is the key iron transport machinery for heme biosynthesis in erythroid precursors. It binds to the transferrin receptor (Tf-R), which clusters in specialized areas of the cell surface, called 'coated pits'. Coated pits, whose assembly is potassium dependent [117], are pooled into the cytosol to rapidly form coated endocytic vesicles. The latter lose the majority of their coat proteins and are then referred to as primary endosomes (Figure 5). Membrane-embedded V-ATPases are ATP-dependent proton (H^+)-pumps. They are present in both endomembrane organelles and cell membrane and lead to alkalinization of the cytoplasm associated with acidification of both extracellular milieu (Figure 2) and intracellular compartments, e.g., endosomal and lysosomal lumen [118–120] (for details, see Figure 5). In the acidified endosome, ferric iron (Fe^{3+}) readily dissociates from Tf [121,122] and is subsequently reduced to ferrous iron (Fe^{2+}) by NADPH-dependent endosomal ferrireductase Steap3 (six-transmembrane epithelial antigen of the prostate) [123], prior to its transport into the cytosol by H^+-coupled endosomal DMT1B-II [124] (see also Figure 5). Cytosolic iron (Fe^{2+}) is the essential substrate for heme biosynthesis in erythroid precursors and erythropoiesis. Heme itself controls the synthesis of globin chains—needed for hemoglobin synthesis—at both transcriptional and translational levels. It is important to note that genetic ablation of Steap3 leads to severe hypochromic [125] and microcytic anemia [126].

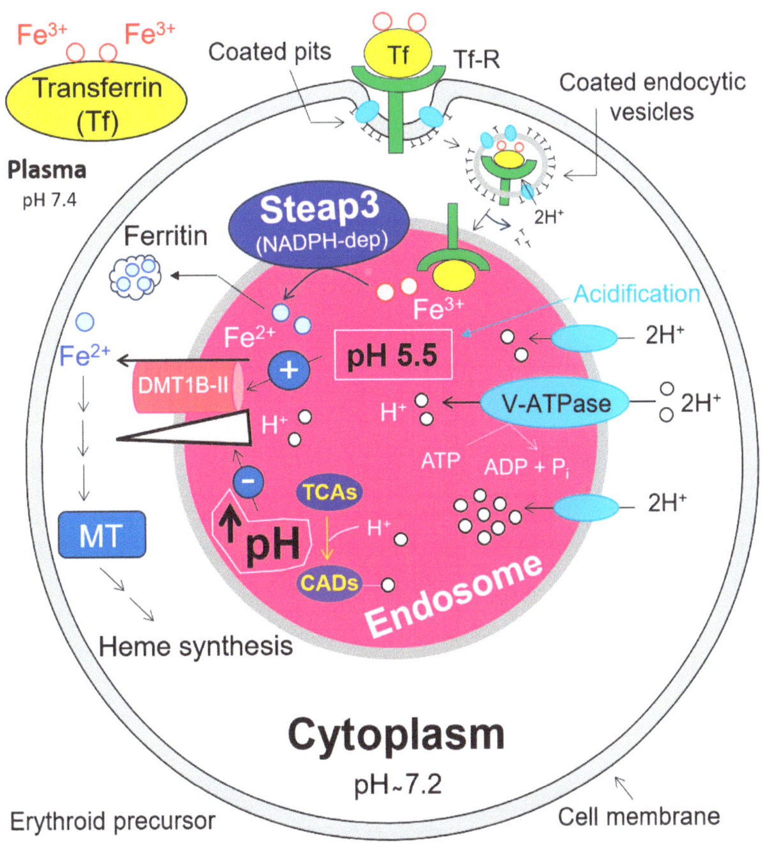

Figure 5. Iron transport into erythroid precursors. This process comprises the endocytosis of Transferrin-bound iron, DMT1B-II-mediated Fe^{2+} export from acidified endosomes into the cytoplasm. MT: mitochondria.

11. Disruption of Lysosomal pH by Tricyclic Antidepressant Desipramine, Its Possible Negative Effect on Endosomal pH, Iron Supply and Erythropoiesis: A High-Risk Drug during Pregnancy?

Tricyclic antidepressants (TCAs) influence norepinephrine (NE) and serotonin (SER) transporters [127]. Desipramine, a representative of TCAs, has two primary targets. On one hand, it preferentially interacts with the NE-transporter and increases NE synaptic transmission by inhibiting NE reuptake, thereby relieving depressive symptoms [128–131]. On the other hand, desipramine has a direct inhibitory effect on lysosomal acid ceramidase [132] and acid sphingomyelinase [133,134]. Both acid ceramidase [135–138] and acid sphingomyelinase [139–141] are aberrantly over-expressed and highly active in patients with dysregulated sphingolipid metabolism. Under acidic conditions, e.g., in lysosomes, endosomes, or in the cytoplasm of glycolytically active cells, TCAs act as cationic amphiphilic drugs (CADs) (TCAs + $H^+ \rightarrow$ CADs), their reactions resembling the formation of ammonium from ammonia and a proton ($NH_3 + H^+ \rightarrow NH_4^+$). The secondary amine and basic lipophilic drug desipramine follow the same principle: it acts as a proton (H^+-ion) acceptor, depleting the free proton and thus increasing the intracellular pH. Elojeimy et al. (2006) showed that in cancer cell lines, desipramine, even at a relatively low dose of 5 μM, neutralizes lysosomal pH [132]. By the same principle, desipramine and other CADs like imipramine, amitriptyline, chlorpromazine, and chloroquine would be able to neutralize the luminal acidification of endosomes in glycolytically active erythroid precursors. Consequently, H^+-coupled Fe^{2+} transport into the cytosol of these cells, and thus the proper heme biosynthesis and heme-dependent erythropoiesis, might be severely affected (see Figure 5).

In addition to this, desipramine inhibits NHE-1 activity [142], a major regulator of intracellular pH. As shown in Figure 2, NHE-3 plays an essential role in H^+-coupled Fe^{2+} transport by DMT1 in enterocytes. To avoid anemia and preserve the naturally increased erythropoietic activity during pregnancy, pregnant women suffering from depression should avoid medications with CAD properties if possible. However, this does not diminish the importance of CAD's clinical applicability regarding cancer [143–145] and anti-viral [146,147] therapies. It is to be noted that the above-mentioned low-dose concentration of desipramine (5 μM) has no significant detectable biological effects on cell organelle-free mature hRBCs [148].

Author Contributions: M.G. designed this project. M.G. mainly and X.P. partly wrote the manuscript. X.P. and M.G. searched the literature and selected the references. M.G. made Figures 1 and 2. M.K. and M.G. made Figure 3. X.P. and M.G. made Figures 4 and 5. All authors have read and agreed to the published version of the manuscript.

Funding: This research received no external funding.

Institutional Review Board Statement: Not applicable.

Informed Consent Statement: Not applicable.

Data Availability Statement: Not applicable.

Acknowledgments: We acknowledge support from the Open Access Publication Fund of the University of Tübingen.

Conflicts of Interest: The authors declare that no competing financial interests or otherwise exist.

References

1. Stipanuk, M.H.; Beck, P.W. Characterization of the enzymic capacity for cysteine desulphhydration in liver and kidney of the rat. *Biochem. J.* **1982**, *206*, 267–277. [CrossRef] [PubMed]
2. Shibuya, N.; Mikami, Y.; Kimura, Y.; Nagahara, N.; Kimura, H. Vascular endothelium expresses 3-mercaptopyruvate sulfurtransferase and produces hydrogen sulfide. *J. Biochem.* **2009**, *146*, 623–626. [CrossRef] [PubMed]
3. Tanizawa, K. Production of H2S by 3-mercaptopyruvate sulphurtransferase. *J. Biochem.* **2011**, *149*, 357–359. [CrossRef] [PubMed]
4. Vitvitsky, V.; Yadav, P.K.; Kurthen, A.; Banerjee, R. Sulfide oxidation by a noncanonical pathway in red blood cells generates thiosulfate and polysulfides. *J. Biol. Chem.* **2015**, *290*, 8310–8320. [CrossRef] [PubMed]

5. Mathai, J.C.; Missner, A.; Kugler, P.; Saparov, S.M.; Zeidel, M.L.; Lee, J.K.; Pohl, P. No facilitator required for membrane transport of hydrogen sulfide. *Proc. Natl. Acad. Sci. USA* **2009**, *106*, 16633–16638. [CrossRef] [PubMed]
6. Allaway, G.P.; Burness, A.T. Site of attachment of encephalomyocarditis virus on human erythrocytes. *J. Virol.* **1986**, *59*, 768–770. [CrossRef] [PubMed]
7. Craig, M.L.; Bankovich, A.J.; Taylor, R.P. Visualization of the transfer reaction: Tracking immune complexes from erythrocyte complement receptor 1 to macrophages. *Clin. Immunol.* **2002**, *105*, 36–47. [CrossRef] [PubMed]
8. Fonseca, A.M.; Pereira, C.F.; Porto, G.; Arosa, F.A. Red blood cells promote survival and cell cycle progression of human peripheral blood T cells independently of CD58/LFA-3 and heme compounds. *Cell Immunol.* **2003**, *224*, 17–28. [CrossRef] [PubMed]
9. Beck, Z.; Brown, B.K.; Wieczorek, L.; Peachman, K.K.; Matyas, G.R.; Polonis, V.R.; Rao, M.; Alving, C.R. Human erythrocytes selectively bind and enrich infectious HIV-1 virions. *PLoS ONE* **2009**, *4*, e8297. [CrossRef]
10. Darbonne, W.C.; Rice, G.C.; Mohler, M.A.; Apple, T.; Hebert, C.A.; Valente, A.J.; Baker, J.B. Red blood cells are a sink for interleukin 8, a leukocyte chemotaxin. *J. Clin. Investig.* **1991**, *88*, 1362–1369. [CrossRef]
11. Horuk, R.; Colby, T.J.; Darbonne, W.C.; Schall, T.J.; Neote, K. The human erythrocyte inflammatory peptide (chemokine) receptor. Biochemical characterization, solubilization, and development of a binding assay for the soluble receptor. *Biochemistry* **1993**, *32*, 5733–5738. [CrossRef]
12. de Winter, R.J.; Manten, A.; de Jong, Y.P.; Adams, R.; van Deventer, S.J.; Lie, K.I. Interleukin 8 released after acute myocardial infarction is mainly bound to erythrocytes. *Heart* **1997**, *78*, 598–602. [CrossRef] [PubMed]
13. Bahl, N.; Du, R.; Winarsih, I.; Ho, B.; Tucker-Kellogg, L.; Tidor, B.; Ding, J.L. Delineation of lipopolysaccharide (LPS)-binding sites on hemoglobin: From in silico predictions to biophysical characterization. *J. Biol. Chem.* **2011**, *286*, 37793–37803. [CrossRef] [PubMed]
14. Liepke, C.; Baxmann, S.; Heine, C.; Breithaupt, N.; Standker, L.; Forssmann, W.G. Human hemoglobin-derived peptides exhibit antimicrobial activity: A class of host defense peptides. *J. Chromatogr. B Analyt. Technol. Biomed. Life Sci.* **2003**, *791*, 345–356. [CrossRef]
15. Ghashghaeinia, M.; Mrowietz, U. Human erythrocytes, nuclear factor kappaB (NFkappaB) and hydrogen sulfide (H(2)S)—From non-genomic to genomic research. *Cell Cycle* **2021**, *20*, 2091–2101. [CrossRef]
16. Perutz, M.F. Submicroscopic structure of the red cell. *Nature* **1948**, *161*, 204. [CrossRef] [PubMed]
17. Su, D.; Asard, H. Three mammalian cytochromes b561 are ascorbate-dependent ferrireductases. *FEBS J.* **2006**, *273*, 3722–3734. [CrossRef]
18. Lloyd, R.V.; Hanna, P.M.; Mason, R.P. The origin of the hydroxyl radical oxygen in the Fenton reaction. *Free Radic. Biol. Med.* **1997**, *22*, 885–888. [CrossRef]
19. Turnbull, A.; Cleton, F.; Finch, C.A. Iron absorption. IV. The absorption of hemoglobin iron. *J. Clin. Investig.* **1962**, *41*, 1897–1907. [CrossRef]
20. Young, G.P.; Rose, I.S.; St John, D.J. Haem in the gut. I. Fate of haemoproteins and the absorption of haem. *J. Gastroenterol. Hepatol.* **1989**, *4*, 537–545. [CrossRef]
21. Raffin, S.B.; Woo, C.H.; Roost, K.T.; Price, D.C.; Schmid, R. Intestinal absorption of hemoglobin iron-heme cleavage by mucosal heme oxygenase. *J. Clin. Investig.* **1974**, *54*, 1344–1352. [CrossRef]
22. Conrad, M.E.; Schade, S.G. Ascorbic acid chelates in iron absorption: A role for hydrochloric acid and bile. *Gastroenterology* **1968**, *55*, 35–45. [CrossRef] [PubMed]
23. Lynch, S.R.; Cook, J.D. Interaction of vitamin C and iron. *Ann. N. Y. Acad. Sci.* **1980**, *355*, 32–44. [CrossRef] [PubMed]
24. May, J.M.; Qu, Z.C.; Whitesell, R.R. Ascorbate is the major electron donor for a transmembrane oxidoreductase of human erythrocytes. *Biochim. Biophys. Acta* **1995**, *1238*, 127–136. [CrossRef] [PubMed]
25. Oakhill, J.S.; Marritt, S.J.; Gareta, E.G.; Cammack, R.; McKie, A.T. Functional characterization of human duodenal cytochrome b (Cybrd1): Redox properties in relation to iron and ascorbate metabolism. *Biochim. Biophys. Acta* **2008**, *1777*, 260–268. [CrossRef] [PubMed]
26. Lane, D.J.; Bae, D.H.; Merlot, A.M.; Sahni, S.; Richardson, D.R. Duodenal cytochrome b (DCYTB) in iron metabolism: An update on function and regulation. *Nutrients* **2015**, *7*, 2274–2296. [CrossRef] [PubMed]
27. Ludwiczek, S.; Rosell, F.I.; Ludwiczek, M.L.; Mauk, A.G. Recombinant expression and initial characterization of the putative human enteric ferric reductase Dcytb. *Biochemistry* **2008**, *47*, 753–761. [CrossRef] [PubMed]
28. McKie, A.T.; Barrow, D.; Latunde-Dada, G.O.; Rolfs, A.; Sager, G.; Mudaly, E.; Mudaly, M.; Richardson, C.; Barlow, D.; Bomford, A.; et al. An iron-regulated ferric reductase associated with the absorption of dietary iron. *Science* **2001**, *291*, 1755–1759. [CrossRef]
29. Gunshin, H.; Mackenzie, B.; Berger, U.V.; Gunshin, Y.; Romero, M.F.; Boron, W.F.; Nussberger, S.; Gollan, J.L.; Hediger, M.A. Cloning and characterization of a mammalian proton-coupled metal-ion transporter. *Nature* **1997**, *388*, 482–488. [CrossRef]
30. Mackenzie, B.; Ujwal, M.L.; Chang, M.H.; Romero, M.F.; Hediger, M.A. Divalent metal-ion transporter DMT1 mediates both H+ -coupled Fe2+ transport and uncoupled fluxes. *Pflugers Arch.* **2006**, *451*, 544–558. [CrossRef]
31. Ehrnstorfer, I.A.; Manatschal, C.; Arnold, F.M.; Laederach, J.; Dutzler, R. Structural and mechanistic basis of proton-coupled metal ion transport in the SLC11/NRAMP family. *Nat. Commun.* **2017**, *8*, 14033. [CrossRef] [PubMed]
32. Newsholme, E.A.; Carrie, A.L. Quantitative aspects of glucose and glutamine metabolism by intestinal cells. *Gut* **1994**, *35*, S13–S17. [CrossRef] [PubMed]

33. Trejdosiewicz, L.K. What is the role of human intestinal intraepithelial lymphocytes? *Clin. Exp. Immunol.* **1993**, *94*, 395–397. [CrossRef] [PubMed]
34. Dahan, S.; Roth-Walter, F.; Arnaboldi, P.; Agarwal, S.; Mayer, L. Epithelia: Lymphocyte interactions in the gut. *Immunol. Rev.* **2007**, *215*, 243–253. [CrossRef] [PubMed]
35. Lutter, L.; Hoytema van Konijnenburg, D.P.; Brand, E.C.; Oldenburg, B.; van Wijk, F. The elusive case of human intraepithelial T cells in gut homeostasis and inflammation. *Nat. Rev. Gastroenterol. Hepatol.* **2018**, *15*, 637–649. [CrossRef] [PubMed]
36. Ma, H.; Tao, W.; Zhu, S. T lymphocytes in the intestinal mucosa: Defense and tolerance. *Cell Mol. Immunol.* **2019**, *16*, 216–224. [CrossRef] [PubMed]
37. Newsholme, E.A.; Crabtree, B.; Ardawi, M.S.M. Glutamine metabolism in lymphocytes: Its biochemical, physiological and clinical importance. *Q. J. Exp. Physiol. Transl. Integr.* **1985**, *70*, 473–489. [CrossRef] [PubMed]
38. Duée, P.-H.; Darcy-Vrillon, B.; Blachier, F.; Morel, M.-T. Fuel selection in intestinal cells. *Proc. Nutr. Soc.* **1995**, *54*, 83–94. [CrossRef]
39. Kight, C.E.; Fleming, S.E. Transamination processes promote incomplete glutamine oxidation in small intestine epithelial cells. *J. Nutr. Biochem.* **1995**, *6*, 27–37. [CrossRef]
40. Fallingborg, J. Intraluminal pH of the human gastrointestinal tract. *Dan. Med. Bull.* **1999**, *46*, 183–196.
41. Fallingborg, J.; Christensen, L.A.; Ingeman-Nielsen, M.; Jacobsen, B.A.; Abildgaard, K.; Rasmussen, H.H. pH-profile and regional transit times of the normal gut measured by a radiotelemetry device. *Aliment. Pharmacol. Ther.* **1989**, *3*, 605–613. [CrossRef] [PubMed]
42. Cham, C.M.; Gajewski, T.F. Glucose availability regulates IFN-gamma production and p70S6 kinase activation in CD8+ effector T cells. *J. Immunol.* **2005**, *174*, 4670–4677. [CrossRef] [PubMed]
43. Kondoh, H.; Lleonart, M.E.; Nakashima, Y.; Yokode, M.; Tanaka, M.; Bernard, D.; Gil, J.; Beach, D. A high glycolytic flux supports the proliferative potential of murine embryonic stem cells. *Antioxid. Redox Signal.* **2007**, *9*, 293–299. [CrossRef]
44. Spencer, T.L.; Lehninger, A.L. L-lactate transport in Ehrlich ascites-tumour cells. *Biochem. J.* **1976**, *154*, 405–414. [CrossRef] [PubMed]
45. Broer, S.; Rahman, B.; Pellegri, G.; Pellerin, L.; Martin, J.L.; Verleysdonk, S.; Hamprecht, B.; Magistretti, P.J. Comparison of lactate transport in astroglial cells and monocarboxylate transporter 1 (MCT 1) expressing Xenopus laevis oocytes. Expression of two different monocarboxylate transporters in astroglial cells and neurons. *J. Biol. Chem.* **1997**, *272*, 30096–30102. [CrossRef] [PubMed]
46. Kozlov, A.M.; Lone, A.; Betts, D.H.; Cumming, R.C. Lactate preconditioning promotes a HIF-1alpha-mediated metabolic shift from OXPHOS to glycolysis in normal human diploid fibroblasts. *Sci. Rep.* **2020**, *10*, 8388. [CrossRef] [PubMed]
47. Sonveaux, P.; Copetti, T.; De Saedeleer, C.J.; Vegran, F.; Verrax, J.; Kennedy, K.M.; Moon, E.J.; Dhup, S.; Danhier, P.; Frerart, F.; et al. Targeting the lactate transporter MCT1 in endothelial cells inhibits lactate-induced HIF-1 activation and tumor angiogenesis. *PLoS ONE* **2012**, *7*, e33418. [CrossRef] [PubMed]
48. Wu, Y.; Wang, M.; Feng, H.; Peng, Y.; Sun, J.; Qu, X.; Li, C. Lactate induces osteoblast differentiation by stabilization of HIF1alpha. *Mol. Cell Endocrinol.* **2017**, *452*, 84–92. [CrossRef] [PubMed]
49. Semenza, G.L. Hypoxia-inducible factor 1: Master regulator of O2 homeostasis. *Curr. Opin. Genet. Dev.* **1998**, *8*, 588–594. [CrossRef]
50. Goldberg, M.A.; Dunning, S.P.; Bunn, H.F. Regulation of the erythropoietin gene: Evidence that the oxygen sensor is a heme protein. *Science* **1988**, *242*, 1412–1415. [CrossRef]
51. Shimoda, L.A.; Fallon, M.; Pisarcik, S.; Wang, J.; Semenza, G.L. HIF-1 regulates hypoxic induction of NHE1 expression and alkalinization of intracellular pH in pulmonary arterial myocytes. *Am. J. Physiol. Lung Cell Mol. Physiol.* **2006**, *291*, L941–L949. [CrossRef] [PubMed]
52. Ghashghaeinia, M.; Koberle, M.; Mrowietz, U.; Bernhardt, I. Proliferating tumor cells mimick glucose metabolism of mature human erythrocytes. *Cell Cycle* **2019**, *18*, 1316–1334. [CrossRef] [PubMed]
53. Donovan, A.; Brownlie, A.; Zhou, Y.; Shepard, J.; Pratt, S.J.; Moynihan, J.; Paw, B.H.; Drejer, A.; Barut, B.; Zapata, A.; et al. Positional cloning of zebrafish ferroportin1 identifies a conserved vertebrate iron exporter. *Nature* **2000**, *403*, 776–781. [CrossRef] [PubMed]
54. Fuqua, B.K.; Lu, Y.; Darshan, D.; Frazer, D.M.; Wilkins, S.J.; Wolkow, N.; Bell, A.G.; Hsu, J.; Yu, C.C.; Chen, H.; et al. The multicopper ferroxidase hephaestin enhances intestinal iron absorption in mice. *PLoS ONE* **2014**, *9*, e98792. [CrossRef] [PubMed]
55. Kuo, Y.M.; Su, T.; Chen, H.; Attieh, Z.; Syed, B.A.; McKie, A.T.; Anderson, G.J.; Gitschier, J.; Vulpe, C.D. Mislocalisation of hephaestin, a multicopper ferroxidase involved in basolateral intestinal iron transport, in the sex linked anaemia mouse. *Gut* **2004**, *53*, 201–206. [CrossRef] [PubMed]
56. Reeves, P.G.; Demars, L.C.; Johnson, W.T.; Lukaski, H.C. Dietary copper deficiency reduces iron absorption and duodenal enterocyte hephaestin protein in male and female rats. *J. Nutr.* **2005**, *135*, 92–98. [CrossRef] [PubMed]
57. Stearman, R.; Yuan, D.S.; Yamaguchi-Iwai, Y.; Klausner, R.D.; Dancis, A. A permease-oxidase complex involved in high-affinity iron uptake in yeast. *Science* **1996**, *271*, 1552–1557. [CrossRef] [PubMed]
58. Doguer, C.; Ha, J.H.; Collins, J.F. Intersection of Iron and Copper Metabolism in the Mammalian Intestine and Liver. *Compr. Physiol.* **2018**, *8*, 1433–1461. [CrossRef] [PubMed]
59. Petrak, J.; Vyoral, D. Hephaestin—A ferroxidase of cellular iron export. *Int. J. Biochem. Cell Biol.* **2005**, *37*, 1173–1178. [CrossRef]
60. Anderson, G.J.; Frazer, D.M.; McKie, A.T.; Vulpe, C.D. The ceruloplasmin homolog hephaestin and the control of intestinal iron absorption. *Blood Cells Mol. Dis.* **2002**, *29*, 367–375. [CrossRef]

61. Helman, S.L.; Zhou, J.; Fuqua, B.K.; Lu, Y.; Collins, J.F.; Chen, H.; Vulpe, C.D.; Anderson, G.J.; Frazer, D.M. The biology of mammalian multi-copper ferroxidases. *Biometals* **2023**, *36*, 263–281. [CrossRef]
62. Vashchenko, G.; MacGillivray, R.T. Multi-copper oxidases and human iron metabolism. *Nutrients* **2013**, *5*, 2289–2313. [CrossRef]
63. Mackenzie, B.; Hediger, M.A. SLC11 family of H+-coupled metal-ion transporters NRAMP1 and DMT1. *Pflugers Arch.* **2004**, *447*, 571–579. [CrossRef]
64. Gkouvatsos, K.; Papanikolaou, G.; Pantopoulos, K. Regulation of iron transport and the role of transferrin. *Biochim. Biophys. Acta* **2012**, *1820*, 188–202. [CrossRef]
65. Nandakumar, S.K.; Ulirsch, J.C.; Sankaran, V.G. Advances in understanding erythropoiesis: Evolving perspectives. *Br. J. Haematol.* **2016**, *173*, 206–218. [CrossRef]
66. Palis, J. Primitive and definitive erythropoiesis in mammals. *Front. Physiol.* **2014**, *5*, 3. [CrossRef]
67. Dzierzak, E.; Philipsen, S. Erythropoiesis: Development and differentiation. *Cold Spring Harb. Perspect. Med.* **2013**, *3*, a011601. [CrossRef]
68. Koury, M.J.; Ponka, P. New insights into erythropoiesis: The roles of folate, vitamin B12, and iron. *Annu. Rev. Nutr.* **2004**, *24*, 105–131. [CrossRef]
69. Visentin, M.; Diop-Bove, N.; Zhao, R.; Goldman, I.D. The intestinal absorption of folates. *Annu. Rev. Physiol.* **2014**, *76*, 251–274. [CrossRef]
70. Baker, H.; Thomson, A.D.; Feingold, S.; Frank, O. Role of the jejunum in the absorption of folic acid and its polyglutamates. *Am. J. Clin. Nutr.* **1969**, *22*, 124–132. [CrossRef]
71. Kozyraki, R.; Cases, O. Vitamin B12 absorption: Mammalian physiology and acquired and inherited disorders. *Biochimie* **2013**, *95*, 1002–1007. [CrossRef]
72. Cooper, B.A. Complex of intrinsic factor and B12 in human ileum during vitamin B12 absorption. *Am. J. Physiol.* **1968**, *214*, 832–835. [CrossRef]
73. Valman, H.B.; Roberts, P.D. Vitamin B12 absorption after resection of ileum in childhood. *Arch. Dis. Child.* **1974**, *49*, 932–935. [CrossRef]
74. Jacobson, L.O.; Goldwasser, E.; Fried, W.; Plzak, L. Role of the kidney in erythropoiesis. *Nature* **1957**, *179*, 633–634. [CrossRef]
75. Bachmann, S.; Le Hir, M.; Eckardt, K.U. Co-localization of erythropoietin mRNA and ecto-5′-nucleotidase immunoreactivity in peritubular cells of rat renal cortex indicates that fibroblasts produce erythropoietin. *J. Histochem. Cytochem.* **1993**, *41*, 335–341. [CrossRef]
76. Koury, S.T.; Bondurant, M.C.; Koury, M.J. Localization of erythropoietin synthesizing cells in murine kidneys by in situ hybridization. *Blood* **1988**, *71*, 524–527. [CrossRef]
77. Lacombe, C.; Da Silva, J.L.; Bruneval, P.; Fournier, J.G.; Wendling, F.; Casadevall, N.; Camilleri, J.P.; Bariety, J.; Varet, B.; Tambourin, P. Peritubular cells are the site of erythropoietin synthesis in the murine hypoxic kidney. *J. Clin. Investig.* **1988**, *81*, 620–623. [CrossRef]
78. Tojo, Y.; Sekine, H.; Hirano, I.; Pan, X.; Souma, T.; Tsujita, T.; Kawaguchi, S.; Takeda, N.; Takeda, K.; Fong, G.H.; et al. Hypoxia Signaling Cascade for Erythropoietin Production in Hepatocytes. *Mol. Cell Biol.* **2015**, *35*, 2658–2672. [CrossRef]
79. Erslev, A.J.; Caro, J.; Kansu, E.; Silver, R. Renal and extrarenal erythropoietin production in anaemic rats. *Br. J. Haematol.* **1980**, *45*, 65–72. [CrossRef]
80. Maxwell, P.H.; Ferguson, D.J.; Osmond, M.K.; Pugh, C.W.; Heryet, A.; Doe, B.G.; Johnson, M.H.; Ratcliffe, P.J. Expression of a homologously recombined erythopoietin-SV40 T antigen fusion gene in mouse liver: Evidence for erythropoietin production by Ito cells. *Blood* **1994**, *84*, 1823–1830. [CrossRef]
81. Suzuki, N.; Yamamoto, M. Roles of renal erythropoietin-producing (REP) cells in the maintenance of systemic oxygen homeostasis. *Pflugers Arch.* **2016**, *468*, 3–12. [CrossRef]
82. Weidemann, A.; Johnson, R.S. Nonrenal regulation of EPO synthesis. *Kidney Int.* **2009**, *75*, 682–688. [CrossRef]
83. Moritz, K.M.; Lim, G.B.; Wintour, E.M. Developmental regulation of erythropoietin and erythropoiesis. *Am. J. Physiol.* **1997**, *273*, R1829–R1844. [CrossRef]
84. Latour, C.; Kautz, L.; Besson-Fournier, C.; Island, M.L.; Canonne-Hergaux, F.; Loreal, O.; Ganz, T.; Coppin, H.; Roth, M.P. Testosterone perturbs systemic iron balance through activation of epidermal growth factor receptor signaling in the liver and repression of hepcidin. *Hepatology* **2014**, *59*, 683–694. [CrossRef]
85. Hennigar, S.R.; Berryman, C.E.; Harris, M.N.; Karl, J.P.; Lieberman, H.R.; McClung, J.P.; Rood, J.C.; Pasiakos, S.M. Testosterone Administration during Energy Deficit Suppresses Hepcidin and Increases Iron Availability for Erythropoiesis. *J. Clin. Endocrinol. Metab.* **2020**, *105*, e1316–e1321. [CrossRef]
86. Coviello, A.D.; Kaplan, B.; Lakshman, K.M.; Chen, T.; Singh, A.B.; Bhasin, S. Effects of graded doses of testosterone on erythropoiesis in healthy young and older men. *J. Clin. Endocrinol. Metab.* **2008**, *93*, 914–919. [CrossRef]
87. Park, C.H.; Valore, E.V.; Waring, A.J.; Ganz, T. Hepcidin, a urinary antimicrobial peptide synthesized in the liver. *J. Biol. Chem.* **2001**, *276*, 7806–7810. [CrossRef]
88. Pigeon, C.; Ilyin, G.; Courselaud, B.; Leroyer, P.; Turlin, B.; Brissot, P.; Loreal, O. A new mouse liver-specific gene, encoding a protein homologous to human antimicrobial peptide hepcidin, is overexpressed during iron overload. *J. Biol. Chem.* **2001**, *276*, 7811–7819. [CrossRef]
89. Katsarou, A.; Pantopoulos, K. Hepcidin Therapeutics. *Pharmaceuticals* **2018**, *11*, 127. [CrossRef]

90. Nemeth, E.; Tuttle, M.S.; Powelson, J.; Vaughn, M.B.; Donovan, A.; Ward, D.M.; Ganz, T.; Kaplan, J. Hepcidin regulates cellular iron efflux by binding to ferroportin and inducing its internalization. *Science* **2004**, *306*, 2090–2093. [CrossRef]
91. De Domenico, I.; Ward, D.M.; Langelier, C.; Vaughn, M.B.; Nemeth, E.; Sundquist, W.I.; Ganz, T.; Musci, G.; Kaplan, J. The molecular mechanism of hepcidin-mediated ferroportin down-regulation. *Mol. Biol. Cell* **2007**, *18*, 2569–2578. [CrossRef]
92. Nicolas, G.; Chauvet, C.; Viatte, L.; Danan, J.L.; Bigard, X.; Devaux, I.; Beaumont, C.; Kahn, A.; Vaulont, S. The gene encoding the iron regulatory peptide hepcidin is regulated by anemia, hypoxia, and inflammation. *J. Clin. Investig.* **2002**, *110*, 1037–1044. [CrossRef]
93. Tsukaguchi, H.; Tokui, T.; Mackenzie, B.; Berger, U.V.; Chen, X.Z.; Wang, Y.; Brubaker, R.F.; Hediger, M.A. A family of mammalian Na+-dependent L-ascorbic acid transporters. *Nature* **1999**, *399*, 70–75. [CrossRef]
94. Ulloa, V.; Garcia-Robles, M.; Martinez, F.; Salazar, K.; Reinicke, K.; Perez, F.; Godoy, D.F.; Godoy, A.S.; Nualart, F. Human choroid plexus papilloma cells efficiently transport glucose and vitamin C. *J. Neurochem.* **2013**, *127*, 403–414. [CrossRef]
95. Amano, A.; Aigaki, T.; Maruyama, N.; Ishigami, A. Ascorbic acid depletion enhances expression of the sodium-dependent vitamin C transporters, SVCT1 and SVCT2, and uptake of ascorbic acid in livers of SMP30/GNL knockout mice. *Arch. Biochem. Biophys.* **2010**, *496*, 38–44. [CrossRef]
96. Kobayashi, T.A.; Shimada, H.; Sano, F.K.; Itoh, Y.; Enoki, S.; Okada, Y.; Kusakizako, T.; Nureki, O. Dimeric transport mechanism of human vitamin C transporter SVCT1. *Nat. Commun.* **2024**, *15*, 5569. [CrossRef]
97. Mann, G.V.; Newton, P. The membrane transport of ascorbic acid. *Ann. N. Y. Acad. Sci.* **1975**, *258*, 243–252. [CrossRef]
98. Ingermann, R.L.; Stankova, L.; Bigley, R.H.; Bissonnette, J.M. Effect of monosaccharide on dehydroascorbic acid uptake by placental membrane vesicles. *J. Clin. Endocrinol. Metab.* **1988**, *67*, 389–394. [CrossRef]
99. Hornung, T.C.; Biesalski, H.K. Glut-1 explains the evolutionary advantage of the loss of endogenous vitamin C-synthesis: The electron transfer hypothesis. *Evol. Med. Public Health* **2019**, *2019*, 221–231. [CrossRef]
100. Goldenberg, H.; Schweinzer, E. Transport of vitamin C in animal and human cells. *J. Bioenerg. Biomembr.* **1994**, *26*, 359–367. [CrossRef]
101. Skou, J.C. Enzymatic Basis for Active Transport of Na^+ and K^+ across Cell Membrane. *Physiol. Rev.* **1965**, *45*, 596–617. [CrossRef]
102. Cereijido, M.; Shoshani, L.; Contreras, R.G. The polarized distribution of Na+, K+-ATPase and active transport across epithelia. *J. Membr. Biol.* **2001**, *184*, 299–304. [CrossRef]
103. Sage, J.M.; Carruthers, A. Human erythrocytes transport dehydroascorbic acid and sugars using the same transporter complex. *Am. J. Physiol. Cell Physiol.* **2014**, *306*, C910–C917. [CrossRef]
104. Montel-Hagen, A.; Kinet, S.; Manel, N.; Mongellaz, C.; Prohaska, R.; Battini, J.L.; Delaunay, J.; Sitbon, M.; Taylor, N. Erythrocyte Glut1 triggers dehydroascorbic acid uptake in mammals unable to synthesize vitamin C. *Cell* **2008**, *132*, 1039–1048. [CrossRef]
105. Montel-Hagen, A.; Sitbon, M.; Taylor, N. Erythroid glucose transporters. *Curr. Opin. Hematol.* **2009**, *16*, 165–172. [CrossRef]
106. May, J.M.; Qu, Z.C.; Whitesell, R.R.; Cobb, C.E. Ascorbate recycling in human erythrocytes: Role of GSH in reducing dehydroascorbate. *Free Radic. Biol. Med.* **1996**, *20*, 543–551. [CrossRef]
107. Hughes, R.E. Reduction of Dehydroasorbic Acid by Animal Tissues. *Nature* **1964**, *203*, 1068–1069. [CrossRef]
108. Dereven'kov, I.A.; Makarov, S.V.; Bui Thi, T.T.; Makarova, A.S.; Koifman, O.I. Studies on the reduction of dehydroascorbic acid by glutathione in the presence of aquahydroxocobinamide. *Eur. J. Inorg. Chem.* **2018**, *2018*, 2987–2992. [CrossRef]
109. Sasaki, H.; Giblin, F.J.; Winkler, B.S.; Chakrapani, B.; Leverenz, V.; Shu, C.C. A protective role for glutathione-dependent reduction of dehydroascorbic acid in lens epithelium. *Investig. Ophthalmol. Vis. Sci.* **1995**, *36*, 1804–1817.
110. Cisternas, P.; Silva-Alvarez, C.; Martinez, F.; Fernandez, E.; Ferrada, L.; Oyarce, K.; Salazar, K.; Bolanos, J.P.; Nualart, F. The oxidized form of vitamin C, dehydroascorbic acid, regulates neuronal energy metabolism. *J. Neurochem.* **2014**, *129*, 663–671. [CrossRef]
111. Puskas, F.; Gergely, P., Jr.; Banki, K.; Perl, A. Stimulation of the pentose phosphate pathway and glutathione levels by dehydroascorbate, the oxidized form of vitamin C. *FASEB J.* **2000**, *14*, 1352–1361. [CrossRef]
112. Orringer, E.P.; Roer, M.E. An ascorbate-mediated transmembrane-reducing system of the human erythrocyte. *J. Clin. Investig.* **1979**, *63*, 53–58. [CrossRef]
113. D'Alessandro, A.; Keele, G.R.; Hay, A.; Nemkov, T.; Earley, E.J.; Stephenson, D.; Vincent, M.; Deng, X.; Stone, M.; Dzieciatkowska, M.; et al. Ferroptosis regulates hemolysis in stored murine and human red blood cells. *bioRxiv* **2024**. [CrossRef]
114. Howie, H.L.; Hay, A.M.; de Wolski, K.; Waterman, H.; Lebedev, J.; Fu, X.; Culp-Hill, R.; D'Alessandro, A.; Gorham, J.D.; Ranson, M.S.; et al. Differences in Steap3 expression are a mechanism of genetic variation of RBC storage and oxidative damage in mice. *Blood Adv.* **2019**, *3*, 2272–2285. [CrossRef]
115. Sender, R.; Fuchs, S.; Milo, R. Revised Estimates for the Number of Human and Bacteria Cells in the Body. *PLoS Biol.* **2016**, *14*, e1002533. [CrossRef]
116. Callender, S.T.; Powell, E.; Witts, L. The life-span of the red cell in man. *J. Pathol. Bacteriol.* **1945**, *57*, 129–139. [CrossRef]
117. Larkin, J.M.; Donzell, W.C.; Anderson, R.G. Potassium-dependent assembly of coated pits: New coated pits form as planar clathrin lattices. *J. Cell Biol.* **1986**, *103*, 2619–2627. [CrossRef]
118. Song, Q.; Meng, B.; Xu, H.; Mao, Z. The emerging roles of vacuolar-type ATPase-dependent Lysosomal acidification in neurodegenerative diseases. *Transl. Neurodegener.* **2020**, *9*, 17. [CrossRef]
119. Harvey, W.R. Physiology of V-ATPases. *J. Exp. Biol.* **1992**, *172*, 1–17. [CrossRef]

120. Touret, N.; Furuya, W.; Forbes, J.; Gros, P.; Grinstein, S. Dynamic traffic through the recycling compartment couples the metal transporter Nramp2 (DMT1) with the transferrin receptor. *J. Biol. Chem.* **2003**, *278*, 25548–25557. [CrossRef]
121. Eckenroth, B.E.; Steere, A.N.; Chasteen, N.D.; Everse, S.J.; Mason, A.B. How the binding of human transferrin primes the transferrin receptor potentiating iron release at endosomal pH. *Proc. Natl. Acad. Sci. USA* **2011**, *108*, 13089–13094. [CrossRef]
122. Dautry-Varsat, A.; Ciechanover, A.; Lodish, H.F. pH and the recycling of transferrin during receptor-mediated endocytosis. *Proc. Natl. Acad. Sci. USA* **1983**, *80*, 2258–2262. [CrossRef]
123. Ohgami, R.S.; Campagna, D.R.; Greer, E.L.; Antiochos, B.; McDonald, A.; Chen, J.; Sharp, J.J.; Fujiwara, Y.; Barker, J.E.; Fleming, M.D. Identification of a ferrireductase required for efficient transferrin-dependent iron uptake in erythroid cells. *Nat. Genet.* **2005**, *37*, 1264–1269. [CrossRef]
124. Yanatori, I.; Kishi, F. DMT1 and iron transport. *Free Radic. Biol. Med.* **2019**, *133*, 55–63. [CrossRef]
125. Grandchamp, B.; Hetet, G.; Kannengiesser, C.; Oudin, C.; Beaumont, C.; Rodrigues-Ferreira, S.; Amson, R.; Telerman, A.; Nielsen, P.; Kohne, E.; et al. A novel type of congenital hypochromic anemia associated with a nonsense mutation in the STEAP3/TSAP6 gene. *Blood* **2011**, *118*, 6660–6666. [CrossRef]
126. Blanc, L.; Papoin, J.; Debnath, G.; Vidal, M.; Amson, R.; Telerman, A.; An, X.; Mohandas, N. Abnormal erythroid maturation leads to microcytic anemia in the TSAP6/Steap3 null mouse model. *Am. J. Hematol.* **2015**, *90*, 235–241. [CrossRef]
127. Zhou, Z.; Zhen, J.; Karpowich, N.K.; Goetz, R.M.; Law, C.J.; Reith, M.E.; Wang, D.N. LeuT-desipramine structure reveals how antidepressants block neurotransmitter reuptake. *Science* **2007**, *317*, 1390–1393. [CrossRef]
128. Ordway, G.A.; Jia, W.; Li, J.; Zhu, M.Y.; Mandela, P.; Pan, J. Norepinephrine transporter function and desipramine: Residual drug effects versus short-term regulation. *J. Neurosci. Methods* **2005**, *143*, 217–225. [CrossRef]
129. Zhu, M.Y.; Kyle, P.B.; Hume, A.S.; Ordway, G.A. The persistent membrane retention of desipramine causes lasting inhibition of norepinephrine transporter function. *Neurochem. Res.* **2004**, *29*, 419–427. [CrossRef]
130. Andersen, J.; Kristensen, A.S.; Bang-Andersen, B.; Stromgaard, K. Recent advances in the understanding of the interaction of antidepressant drugs with serotonin and norepinephrine transporters. *Chem. Commun.* **2009**, *15*, 3677–3692. [CrossRef]
131. Hyman, S.E.; Nestler, E.J. Initiation and adaptation: A paradigm for understanding psychotropic drug action. *Am. J. Psychiatry* **1996**, *153*, 151–162. [CrossRef] [PubMed]
132. Elojeimy, S.; Holman, D.H.; Liu, X.; El-Zawahry, A.; Villani, M.; Cheng, J.C.; Mahdy, A.; Zeidan, Y.; Bielwaska, A.; Hannun, Y.A.; et al. New insights on the use of desipramine as an inhibitor for acid ceramidase. *FEBS Lett.* **2006**, *580*, 4751–4756. [CrossRef] [PubMed]
133. Hurwitz, R.; Ferlinz, K.; Sandhoff, K. The tricyclic antidepressant desipramine causes proteolytic degradation of lysosomal sphingomyelinase in human fibroblasts. *Biol. Chem. Hoppe Seyler* **1994**, *375*, 447–450. [CrossRef] [PubMed]
134. Kolzer, M.; Werth, N.; Sandhoff, K. Interactions of acid sphingomyelinase and lipid bilayers in the presence of the tricyclic antidepressant desipramine. *FEBS Lett.* **2004**, *559*, 96–98. [CrossRef] [PubMed]
135. Lai, M.; Realini, N.; La Ferla, M.; Passalacqua, I.; Matteoli, G.; Ganesan, A.; Pistello, M.; Mazzanti, C.M.; Piomelli, D. Complete Acid Ceramidase ablation prevents cancer-initiating cell formation in melanoma cells. *Sci. Rep.* **2017**, *7*, 7411. [CrossRef] [PubMed]
136. Saad, A.F.; Meacham, W.D.; Bai, A.; Anelli, V.; Elojeimy, S.; Mahdy, A.E.; Turner, L.S.; Cheng, J.; Bielawska, A.; Bielawski, J.; et al. The functional effects of acid ceramidase overexpression in prostate cancer progression and resistance to chemotherapy. *Cancer Biol. Ther.* **2007**, *6*, 1455–1460. [CrossRef] [PubMed]
137. Seelan, R.S.; Qian, C.; Yokomizo, A.; Bostwick, D.G.; Smith, D.I.; Liu, W. Human acid ceramidase is overexpressed but not mutated in prostate cancer. *Genes. Chromosomes Cancer* **2000**, *29*, 137–146. [CrossRef] [PubMed]
138. Beckham, T.H.; Lu, P.; Cheng, J.C.; Zhao, D.; Turner, L.S.; Zhang, X.; Hoffman, S.; Armeson, K.E.; Liu, A.; Marrison, T.; et al. Acid ceramidase-mediated production of sphingosine 1-phosphate promotes prostate cancer invasion through upregulation of cathepsin B. *Int. J. Cancer* **2012**, *131*, 2034–2043. [CrossRef] [PubMed]
139. Awojoodu, A.O.; Keegan, P.M.; Lane, A.R.; Zhang, Y.; Lynch, K.R.; Platt, M.O.; Botchwey, E.A. Acid sphingomyelinase is activated in sickle cell erythrocytes and contributes to inflammatory microparticle generation in SCD. *Blood* **2014**, *124*, 1941–1950. [CrossRef]
140. Lang, P.A.; Schenck, M.; Nicolay, J.P.; Becker, J.U.; Kempe, D.S.; Lupescu, A.; Koka, S.; Eisele, K.; Klarl, B.A.; Rubben, H.; et al. Liver cell death and anemia in Wilson disease involve acid sphingomyelinase and ceramide. *Nat. Med.* **2007**, *13*, 164–170. [CrossRef]
141. Momchilova, A.; Pankov, R.; Alexandrov, A.; Markovska, T.; Pankov, S.; Krastev, P.; Staneva, G.; Vassileva, E.; Krastev, N.; Pinkas, A. Sphingolipid Catabolism and Glycerophospholipid Levels Are Altered in Erythrocytes and Plasma from Multiple Sclerosis Patients. *Int. J. Mol. Sci.* **2022**, *23*, 7592. [CrossRef] [PubMed]
142. Choi, S.Y.; Li, J.; Jo, S.H.; Lee, S.J.; Oh, S.B.; Kim, J.S.; Lee, J.H.; Park, K. Desipramine inhibits Na+/H+ exchanger in human submandibular cells. *J. Dent. Res.* **2006**, *85*, 839–843. [CrossRef] [PubMed]
143. Ellegaard, A.M.; Dehlendorff, C.; Vind, A.C.; Anand, A.; Cederkvist, L.; Petersen, N.H.T.; Nylandsted, J.; Stenvang, J.; Mellemgaard, A.; Osterlind, K.; et al. Repurposing Cationic Amphiphilic Antihistamines for Cancer Treatment. *EBioMedicine* **2016**, *9*, 130–139. [CrossRef] [PubMed]
144. Petersen, N.H.; Olsen, O.D.; Groth-Pedersen, L.; Ellegaard, A.M.; Bilgin, M.; Redmer, S.; Ostenfeld, M.S.; Ulanet, D.; Dovmark, T.H.; Lonborg, A.; et al. Transformation-associated changes in sphingolipid metabolism sensitize cells to lysosomal cell death induced by inhibitors of acid sphingomyelinase. *Cancer Cell* **2013**, *24*, 379–393. [CrossRef] [PubMed]

145. Arimochi, H.; Morita, K. Desipramine induces apoptotic cell death through nonmitochondrial and mitochondrial pathways in different types of human colon carcinoma cells. *Pharmacology* **2008**, *81*, 164–172. [CrossRef] [PubMed]
146. Pakkanen, K.; Salonen, E.; Makela, A.R.; Oker-Blom, C.; Vattulainen, I.; Vuento, M. Desipramine induces disorder in cholesterol-rich membranes: Implications for viral trafficking. *Phys. Biol.* **2009**, *6*, 046004. [CrossRef] [PubMed]
147. Salata, C.; Calistri, A.; Parolin, C.; Baritussio, A.; Palu, G. Antiviral activity of cationic amphiphilic drugs. *Expert. Rev. Anti Infect. Ther.* **2017**, *15*, 483–492. [CrossRef]
148. Pan, X.; Giustarini, D.; Lang, F.; Rossi, R.; Wieder, T.; Koberle, M.; Ghashghaeinia, M. Desipramine induces eryptosis in human erythrocytes, an effect blunted by nitric oxide donor sodium nitroprusside and N-acetyl-L-cysteine but enhanced by Calcium depletion. *Cell Cycle* **2023**, *22*, 1827–1853. [CrossRef]

antioxidants

MDPI

Article

Molecular Study on Twin Cohort with Discordant Birth Weight

Payal Chakraborty [1,2], Hajnalka Orvos [3] and Edit Hermesz [1,*]

[1] Department of Biochemistry and Molecular Biology, Faculty of Science and Informatics, University of Szeged, P.O. Box 533, H-6701 Szeged, Hungary; payal.chakraborty@jisuniversity.ac.in
[2] Department of Pharmaceutical Technology, JIS University, 81, Nilgunj Road, Kolkata 700109, India
[3] Department of Obstetrics and Gynaecology, Faculty of Medicine, University of Szeged, Semmelweis u. 1, H-6725 Szeged, Hungary; orvosh@obgyn.szote.u-szeged.hu
* Correspondence: hermesz@bio.u-szeged.hu

Abstract: The increased rate of twinning has pointed out newer challenges in clinical practices related to gestational complications, intrauterine growth restriction, perinatal mortality, and comorbidities. As a twin pregnancy progresses, the increased demand for oxygen supply can easily disrupt the redox homeostasis balance and further impose a greater challenge for the developing fetuses. A substantial birth-weight difference acts as an indicator of a deficit in oxygenation or blood flow to one of the fetuses, which might be related to a low bioavailable nitric oxide level. Therefore, in this study, we focused on networks involved in the adjustment of oxygen supply, like the activation of inducible and endothelial nitric oxide synthase (NOS3) along with free radical and lipid peroxide formation in mature twin pairs with high birth-weight differences. The selected parameters were followed by immunofluorescence staining, fluorescence-activated cell sorting analysis, and biochemical measurements in the umbilical cord vessels and fetal red blood cells. Based on our data set, it is clear that the lower-weight siblings are markedly exposed to persistent intrauterine hypoxic conditions, which are connected to a decreased level in NOS3 activation. Furthermore, the increased level of peroxynitrite aggravates lipid peroxidation and induces morphological and functional damage and loss in redox homeostasis.

Keywords: birth-weight discordance; free radicals; hypoxia; nitric oxide synthases; oxidative stress; peroxynitrite; red blood cells; umbilical cord vessels

Citation: Chakraborty, P.; Orvos, H.; Hermesz, E. Molecular Study on Twin Cohort with Discordant Birth Weight. *Antioxidants* **2023**, *12*, 1370. https://doi.org/10.3390/antiox12071370

Academic Editors: Angelo D'Alessandro and Alkmini T. Anastasiadi

Received: 23 May 2023
Revised: 25 June 2023
Accepted: 28 June 2023
Published: 30 June 2023

1. Introduction

Multiple/twin pregnancies have significantly increased in the last few decades, which can be partly traced back to the applied expansion of assisted reproduction techniques and a steep surge in the average maternal age during the first pregnancy [1,2]. The increased rate of twinning has pointed out newer challenges in clinical practices related to gestational complications [3]. The appropriate fetal development critically depends on the efficiency of the fetoplacental circulation through the umbilical cord (UC) vessels. The activation of molecular parameters involved in the regulation of the oxygen supply of the developing fetus, like endothelial nitric oxide synthase (NOS3), its phosphorylation status at the serine1177 residue (pNOS3), and the upregulation of the inducible nitric oxide synthase (NOS2), is crucial. The UC vein carries oxygenated blood, while arteries carry the deoxygenated blood in the vascular system. With any UC disorder, structural and/or functional alterations would reflect the *in-utero* condition and can be directly or indirectly connected to intrauterine hypoxia, impaired blood flow to the developing fetus, and retardation in fetal growth. Since UC vessels lack innervation, the regulation of nitric oxide (NO) production by NOS3 expression and activation in the vessels counts as determining factors in the control of blood flow and maintenance of the vascular tone [4,5]. Alteration in the NOS3 expression and activation and/or NO degradation due to the reactive oxygen species (ROS) may serve as markers for an unfavorable cardiovascular condition. A low amount of bioavailable NO in the vessel walls can lead to the formation of endothelial dysfunction (ED), which may cause cardiovascular complications during fetal and post-natal life [6–8]. As an alternative to the NO-producing pathway, the upregulation of inducible nitric oxide synthase (NOS2) expression may serve as a rescue mechanism to increase the NO level in vascular circulation [9–13]. Apart from the upregulation of NOS2, hypoxic condition and the deficiency of bioavailable NO can be sensed by healthy fetal red blood cells (RBCs). Therefore, under stressful conditions, RBCs themselves are able to synthesize NOS3 (RBC-NOS3) and can serve as an emergency NO-producing entity [14–16].

Per the study design, neonates are categorized based on their birth weights as appropriate, large, and small for the gestational age. In addition, especially for twin/multiple pregnancies, significant birth-weight differences may occur between the siblings. Based on the consensus laid down by the American College of Obstetricians and Gynecologists (2016), the threshold value of birth-weight discordance was considered to be between 20–25% difference in the actual weight between the twin siblings.

In our previous publications, we followed the NO-producing pathways in singletons and birth weight–non-discordant, mature and pre-mature twin pairs [13,17]. We presented significant evidence for the altered NOS3 expression and its activation capacity in the UC vessels depending on the gestational age and birth weights of the twin neonates. Additionally, we presented evidence that besides the vascular walls, the RBC-NOS3-NO pathway also gets activated during the pre-term pregnancy status. However, as the pregnancy progresses (37–39 weeks), there arises a higher demand for NO production because of the increasing hypoxic conditions with an elevated NO degradation rate due to ROS. In mature twin populations, we have seen that there lies a significant impairment in the NO-producing capacity via the NOS3 pathway, along with an increase in birth weight [13,17].

ED refers to several pathological conditions and impairment in endothelium-dependent vasorelaxation due to the loss of NO bioactivity [18,19]. In recent years, special attention has been paid to the investigation of NO-mediated signaling pathways not only in the vessel endothelium but also in cooperation with the RBCs. Here, we point out the relationship between the birth weight and UC endothelium NOS3-, NOS2-, and the circulating RBC-NO-producing capacity to further support their importance in *in-utero* fetal development. Moreover, a deeper knowledge and understanding of such complementary relationships between these complex systems will allow them to emerge as early biomarkers associated with the risk of cardiovascular events.

2. Materials and Methods

2.1. Clinical Samples

Umbilical cord samples along with UC arterial blood were collected at the Department of Obstetrics and Gynecology at the University of Szeged, Hungary, in the time period of 2014–2022. Samples from informed volunteers with twin pregnancies were handled according to the Declaration of Helsinki. The study protocol was approved by the Institutional Ethics Committee (16/2014 and 20/2016). In the requisition process, we excluded clinical factors like (i) maternal age below 18 years; (ii) gestational age less than 37 weeks; (iii) gestational diabetes, infection, and inflammatory conditions or disorders such as cardiovascular diseases; (iv) complications or difficulty during delivery, (v) malformations or evidence of genetic disorders; and (vi) neonates from mothers addicted to alcohol or with smoking habits. Under any circumstances, the nutritional status of the women during pregnancy was kept satisfactory, and no cases of malnutrition had been reported. Here, in total, we considered 18 pairs of mature, birth weight–discordant siblings (with a minimum of 25% birth-weight difference) and grouped them according to their birth-weight and gestational-percentile ranges; likewise, high weight–discordant (D-Hwt) and low weight–discordant (D-Lwt) twin siblings have birth-weight ranges of 2800–3500 g and 2100–2600 g, respectively, and gestational percentiles of ~75–95 percent and ~10–50 percent, respectively. The average timing between the twin delivery was 1–10 min, depending on the mode of delivery (vaginal deliveries, on average, had a maximum difference of about 10 min; in the case of cesarean section, it was only 1–3 min). In detail, the clinical parameters of the above-mentioned data sets are demonstrated in Table S1 (Supplementary Material).

2.2. Immunohistochemistry in the Umbilical Cord Sections

Sample processing: Small segments of the umbilical cords of 2–3 cm in length were immediately fixed in 4% (*w*/*v*) paraformaldehyde in 0.05 M phosphate buffer (PB) upon arrival and cryopreserved with 30% (*w*/*v*) sucrose in PB added with 0.1% (*v*/*w*) Na-azide. Samples were kept at 4 °C until the further process of embedding. Fixed UC segments were embedded in Tissue-Tek® O.C.T.™ (4583) from Sakura Finetek Europe (Alphen aan den Rijn, The Netherlands) for the purpose of cryosectioning slice thicknesses of 16 µm. Finally, the UC sections were mounted on Superfrost™ ultra plus® microscope slides (J3800AMNZ) from Thermo Fisher Scientific (Waltham, MA, USA) and stored at −80 °C until the following steps [20].

Immunohistochemistry on the specimen samples: The slides with the UC sections were thawed, dried at room temperature (RT), and permeabilized with 0.1% Triton X-100 for 20 min, which was followed by the blocking of a nonspecific binding using 4% (*w*/*v*) bovine serum albumin (BSA) and 5% (*v*/*v*) normal goat serum (NGS) in PB. Using the process of immunohistochemistry technique, UC sections were immunolabelled with the primary antibodies at 4 °C overnight. The slides, after consecutive washing, were further incubated with the secondary antibodies, goat anti-mouse Alexa® 647- and/or goat anti-rabbit Alexa® 488-conjugated in 1:2000 dilution, in the dark for 2 h at RT. Finally, for the nuclei staining, slides were washed with 0.05 M PB 2–3 times and counterstained with 4′,6-diamidino-2-phenylindole (D9542) from Sigma-Aldrich (St. Louis, MO, USA), using a concentration of 1 µg/cm^3, in the dark for at least 5 min [21].

Visualizing using Fluorescence Microscope: To check and analyze under an epifluorescence microscope (Nikon Eclipse 80i, 100× and 50× immersion objective; Nikon Zeiss Microscopy GmbH, Jena, Germany), the immunolabelled slides were semi-dried and mounted using Antifading, Bright Mount/Plus aqueous mounting medium (ab103748) from Abcam (Cambridge, UK). Images were captured using a QImaging RETIGA 4000R camera with Capture Pro 6.0 software (QImaging, Surrey, BC, Canada). The acquired images with 100× or 50× magnification were semi-quantitatively analyzed using the ImageJ® 1.50i software (National Institutes of Health, Bethesda, MD, USA).

2.3. Immunocytochemistry and Fluorescence-Activated Cell Sorting (FACS) on the Fetal RBCs

The whole blood was immediately processed upon arrival for the RBC fraction isolation; the whole blood was subjected to centrifugation at 200× *g* for 10 min at an optimum temperature of 22 °C, and the lower 2/3 of the RBC populations were collected. The purity of the isolated RBC was checked by staining with the RBC-specific marker anti-Glycophorin A (CD235a) from Thermo Fisher Scientific (Waltham, MA, USA), where the sample purity was >95% [21]. The freshly isolated RBCs were fixed using 4% (*w*/*v*) paraformaldehyde in 0.05 M PB at 4 °C in the neutral pH for 1 h. Next, using 0.1% Triton X-100 for 30 min, the RBCs were permeabilized at RT. After permeabilization, the RBCs were blocked for nonspecific antibody binding for 1 h in PB containing 4% (*w*/*v*) BSA and 5% (*v*/*v*) NGS. After blocking, the RBCs were immunolabelled with the primary antibodies at 4 °C overnight. This was followed by consecutive washing and incubation with the goat anti-mouse Alexa® 647 and goat anti-rabbit Alexa® 488-conjugated secondary antibodies for 2 h at RT (see Table S2 for the detailed list of antibodies). In the final step, the RBCs were thoroughly washed and prepared for quantification through FACS analysis (FACS, BD FACSCalibur™; BD Biosciences, Franklin Lakes, NJ, USA) using the software FlowJo™ (FlowJo™ Software for Windows Version 10; Ashland, OR, USA).

2.4. Determination of Peroxynitrite ($ONOO^-$) Level

An isolated RBC fraction was divided into aliquots and stored at −20 °C until further processing for biochemical measurements. To determine the $ONOO^-$ levels of the RBC populations, spectrophotometric measurements were performed at 302 nm using a GENESYS 10S UV-Vis spectrophotometer. Hemolyzed samples were mixed with 1 M NaOH solution in a ratio of 1:250, and the increase in absorbance was followed until it reached a stable equilibrium. Then, the samples were mixed with 100 mM PB (pH = 7.4) in 1:250 ratios. On this neutral pH, the $ONOO^-$ decomposed, and a decrease in the absorbance was observed until the equilibrium point [22].

The $ONOO^-$ concentration was calculated as a result of the different absorbance at the two distinct pH values according to the Lambert–Beer law (ε_{ONOO^-} = 1670 M^{-1} cm^{-1}). The final results were calculated relative to the protein concentration (μmol/mg protein).

2.5. Blood Smear Image Processing and Data Analysis

The total blood originating from the siblings with substantial birth-weight differences (n = 18 pairs) was used upon arrival for smear preparation. Smears were immunolabelled for anti-NOS3/Alexa® 647-conjugated antibodies and mounted in Antifading, Bright Mount/Plus aqueous mounting medium (Abcam ab103748, Cambridge, UK). Blood smears were examined under the epifluorescence microscope (Nikon Eclipse 80i, 100× immersion objective; Nikon Zeiss Microscopy GmbH, Jena, Germany) with a QImaging RETIGA 4000R camera using Capture Pro 6.0 software (QImaging, Surrey, BC, Canada). Smear images were processed and analyzed using the scientific software MatLab. Different phenotypic variants were calculated using the automated bio-image analysis tool CellProfiler™ (available at http://cellprofiler.org and accessed on 15 November 2018) [23].

2.6. Statistical Analysis

Statistical significance was accepted at values ** $p \leq 0.01$ and *** $p \leq 0.001$. All calculations were performed through one-way analysis of variance (ANOVA) and complied with the Newman–Keuls multiple comparison test using the GraphPad Prism Statistical Software version 6.0.

3. Results

3.1. Expression of Endothelial and Inducible NOS in the Birth Weight–Discordant Populations

Molecular parameters of UC arteries and veins originating from twin pairs with high birth-weight differences were compared. First, sections of the vessels were double immunolabelled for NOS3/pNOS3 and subjected to Image J© (1.50i) evaluation.

Considering the high- versus low-weight siblings, there is a significant difference between their NOS3 expressions/activations, both in the vessels. In the arteries, a ~60% lower value was measured for both the NOS3 and pNOS3 intensities. The vein was also affected, but the level of alteration was remarkably smaller, ~40% in both the NOS3 expression and the Ser1177 phosphorylation intensity in the lower-weight siblings (Figures 1, S1 and S3). In the case of impaired NOS3 expression/activation, an increased NOS2 level could serve as a rescue mechanism to increase the bioavailable NO level. However, in both vessels, the NOS2 level was found to be similar between the siblings (Figures 1, S1 and S3); there was only a 5–8% difference in the measured intensity values.

Comparing the mature, birth weight–discordant and non-discordant populations, in the case of high-weight neonates, the measured parameters were comparable for both the vessels and RBCs (Figure S4). The discordant population was "even in a better condition" in terms of NOS3 expression and activation; the efficacy increase was about 20–40% in both the veins and the RBCs. However, the values of the low-weight siblings were far below those of their age-matched, low birth-weight, non-discordant twin pairs, especially in the case of arteries and RBCs. In the arteries, there were ~40% and 30% differences in the NOS3 expression and its activation, respectively, and neither the NOS2 activity in the vessels nor the circulating RBC populations were able to compensate for the impaired NO production level (Figure S5).

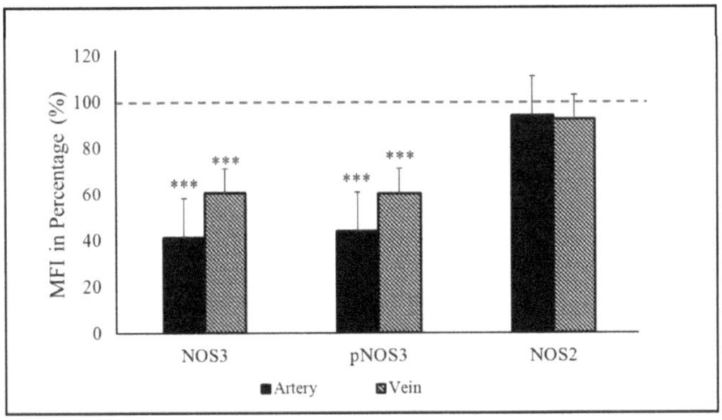

Figure 1. Endothelial (NOS3) and inducible nitric oxide synthase (NOS2) levels in the umbilical cord vessels. Quantifying the mean fluorescence intensity (MFI) in percentage determines the level of NOS3 and its phosphorylated status at Ser1177 residue (pNOS3) along with the expression of NOS2 of the low-weight twin neonates (D-Lwt), where the values of their high-weight siblings (D-Hwt) were taken as 100% (red dotted line) (n = 18 pairs). Statistical significance was accepted at *** $p < 0.001$ based on one-way ANOVA using the Newman–Keuls multiple comparison test. Statistical analysis was performed on the measured MFI data set.

3.2. NOS3 Expression and Activation Level in the Red Blood Cell of the Birth Weight–Discordant Study Populations

NOS3 protein level and its activation by the phosphorylation at the Serine1177 position were immunostained for NOS3/Glycophorin and pNOS3/Glycophorin and analyzed using FACS (Figure 2). Representative histograms indicate that, in the low-weight neonates, both the NOS3 expression and activation (pNOS3) were far behind the higher-weight siblings' values (an arbitrary borderline at 10^2 was considered along the x-axis dividing the total RBC population into basal- and high-intensity groups). In the higher-weight neonates, ~89% and 98% of the Glycophorin-positive cells showed high NOS3 and pNOS3 intensity levels, respectively, whereas, in the case of their smaller siblings, these numbers reached only 12% and 18% (Figure 2A,B). Taking into account the total data set (n = 18), it is clear that there are differences of 35–38% and 55–62% in NOS3 and pNOS3 intensity levels, respectively, between the siblings (Figure 2C).

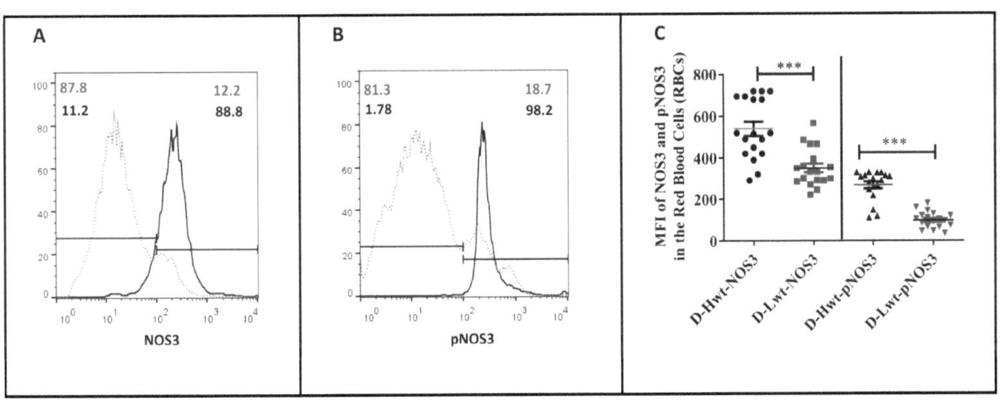

Figure 2. Fluorescence-activated cell sorting analysis of the endothelial nitric oxide synthase (NOS3) expression and its activation status on the immunolabelled fetal red blood cells (RBCs). Panels (**A**,**B**) show representative histogram plots indicating NOS3 and pNOS3 intensity levels in the mature, discordant

high-weight (D-Hwt—black line) and low-weight (D-Lwt—gray dots) twin siblings. According to the blank sample, an arbitrary borderline at 10^2 was considered dividing the total RBC population into basal and high intensity levels. Graph (**C**) quantifies the mean fluorescence intensity (MFI) values of the total NOS3 expression and its phosphorylation status on the immunolabelled UC arterial RBCs., as derived from both the Hwt and Lwt mature discordant (n = 18 pairs) twin siblings. The statistical significance was accepted at *** $p < 0.001$ based on one-way ANOVA using the Newman–Keuls multiple comparison test.

3.3. Membrane Damage Verification in the UC Vessels' Endothelium and Red Blood Cells; Measurement of 4-Hydroxynonenal Level as a Lipid Peroxidation Marker

During the period of twin pregnancies, there exists a high intra-uterine hypoxic environment. Subsequently, there occurs an excessive production of ROS, which induces the formation of oxygenated α, β-unsaturated aldehydes, such as 4-hydroxynonenal (4-HNE). The lipid peroxidation level was followed by immunolabelling the aldehyde-protein adduct using an anti-4-HNE antibody. In the vessels, though, there were significant differences between the high and low birth-weight siblings in the study population (~35–40% of the higher value was indicated in the arteries, and a 15% rise occurred in the vein) (Figure 3A); the major differences were measured in the circulating RBCs. The intensity level (MFI value) of 4-HNE doubled in the smaller-weight siblings (Figure 3B), and ~53% of the RBCs showed extremely high intensity values (i.e., above 10^3) based on the FACS analyses data set (Figure 3C).

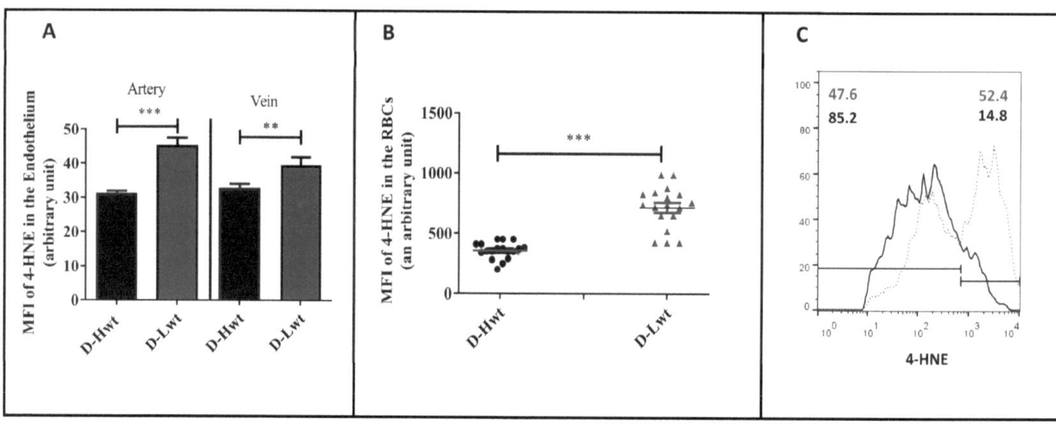

Figure 3. 4-hydroxynonenal (4-HNE) level in the immunolabelled umbilical cord (UC) vessels and red blood cells (RBCs) of the mature, birth weight–discordant twin neonates. Graph (**A**,**B**) determines the mean fluorescence intensity (MFI) values to measure the 4-HNE level in the immunolabelled UC vessel sections and the UC arterial RBCs, respectively, between high-weight (D-Hwt -black) and low-weight (D-Lwt—gray) siblings (n = 18 pairs). Panel (**C**) shows representative histograms as measured from the fluorescence-activated cell sorting analysis using anti-4 HNE primary antibodies on the UC arterial RBCs, (D-Hwt—black line) and low-weight (D-Lwt—gray dots). According to the blank sample, an arbitrary borderline at 10^3 was considered dividing the total RBC population into basal and high intensity levels. All the measured statistical significance was accepted at ** $p < 0.01$ and *** $p < 0.001$ based on one-way ANOVA using the Newman–Keuls multiple comparison test.

3.4. Peroxynitrite ($ONOO^-$) Accumulation in Red Blood Cell Populations

One of the major culprits for the elevated level of lipid peroxidation is the increased formation of $ONOO^-$ via the spontaneous reaction between superoxide anion ($O_2^{\bullet-}$) and NO. More than 50% of the RBC samples derived from the low-weight population showed a highly increased level of $ONOO^-$ (40–60% differences between siblings), while the high-weight siblings' $ONOO^-$ levels were quite comparable to their age- and weight-matched, birth weight–non-discordant twin neonates (Figures 4 and S6).

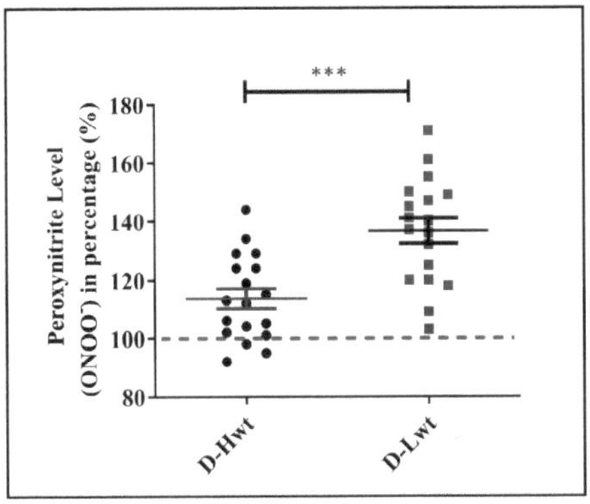

Figure 4. Peroxynitrite level in the mature, birth weight–discordant twin population. Peroxynitrite levels of RBC populations derived from mature (D-Hwt—black and D-Lwt—gray), birth weight–discordant siblings (n = 18 pairs). Values were expressed in percentages of age- and weight-matched, mature, birth weight–non-discordant twin neonates, taken as control (100%). Statistical significance was accepted at *** $p < 0.001$ by one-way ANOVA using the Newman–Keuls multiple comparison test.

3.5. Accumulation of Morphological Variants in the Birth Weight–Discordant Red Blood Cell Populations

As a further consequence of the substantial stress condition, significant alterations in the phenotypical variants were detected, and as a result, RBCs were losing their characteristic biconcave disk-shaped structures. Altogether, ~17,000 cells were counted from the n = 18 pairs of siblings and analyzed using the bio-image analysis tool Cell Profiler™. In the Lwt siblings, the Burr cell population was found to be ~30% of the total RBC population (Figures 5 and S7), and even besides that, a formation of the Rouleaux phenotype was also detectable within ~4–5%.

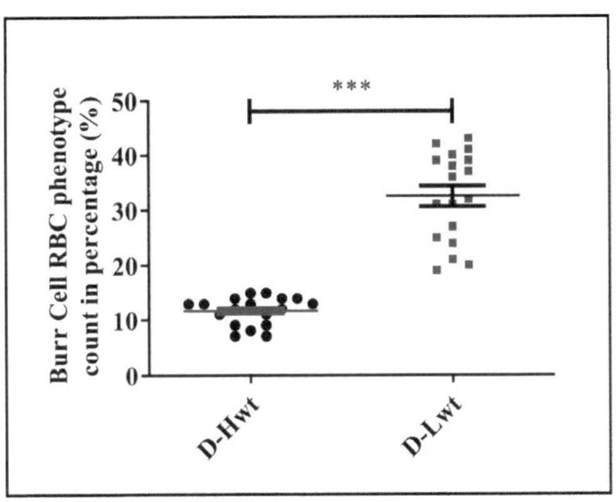

Figure 5. Distribution of the phenotypical variations in the fetal red blood cell population. Frequency of the Burr cell count was obtained as percentage with respect to healthy biconcave-shaped cells taken as control, where the total cell count of approximately ~17,000 cells was considered from the (n = 18 pairs) mature, birth weight–discordant twin siblings. Statistical significance was accepted at *** $p < 0.001$ by one-way ANOVA using the Newman–Keuls multiple comparison test.

4. Discussion

Substantial birth-weight differences between the twin siblings indicate a clinical condition of insufficient oxygenation and reduced blood flow to one of the fetuses. Birth-weight discordance may be well-connected to long-term intrauterine hypoxia and low bioavailable NO levels. In this study, mature, birth weight–discordant twin pairs were analyzed, mostly focusing on the NO-producing capacity. In the vascular system, the bioavailability of NO is crucial for the regulation of redox balance and functionality, most particularly in the case of the UC with no innervation. UC vessel-related complications can be a useful device for the study of various pathophysiological mechanisms related to neonatal cardiovascular disorders. Additionally, impairment in the primary source of NO production via the NOS3 signaling pathway offers an opportunity to have an insight into the potential compensatory mechanisms, like the NOS2 upregulation and the activation of the RBC-NOS3-NO pathway.

First, we checked the values of higher-weight siblings, considered them as a reference, and found that the NOS3 system and its activation status are highly undermined in siblings with low percentile values. In parallel, the NOS2 isoform might play an active role in order to maintain adequate vascular tone since its catalytic activity is 100–1000-fold that of NOS3 [11]. However, in the birth weight–discordant siblings, there were no signs of NOS2 upregulation in any of the vessels. Additionally, in the low–birth weight neonates, the RBC-NOS3 system was also not available to further increase the level of bioavailable NO. Impairment in the RBC-NOS3 system might be associated with an elevated level of $ONOO^-$ formation (~20–60% increase). Among the other factors, NOS3 was tightly regulated by its substrate and cofactors availability [24]. The process of NOS3 coupling is the most crucial step toward its activation. So, the oxidative milieu–induced deficiency of cofactors, like tetrahydrobiopterin, is one of the major limitations of NOS3 homodimer formation, resulting in the generation of $O_2^{\bullet -}$ instead of NO [25,26]. Thus, here, the NOS3 itself becomes a source of free radicals by creating a loop of spontaneous reaction, where the elevated level of $O_2^{\bullet -}$ scavenges the limited bioavailable NO by forming the deleterious $ONOO^-$ entity [27]. In our study, nearly 45% of the lower-weight siblings showed a 30–60% increase in the $ONOO^-$ levels. The NOS3-$O_2^{\bullet -}$ activation phenomenon in the RBCs, especially in the neonatal system, was so far without any example, but currently, endothelial cell lines have shown quite a few supportive results [27–31]. Accumulation of the highly reactive $ONOO^-$ is toxic for the cell not only because of its impact on the NO production but also because it can even initiate and aggravate the rate of lipid peroxidation in the cellular membranes, resulting in the excessive accumulation of the aldehyde, 4-HNE. The 4-HNE is highly reactive and can form adducts with nucleic acids [32], phospholipids, and nucleophilic amino acids, such as cysteine, histidine, and lysine residues. Moreover, the 4-HNE cross-linked conjugates propagate lipid peroxidation with subsequent loss in redox homeostasis and lead to pathological disorders [33–36]. In our data set, it was quite striking that the formation of 4-HNE was exceptionally higher in the case of RBCs derived from the low-weight siblings in comparison to the vascular endothelial cells. Again, the high $ONOO^-$ level in the circulating adult RBCs was somehow also responsible for the increase of Arginase1 level both in the RBCs, as well as in the vessel endothelium. The increased level of Arginase1 might play a pivotal role in the formation of ED [37].

In addition to the functional impairment of the RBCs, their morphological changes indicate a massive response to stress exposure in siblings with lower birth weights. There is not much information about the frequency of fetal/neonatal morphological variants of RBCs. Based on Zipursky and coworkers' study on mature and pre-mature neonatal blood samples, only 43–39% of the RBCs showed regular discoid shapes, while in the healthy adult population, this value reached ~78% [38]. The exact mechanism responsible for the Burr cell formation is unknown, but this type of shape transformation might be associated with the decrease in the concentration of ATP within the cell [39]. The depletion of the ATP pool affects the oxygen affinity of hemoglobin, and the release of oxygen may be less effective from the RBCs with this kind of phenotype [40]. Ruef and Linderkamp published a detailed analysis of the Burr cell population for mature and pre-mature singletons with values of 1.4 and 1.7%, respectively [41]. In our practice, we found that an average of 3% of Burr cell variants was present in the neonates born from healthy single pregnancies. On the other hand, in the case of singletons born to smoking mothers [16], this value has reached 15–18%. Here, appearances of the Burr cell population were connected to the highly elevated level of $ONOO^-$ formation and 4-HNE production. In the case of mature or pre-mature twins, we measured ~10–12% of this type of phenotypical variant, with a clear indication of elevated oxidative stress. It has been documented that an increase in the formation of Burr cells population can be well-connected to some pathological events, like that of kidney failure, pyruvate kinase deficiency, and certain liver diseases [42–44].

Our data set clearly indicates that there lie significant differences between the arteries and veins. It is always the artery where the impairment in the NO-producing capacity is more prominent, especially in the case of the low-weight siblings. Summarizing this fact with the highly impaired RBC functionality, it seems like the fetoplacental circulation gets remarkably hindered in comparison to the uteroplacental circulation in the lower-weight population.

The guidelines for uncomplicated twin pregnancies and the suggested protocol for a healthy gestational period, as described by the American College of Obstetricians and Gynecologists and the Society for Maternal-Fetal Medicine, respectively, is between 38 and 39 weeks. Despite this, from the clinician's point of view, a clause often arises whether an earlier timing, even at the late pre-mature period, may be more favorable for neonatal outcome [45,46]. Based on our present and previous studies on twins, we assume that an earlier delivery time can be considered, especially in the case of mature, birth weight–discordant pregnancy. It seems like, in this case, neither the NOS3-NO-producing capacity nor the "built in" rescue mechanisms, like NOS2 and RBC-NOS3, are sufficient enough to support adequate blood flow to the lower-weight fetuses. Contrastingly, in our recent publications on mature and pre-mature, birth weight–non-discordant and pre-mature, birth weight–discordant twin neonates, NOS2 was highly upregulated in the arteries in the case of impaired NOS3 performance, which showcased a compensatory mechanism [13,17]. Similarly, we also demonstrated the pivotal role of fetal RBCs with an export mechanism of NO bioactivity for the lower-weight siblings, regardless of maturity. The RBC populations in all such cases still had the capacity to increase the NO production via upregulation of the RBC-NOS3-NO pathway. This phenomenon suggested an efficient crosstalk mechanism between the RBCs and the impaired vascular system [13,17].

In spite of the major differences between the birth weight–discordant siblings, the lower-weight siblings still come under the category of appropriate for gestational age. Moreover, as examined in our previous study, this weight range is similar to the low-weight, mature, non-discordant twin pairs [13]. Based on the comparison of our data sets (Figure S5), it might be clarified that, in the D-Lwt neonates, the NO production capacity of the arterial vessels, in addition to the RBC populations, remains far behind the weight- and age-matched birth weight–non-discordant twins. This fact further supports our earlier hypothesis that, in the case of the low weight–discordant neonates, it is the fetoplacental circulation that gets hampered and not the uteroplacental one. The UC is fully embryonic in origin, and the vessels in it are considered to be a direct elongation of the vascular system for the developing fetus. This is the reason it is important to highlight that not only the birth weight but also the overall status of the endothelial layer in the UC vessels, along with the circulating fetal RBCs, form the determining factors for the healthy condition of neonates. Our results clearly indicate that the accumulation of strong pro-oxidant entities, like the toxic $ONOO^-$, in the circulation can significantly deteriorate the physiological status of the vascular system.

5. Conclusions

In the case of the birth weight–discordant siblings, even though their birth weights matched the appropriate weights for gestational age groups, the molecular parameters suggest that the lower-weight siblings are markedly exposed to persistent, intensive intrauterine hypoxic stress conditions. The RBC population of the same may act as a biosensor along with all the detectable phenotypic alterations and functional impairments, which can be taken as blueprints of such pathological conditions. It is quite probable this may have long-term consequences during the *in-utero* fetal development and even in their post-natal/adult lives.

Conclusively, to mitigate the long-term consequences of an extreme hypoxic condition of the vascular system, an earlier timing of delivery may be beneficial for a better clinical outcome of the lower-weight siblings.

Supplementary Materials: The following supporting information can be downloaded from https://www.mdpi.com/article/10.3390/antiox12071370/s1: Figure S1: Epifluorescent images of the umbilical cord artery and vein endothelium with relation to endothelial nitric oxide synthase (NOS3) expression and its activation (pNOS3) pattern; Figure S2: Representation of epifluorescent images of the umbilical cord vessels endothelium immunolabelled for inducible nitric oxide synthase (NOS2); Figure S3: Endothelial (NOS3) and inducible nitric oxide synthase (NOS2) levels in the umbilical cord vessels; Figure S4: Evaluation of the expression pattern of endothelial (NOS3) and inducible nitric oxide synthase (NOS2) and the phosphorylation status of NOS3 (pNOS3) at the Ser1177 residues in between the mature, birth weight–discordant versus non-discordant high-weight twin groups; Figure S5: Evaluation of the expression pattern of endothelial (NOS3) and inducible nitric oxide synthases (NOS2) and the phosphorylation status of NOS3 (pNOS3) at the Ser1177 residues in between the mature, birth weight–discordant versus non-discordant low-weight twin groups; Figure S6: Distribution pattern of the peroxynitrite level between the mature, birth weight–discordant twin siblings' Figure S7: Representative epifluorescent images of red blood cell variants originated from high-weight (D-Hwt) and low-weight (D-Lwt) siblings; Table S1: Clinical parameters of the study groups and the maternal age; and Table S2: List of the sources and dilutions of primary and secondary antibodies.

Author Contributions: P.C. performed the experiments. H.O. organized the sample collections. E.H. designed and supervised the study and wrote the manuscript with support from P.C. All authors have read and agreed to the published version of the manuscript.

Funding: This research was funded by the European Union and the Hungarian Government in the framework of the GINOP-2.3.2-15-2016-00035 project.

Institutional Review Board Statement: The study was conducted according to the guidelines of the Declaration of Helsinki and approved by the Ethics Committee of the Department of Obstetrics and Gynecology (16/2016).

Informed Consent Statement: For the sample collections, written informed consent was obtained from pregnant mothers.

Data Availability Statement: The data will be available upon request.

Acknowledgments: The authors are grateful to Anirban Lahiri and Arka Bhattacharya for their contributions to the preparation of the figures.

Conflicts of Interest: The authors declare no conflict of interest.

References

1. Ochoa, J.J.; Ramirez-Tortosa, M.C.; Quiles, J.L.; Palomino, N.; Robles, R.; Mataix, J.; Huertas, J.R. Oxidative stress in erythrocytes from premature and full-term infants during their first 72 h of life. *Free Radic. Res.* **2003**, *37*, 317–322. [CrossRef] [PubMed]
2. Umranikar, A.; Parmar, D.; Davies, S.; Fountain, S. Multiple births following in vitro fertilization treatment: Redefining success. *Eur. J. Obstet. Gynecol. Reprod. Biol.* **2013**, *170*, 299–304. [CrossRef] [PubMed]
3. Puccio, G.; Giuffré, M.; Piccione, M.; Piro, E.; Malerba, V.; Corsello, G. Intrauterine growth pattern and birthweight discordance in twin pregnancies: A retrospective study. *Ital. J. Pediatr.* **2014**, *40*, 43. [CrossRef] [PubMed]
4. Fox, S.B.; Khong, T.Y. Lack of innervation of human umbilical cord. An immunohistological and histochemical study. *Placenta* **1990**, *11*, 59–62. [CrossRef] [PubMed]
5. Vanhoutte, P.M.; Shimokawa, H.; Feletou, M.; Tang, E.H.C. Endothelial dysfunction and vascular disease—A 30th anniversary update. *Acta Physiol.* **2017**, *219*, 22–96. [CrossRef]
6. Berk, B.C.; Haendeler, J.; Sottile, J. Angiotensin II, atherosclerosis, and aortic aneurysms. *J. Clin. Investig.* **2000**, *105*, 1525–1526. [CrossRef]

7. Granger, J.P.; Alexander, B.T.; Llinas, M.T.; Bennett, W.A.; Khalil, R.A. Pathophysiology of hypertension during preeclampsia linking placental ischemia with endothelial dysfunction. *Hypertension* **2001**, *38*, 718–722. [CrossRef]
8. Nedeljkovic, Z.S.; Gokce, N.; Loscalzo, J. Mechanisms of oxidative stress and vascular dysfunction. *Postgrad. Med. J.* **2003**, *79*, 195–200. [CrossRef]
9. Selye, H. Stress and disease. *Science* **1955**, *122*, 625–631. [CrossRef]
10. Geng, Y.J.; Wu, Q.; Muszynski, M.; Hansson, G.K.; Libby, P. Apoptosis of vascular smooth muscle cells induced by in vitro stimulation with interferon-γ, tumor necrosis factor-α, and interleukin-1β. *Arter. Thromb. Vasc. Biol.* **1996**, *16*, 19–27. [CrossRef]
11. Farah, C.; Michel, L.Y.M.; Balligand, J.L. Nitric oxide signalling in cardiovascular health and disease. *Nat. Rev. Cardiol.* **2018**, *15*, 292–316. [CrossRef] [PubMed]
12. Liu, X.M.; Chapman, G.B.; Peyton, K.J.; Schafer, A.I.; Durante, W. Carbon monoxide inhibits apoptosis in vascular smooth muscle cells. *Cardiovasc. Res.* **2002**, *55*, 396–405. [CrossRef] [PubMed]
13. Chakraborty, P.; Dugmonits, K.N.; Orvos, H.; Hermesz, E. Mature twin neonates exhibit oxidative stress via nitric oxide synthase dysfunctionality: A prognostic stress marker in the red blood cells and umbilical cord vessels. *Antioxidants* **2020**, *9*, 845. [CrossRef] [PubMed]
14. Cortese-Krott, M.; Rodriguez-Mateos, A.; Sansone, R.; Kuhnle, G.; Thasian-Sivarajah, S.; Krenz, T.; Horn, P.; Krisp, C.; Wolters, D.; Heiss, C.; et al. Human red blood cells at work: Identification and visualization of erythrocytic eNOS activity in health and disease. *Blood* **2012**, *120*, 4229–4237. [CrossRef]
15. Cortese-Krott, M.M.; Kelm, M. Endothelial nitric oxide synthase in red blood cells: Key to a new erythrocrine function? *Redox Biol.* **2014**, *2*, 251–258. [CrossRef]
16. Dugmonits, K.N.; Chakraborty, P.; Hollandi, R.; Zahorán, S.; Pankotai-Bodó, G.; Horváth, P.; Orvos, H.; Hermesz, E. Maternal Smoking Highly Affects the Function, Membrane Integrity, and Rheological Properties in Fetal Red Blood Cells. *Oxid. Med. Cell. Longev.* **2019**, *2019*, 1509798. [CrossRef]
17. Chakraborty, P.; Khamit, A.; Hermesz, E. Fetal oxygen supply can be improved by an effective cross-talk between fetal erythrocytes and vascular endothelium. *Biochim. Biophys. Acta-Mol. Basis Dis.* **2021**, *1867*, 166243. [CrossRef]
18. Ding, H.; Triggle, C.R. Endothelial cell dysfunction and the vascular complications associated with type 2 diabetes: Assessing the health of the endothelium. *Vasc. Health Risk Manag.* **2005**, *1*, 55–71. [CrossRef]
19. Barthelmes, J.; Nägele, M.P.; Ludovici, V.; Ruschitzka, F.; Sudano, I.; Flammer, A.J. Endothelial dysfunction in cardiovascular disease and Flammer syndrome-similarities and differences. *EPMA J.* **2017**, *8*, 99–109. [CrossRef]
20. Nishikawa, E.; Matsumoto, T.; Isige, M.; Tsuji, T.; Mugisima, H.; Takahashi, S. Comparison of capacities to maintain hematopoietic stem cells among different types of stem cells derived from the placenta and umbilical cord. *Regen. Ther.* **2016**, *4*, 48–61. [CrossRef]
21. Chakraborty, P.; Dugmonits, K.N.; Végh, A.G.; Hollandi, R.; Horváth, P.; Maléth, J.; Hegyi, P.; Németh, G.; Hermesz, E. Failure in the compensatory mechanism in red blood cells due to sustained smoking during pregnancy. *Chem. Biol. Interact.* **2019**, *313*, 108821. [CrossRef] [PubMed]
22. Huie, R.E.; Padmaja, S. The reaction of no with superoxide. *Free Radic. Res.* **1993**, *18*, 195–199. [CrossRef] [PubMed]
23. Carpenter, A.E.; Jones, T.R.; Lamprecht, M.R.; Clarke, C.; Kang, I.H.; Friman, O.; Guertin, D.A.; Chang, J.H.; Lindquist, R.A.; Moffat, J.; et al. CellProfiler: Image analysis software for identifying and quantifying cell phenotypes. *Genome. Biol.* **2006**, *7*, R100. [CrossRef] [PubMed]
24. Michel, T.; Feron, O. Nitric oxide synthases: Which, where, how, and why? *J. Clin. Investig.* **1997**, *100*, 2146–2152. [CrossRef] [PubMed]
25. Förstermann, U.; Li, H. Therapeutic effect of enhancing endothelial nitric oxide synthase (eNOS) expression and preventing eNOS uncoupling. *Br. J. Pharmacol.* **2011**, *164*, 213–223. [CrossRef]
26. Förstermann, U.; Sessa, W.C. Nitric oxide synthases: Regulation and function. *Eur. Heart J.* **2012**, *33*, 829–837. [CrossRef] [PubMed]
27. Cau, S.B.A.; Carneiro, F.S.; Tostes, R.C. Differential modulation of nitric oxide synthases in aging: Therapeutic opportunities. *Front. Physiol.* **2012**, *3*, 218. [CrossRef]
28. Cosentino, F.; Lüscher, T.F. Tetrahydrobiopterin and endothelial nitric oxide synthase activity. *Cardiovasc. Res.* **1999**, *43*, 274–278. [CrossRef]
29. Pritchard, K.A., Jr.; Groszek, L.; Smalley, D.M.; Sessa, W.C.; Wu, M.; Villalon, P.; Wolin, M.S.; Stemerman, M.B. Native low-density lipoprotein increases endothelial cell nitric oxide synthase generation of superoxide anion. *Circ. Res.* **1995**, *77*, 510–518. [CrossRef]
30. Landmesser, U.; Dikalov, S.; Price, S.R.; McCann, L.; Fukai, T.; Holland, S.M.; Mitch, W.E.; Harrison, D.G. Oxidation of tetrahydrobiopterin leads to uncoupling of endothelial cell nitric oxide synthase in hypertension. *J. Clin. Investig.* **2003**, *111*, 1201–1209. [CrossRef]
31. Laursen, J.B.; Somers, M.; Kurz, S.; McCann, L.; Warnholtz, A.; Freeman, B.A.; Tarpey, M.; Fukai, T.; Harrison, D.G. Endothelial regulation of vasomotion in apoE-deficient mice: Implications for interactions between peroxynitrite and tetrahydrobiopterin. *Circulation* **2001**, *103*, 1282–1288. [CrossRef] [PubMed]
32. Nair, J.; De Flora, S.; Izzotti, A.; Bartsch, H. Lipid peroxidation-derived etheno-DNA adducts in human atherosclerotic lesions. *Mutat. Res.* **2007**, *621*, 95–105. [CrossRef] [PubMed]
33. Doorn, J.A.; Petersen, D.R. Covalent adduction of nucleophilic amino acids by 4-hydroxynonenal and 4-oxononenal. *Chem. Biol. Interact.* **2003**, *143–144*, 93–100. [CrossRef] [PubMed]

34. Benedetti, A.; Comporti, M.; Esterbauer, H. Identification of 4-hydroxynonenal as a cytotoxic product originating from the peroxidation of liver microsomal lipids. *Biochim. Biophys. Acta* **1980**, *620*, 281–296. [CrossRef] [PubMed]
35. Esterbauer, H.; Schaur, R.J.; Zollner, H. Chemistry and biochemistry of 4-hydroxynonenal, malonaldehyde and related aldehydes. *Free Radic. Biol. Med.* **1991**, *11*, 81–128. [CrossRef]
36. Chapple, S.J.; Cheng, X.; Mann, G.E. Effects of 4-hydroxynonenal on vascular endothelial and smooth muscle cell redox signaling and function in health and disease. *Redox Biol.* **2013**, *1*, 319–331. [CrossRef]
37. Mahdi, A.; Tengbom, J.; Alvarsson, M.; Wernly, B.; Zhou, Z.; Pernow, J. Red Blood Cell Peroxynitrite Causes Endothelial Dysfunction in Type 2 Diabetes Mellitus via Arginase. *Cells* **2020**, *9*, 1712. [CrossRef]
38. Zipursky, A.; Brown, E.; Palko, J.; Brown, E.J. The erythrocyte differential count in newborn infants. *Am. J. Pediatr. Hematol. Oncol.* **1983**, *5*, 45–51.
39. Elgsaeter, A.; Mikkelsen, A. Shapes and shape changes in vitro in normal red blood cells. *Biochim. Biophys. Acta* **1991**, *1071*, 273–290. [CrossRef]
40. Chowdhury, A.; Dasgupta, R.; Majumder, S.K. Changes in hemoglobin-oxygen affinity with shape variations of red blood cells. *J. Biomed. Opt.* **2017**, *22*, 105006. [CrossRef]
41. Ruef, P.; Linderkamp, O. Deformability and Geometry of Neonatal Erythrocytes with Irregular Shapes. *Pediatr. Res.* **1999**, *45*, 114–119. [CrossRef] [PubMed]
42. Mandal, A.K.; Taylor, C.A.; Bell, R.D.; Hillman, N.M.; Jarnot, M.D.; Cunningham, J.D.; Phillips, L.G. Erythrocyte deformation in ischemic acute tubular necrosis and amelioration by splenectomy in the dog. *Lab. Investig.* **1991**, *65*, 566–576. [PubMed]
43. Turchetti, V.; De Matteis, C.; Leoncini, F.; Trabalzini, L.; Guerrini, M.; Forconi, S. Variations of erythrocyte morphology in different pathologies. *Clin. Hemorheol. Microcirc.* **1997**, *17*, 209–215. [PubMed]
44. Suljević, D.; Mitrašinović-Brulić, M.; Fočak, M. L-cysteine protective effects against platelet disaggregation and echinocyte occurrence in gentamicin-induced kidney injury. *Mol. Cell. Biochem.* **2023**, *478*, 13–22. [CrossRef]
45. Dodd, J.M.; Crowther, C.A.; Haslam, R.R.; Robinson, J.S. Elective birth at 37 weeks of gestation versus standard care for women with an uncomplicated twin pregnancy at term: The Twins Timing of Birth Randomised Trial. *BJOG* **2012**, *119*, 964–973. [CrossRef] [PubMed]
46. Cheong-See, F.; Schuit, E.; Arroyo-Manzano, D.; Khalil, A.; Barrett, J.; Joseph, K.S.; Asztalos, E.; Hack, K.; Lewi, L.; Lim, A.; et al. Prospective risk of stillbirth and neonatal complications in twin pregnancies: Systematic review and meta-analysis. *BMJ* **2016**, *354*, i4353. [CrossRef]

antioxidants

MDPI

Article

Alterations in Antioxidant Status and Erythrocyte Properties in Children with Autism Spectrum Disorder

Tomas Jasenovec [1], Dominika Radosinska [2], Katarina Jansakova [1], Maria Kopcikova [1], Aleksandra Tomova [1], Denisa Snurikova [3], Norbert Vrbjar [3] and Jana Radosinska [1,3,*]

1 Institute of Physiology, Faculty of Medicine, Comenius University in Bratislava, Sasinkova 2, 811 08 Bratislava, Slovakia; tomas.jasenovec@fmed.uniba.sk (T.J.); katarina.jansakova@fmed.uniba.sk (K.J.); maria.kopcikova@fmed.uniba.sk (M.K.); aleksandra.tomova@fmed.uniba.sk (A.T.)

2 Institute of Medical Biology, Genetics and Clinical Genetics, Faculty of Medicine, Comenius University in Bratislava, Sasinkova 4, 811 08 Bratislava, Slovakia; dominika.radosinska@fmed.uniba.sk

3 Centre of Experimental Medicine, Slovak Academy of Sciences, Institute for Heart Research, Dúbravská Cesta 9, 841 04 Bratislava, Slovakia; usrddena@savba.sk (D.S.); norbert.vrbjar@savba.sk (N.V.)

* Correspondence: jana.radosinska@fmed.uniba.sk

Abstract: Erythrocytes are responsible for the transport of oxygen within the organism, which is particularly important for nerve tissues. Erythrocyte quality has been shown to be deteriorated in oxidative stress conditions. In this study, we measured the same series of oxidative stress markers in plasma and erythrocytes to compare the differences between neurotypical children (controls) and children with autism spectrum disorder (ASD). We also focused on erythrocyte properties including their deformability, osmotic resistance, Na,K-ATPase activity, nitric oxide levels and free radical levels in children with ASD and controls. Greater oxidative damage to proteins and lipids was observed in the erythrocytes than in the plasma of ASD subjects. Additionally, antioxidant enzymes were more active in plasma samples from ASD children than in their erythrocytes. Significantly higher nitric oxide level and Na,K-ATPase enzyme activity were detected in erythrocytes of ASD individuals in comparison with the controls. Changes in oxidative status could at least partially contribute to the deterioration of erythrocyte morphology, as more frequent echinocyte formation was detected in ASD individuals. These alterations are most probably responsible for worsening the erythrocyte deformability observed in children with ASD. We can conclude that abnormalities in antioxidant status and erythrocyte properties could be involved in the pathomechanisms of ASD and eventually contribute to its clinical manifestations.

Keywords: autism; oxidative stress; erythrocytes; deformability; nitric oxide; Na,K-ATPase; erythrocyte morphology

Citation: Jasenovec, T.; Radosinska, D.; Jansakova, K.; Kopcikova, M.; Tomova, A.; Snurikova, D.; Vrbjar, N.; Radosinska, J. Alterations in Antioxidant Status and Erythrocyte Properties in Children with Autism Spectrum Disorder. *Antioxidants* **2023**, *12*, 2054. https://doi.org/10.3390/antiox12122054

Academic Editors: Angelo D'Alessandro and Alkmini T. Anastasiadi

Received: 31 October 2023
Revised: 21 November 2023
Accepted: 27 November 2023
Published: 28 November 2023

1. Introduction

Oxidative stress is considered a typical feature of most diseases. The imbalance between the production and elimination of free radicals leads to an elevation of reactive molecules that circulate in the blood. Circulating radicals come into contact with red blood cells (RBCs), which are responsible for the transport of oxygen within the organism. Exposure to oxygen, as well as a limited ability to resist oxidative damage (as RBCs are not real cells), can lead to the deterioration of RBC quality in a variety of pathological conditions, as shown by human studies [1]. RBC functionality is particularly important for nerve tissue because of the continuous and high oxygen requirements of nerve cells. To accomplish this oxygen-delivery function, RBCs are supposed to deform repeatedly during their transport through the capillaries, as the smallest ones are narrower than the diameter of RBCs. The deformability of RBCs is thus crucial for functional microcirculation. In addition to oxidative damage, the ability of RBCs to reversibly deform is influenced and conditioned by proper ion homeostasis, which is mostly ensured by the enzyme Na,K-ATPase, as well as by the production of nitric oxide (NO) and the mechanical properties of the RBC membrane [2].

Autism spectrum disorder (ASD) represents a complex of neurodevelopmental disorders with increasing incidence. ASD remains the focus of many research teams searching for biomarkers, indicating the severity and progression of this disorder or helping to clarify its etiopathogenesis, which is still unclear. ASD is characterized by core symptoms that include problems in social communication and interaction, as well as restricted repetitive patterns of behavior, activities or interests [3]. In addition to these behavioral characteristics, children with ASD typically suffer from a wide range of somatic problems at a higher rate than their neurotypical counterparts, e.g., feeding and gastrointestinal disorders [4,5], immunological disturbances [6], sleep disorders [7], neurological diseases and a variety of others [8,9]. Oxidative stress, implicated in the pathophysiology of multiple disorders, has been shown to be a common denominator of neurodevelopmental diseases including ASD [10–12]. It could represent a consequence of the insufficient action of cellular antioxidants [13] or the enhanced production of reactive substances, e.g., due to mitochondrial dysfunction [14,15] in ASD conditions. As mentioned previously, oxidative stress generates unfavorable conditions for the optimal functioning of RBCs, leading to the deterioration of their properties [1]. A decrease in the quality of RBCs could have a significant impact on hemorheology [16], and thus could aggravate the clinical symptoms of multiple disorders due to the suboptimal delivery of oxygen to metabolically active tissues [17]. We have previously proposed that abnormalities in RBC deformability may be involved in the pathomechanisms of ASD and contribute to its clinical manifestations [18]. The aim of this study was to determine a series of oxidative stress markers in blood plasma and in RBCs, as well as to focus on RBC properties including the deformability, Na,K-ATPase activity, and NO and free radical production by RBCs in ASD and neurotypical children. The reasoning for such an approach was based on previously observed indices of oxidative stress and deteriorations of RBC quality in ASD, e.g., see [19–21]. However, relatively few studies have focused on the simultaneous determination of the oxidative state in blood plasma and in RBCs, together with parameters affecting RBC functions. Furthermore, not all data that are available in databases show consistent results.

2. Materials and Methods

2.1. Study Design

We enrolled 36 non-verbal children diagnosed with ASD (32 boys and 4 girls) and 17 neurotypical children (12 boys and 5 girls) in this study. The median age of children with ASD was 3.3 years, ranging from a minimum of 2.7 years to a maximum of 7.6 years, while the neurotypical children had ages ranging from 2.4 to 6.3 years, with a median value of 5.4 years, without significant differences between both groups. ASD diagnosis followed the criteria outlined in the DSM-V (American Psychiatric Association 2013). We administered the Autism Diagnostic Interview-Revised (ADI-R) and the Autism Diagnostic Observation Schedule–second edition (ADOS-2) as our diagnostic tools, as described previously [18]. Children with suspected ASD diagnosis were administered Module 1 of ADOS-2, specifically designed for children over 30 months of age who do not consistently use phrase speech. This module consists of 10 structured tasks to assess social interactions and communication deficits. Each type of behavior has its own score that is needed for the diagnostic algorithm. Participants assigned to the control group were selected from a group of children with no known history of ASD. We also administered the M-CHAT screening questionnaire with their parents. There were no significant scores indicative of ASD. All evaluations and ASD diagnoses were conducted at the Academic Research Center for Autism, Institute of Physiology, Faculty of Medicine, Comenius University in Bratislava. This study received approval from the Ethics Committee of the Faculty of Medicine, Comenius University and University Hospital in Bratislava, Slovak Republic.

Informed consent was obtained from all caregivers of the children. The study adhered to the ethical principles outlined in the Declaration of Helsinki.

2.2. Blood Processing

Venous blood was drawn into EDTA-containing vacutainer tubes right after the diagnostic procedures. The blood was immediately centrifuged at 1150× *g* for 5 min at 4 °C. The plasma was collected, while the buffy coat and the upper 20% of RBCs were removed. The remaining RBCs were subjected to three washes with saline solution. A portion of the RBCs were subjected to osmotic hemolysis in distilled water (*v:v*, 1:19). Plasma and hemolyzed RBCs were stored at −80 °C until further analysis.

2.3. Parameters of Oxidative Stress and Antioxidant Status

Oxidative stress markers, antioxidant status and carbonyl stress were detected in both plasma and hemolyzed RBCs. All chemicals were purchased from Merck Sigma-Aldrich (Merck KGaA, Darmstadt, Germany), and analyses were performed by using a Synergy H1 Hybrid Multi-mode Reader machine (Agilent, Santa Clara, CA, USA).

Measurement of oxidative stress markers included the detection of thiobarbituric acid reactive substances (TBARS) as a marker of lipid peroxidation and advanced oxidation protein products (AOPP) as a marker or protein oxidation. For the assessment of TBARS, 20 μL of the samples and 1,1,3,3-tetraethoxypropane used for the calibration curve were mixed in a PCR plate with 30 μL of distilled water, 20 μL of 0.67% thiobarbituric acid prepared with dimethyl sulphoxide and 20 μL of glacial acetic acid. The plate was then incubated for 45 min at 95 °C. After incubation, 100 μL of n-butanol was added into every well, briefly mixed and subsequently centrifuged at 2000× *g* and 4 °C for 10 min. The fluorescence was detected in the upper organic phase (70 μL) transferred into a dark microtiter plate at λ_{ex} = 515 nm and λ_{em} = 553 nm. To measure AOPP, 200 μL of chloramine T used as a standard was incubated for 6 min with potassium iodide and samples were added into a transparent microtiter plate. Twenty microliters of glacial acetic acid was added into both samples and standards and mixed for 2 min. The absorbance was then read at λ = 340 nm [22].

As a marker of antioxidant status, the assessment of ferric reducing antioxidant power (FRAP) was applied according to the method developed by Benzie and Strain [23]. The absorbance of 200 μL of freshly prepared and warmed FRAP reagent composed of acetate buffer, iron (III) chloride hexahydrate solution and distilled water was read as a blank at λ = 593 nm. Then, we added a 20 μL mixture of the samples and iron (II) sulphate heptahydrate used as a standard into a transparent microtiter plate and measured the absorbance at λ = 593 nm.

Carbonyl stress was assessed via the measurement of fructosamine, representing a marker of advanced glycation of proteins [24]. A 20 μL mixture of the samples and standards (1-deoxy-morpholino-D-fructose) was incubated with 100 μL of a 0.25 mmol/L nitroblue tetrazolium (NBT) solution composed of sodium carbonate and 1 mmol/L NBT at 37 °C for 15 min. The whole mixture in a plate was then measured at λ = 530 nm.

The measurement of reduced and oxidized glutathione (GSH/GSSG), and the subsequent calculation of their ratio, was used as an indicator of oxidative stress. A 10 μL mixture of the samples and reduced L-glutathione as a standard was added into a dark microtiter plate, mixed with 10 μL of O-phtalaldehyde solution and PBS with 2.5 mmol EDTA-Na_2 and incubated at room temperature for 15 min. After incubation, fluorescence was measured at λ_{ex} = 350 nm and λ_{em} = 460 nm. The GSSG detection procedure included the mixing of 25 μL samples and oxidized (-)-glutathione used as a standard with 10 μL of N-ethylmaleimide and incubating the mixture at room temperature for 40 min. Thereafter, 10 μL of the sample/standard mixture was transferred into a dark plate and incubated for 15 min with a 10 μL O-phtalaldehyde solution and 180 μL of 0.1 mmol/L NaOH. Fluorescence was measured at λ_{ex} = 350 nm and λ_{em} = 460 nm.

Additionally, the activity of three important enzymes of cellular antioxidant defense, namely superoxide dismutase (SOD), catalase (CAT) and glutathione peroxidase (GPx), was determined in plasma samples and hemolysates. For CAT, the method presented by Aebi [25] was applied. The sample (10 μL) was incubated for 3 min either with hydrogen peroxide solution or distilled water (100 μL). Then, ammonium molybdate (240 μL, 32.4 mM) was added and absorbance was measured at λ = 374 nm. For SOD and GPx activities, commercial kits were used (CS0009 and MAK437, Merck Sigma-Aldrich, Merck KGaA, Darmstadt, Germany), following the manufacturer's instructions.

2.4. *Erythrocyte Parameters Assessment*

To evaluate the RBC deformability, we employed a filtration technique, consistent with our previous experiments [26]. In short, RBC deformability was determined as the ratio of the count of RBCs that passed through a membrane filter with pores that were 5 μm in diameter (Ultrafree-MC SV Centrifugal Filter; Merck Millipore Ltd., Tullagreen Carrigtwohill, Ireland) to the initial RBC count before centrifugation. The Sysmex hematological analyzer (Sysmex F-820, Sysmex Corp, Tokyo, Japan) was used to count RBCs as well as to determine the mean cell volume (MCV) and RBC distribution width (RDW-SD). The MCV parameter reflects the size of a single RBC, and RDW indicates the differences in the volume of RBCs. The morphology of RBCs was observed using light microscopy. Ten microliters of whole blood was mixed with 90 μL of physiological solution, and blood smears were prepared from the mixture and processed immediately. RBCs were captured using a microscope (Axiolab 5, Zeiss, Jena, Germany) and classified into three categories, namely normal disc cells, echinocytes I (irregularly shaped RBCs), and echinocytes II (RBCs with spikes), as described previously [27]. On average, 750 RBCs per sample were analyzed.

2.5. *Fluorescent Microscope Techniques*

For the purposes of assessing RBC nitric oxide (NO) production, RBC calcium ion content, as well as reactive oxygen species presence, we utilized the following fluorescent probes: 4,5-diaminofluorescein diacetate (DAF-2 DA, 25 μmol/L, ab145283, Abcam, Cambridge, UK), Fluo-3 AM (4 μmol/L, ENZ52004 Enzo Life Sciences, Inc., Farmingdale, NY, USA) and dichlofluorescein (DCF, 50 μmol/L, D6883, Merck, Darmstadt, Germany). Before incubation with the corresponding probes, washed RBCs were diluted in modified physiological solution (*v:v*, 1:19; in mmol/L: NaCl 119, KCl 4.7, $NaHCO_3$ 25, $MgSO_4{\cdot}7H_2O$ 1.17, KH_2PO_4 1.18, $CaCl_2{\cdot}2H_2O$ 2.5, Na_2EDTA 0.03, glucose 5.5, and pH 7.4), whereas for the Fluo 3 AM probe, the solution was adjusted to a slightly higher calcium level, as described previously [28] (in mmol/L: NaCl 116.1, KCl 4.6, $NaHCO_3$ 24.4, $MgSO_4{\cdot}7H_2O$ 1.14, KH_2PO_4 1.15, $CaCl_2{\cdot}2H_2O$ 5, Na_2EDTA 0.03, glucose 5.37, and pH 7.4). The suspension was subsequently treated with a particular fluorescent probe, and the samples were incubated in the dark at 37 °C. After 30 min of incubation, the samples were washed and resuspended in the corresponding solutions. The fluorescence signal was captured using Zeiss Microscope Fluorescence Filter Cube Set 38 HE 489038-9901 (λ_{ex} = 470/40 nm, λ_{em} = 525/50 nm) and the fluorescence microscope Axiolab 5 (Zeiss, Jena, Germany). Quantification of the fluorescent signal intensity was performed using the ZEN 3.4 Blue (Carl Zeiss Microscopy GmbH, Jena, Germany) and ImageJ 1.53e software (National Institutes of Health, Bethesda, MD, USA), while it is expressed in arbitrary units per individual RBC.

2.6. *Na,K-ATPase Enzyme Kinetics Assessment*

The isolation of RBC membranes and the determination of Na,K-ATPase kinetic parameters were performed as described previously [27], while chemicals were purchased from Sigma-Aldrich (St. Louis, MO, USA). Washed RBCs were homogenized in 50 mmol/L TRIS, followed by centrifugation at 13,000× *g* for 30 min at 4 °C. The supernatant was then discarded, and the membrane-containing pellets were subjected to repeated homogenization and centrifugation in 30, 20, and 10 mmol/L TRIS solutions to eliminate residual hemoglobin. Na,K-ATPase activities were measured across a range of Na^+ concentrations

(2–100 mmol/L). RBC membrane proteins (50 μg) were preincubated for 20 min at 37 °C, and ATP (final concentration 8 mmol/L) was subsequently added. The chemical reaction was terminated after 20 min with 12% trichloroacetic acid. Inorganic phosphate generated via ATP hydrolysis was spectrophotometrically determined at λ = 700 nm. The obtained data on Na,K-ATPase activity at each NaCl concentration were used to generate kinetic curves and to derive the kinetic parameters of the Na,K-ATPase enzyme: V_{max} (maximum reaction velocity, i.e., the rate of the reaction at which the enzyme shows the highest turnover) and K_{Na} (the concentration of Na^+ required for half-maximum activation of the enzyme).

2.7. Assessment of Red Blood Cell Osmotic Resistance

The RBC osmotic resistance was determined using the established procedure [29]. A set of NaCl solutions ranging from 0% (i.e., distilled water) to 0.9% were prepared. Washed RBCs were exposed to each NaCl solution for 30 min, followed by centrifugation. In supernatants, the level of hemolysis was determined spectrophotometrically at λ = 540 nm. Supernatants from 0.9% NaCl were considered nonhemolytic, while those from distilled water were used as a reference for 100% hemolysis. Subsequently, the concentration of NaCl resulting in 50% hemolysis (IC_{50}) was calculated. A decrease in the IC_{50} value corresponds to an increase in the RBC osmotic resistance.

2.8. Statistical Analyses

Outliers were identified via the standardized Grubbs test and subsequently excluded from further analyses. The D'Agostino–Pearson test was employed to assess the data normality. Data are presented either as means ± standard deviations or as medians with interquartile ranges when a non-Gaussian distribution was observed. Statistical significance between groups was assessed using the unpaired t-test when data were normally distributed or Mann–Whitney test when not. The occurrence of different morphologies of RBCs was analyzed using the chi-square test. Significance was set at $p < 0.05$. Data analysis was performed using GraphPad Prism 7.02 software (GraphPad Software, San Diego, CA, USA).

3. Results

3.1. Parameters of Oxidative Stress and Antioxidant Status

Focusing on blood plasma, higher concentrations of fructosamine, a marker of early protein glycation, were observed in children with ASD compared with neurotypical children. The investigation of glycation parameters has been suggested as clinically relevant not only for the management of diabetes mellitus, but also for the panel of metabolic markers for the diagnosis of ASD [30]. In addition, higher activities of antioxidant enzymes-catalase and glutathione peroxidase were observed in children with ASD compared with neurotypical children.

In RBCs, higher levels of markers indicating oxidative damage to proteins (AOPP) and lipid peroxidation (TBARS) were found in children with ASD when compared with the controls.

The decrease in the plasma GSH/GSSG ratio (general marker of oxidative stress indicating the thiol–disulfide balance) and the increase in catalase activity in RBCs did not reach the level of statistical significance ($p = 0.06$ for both). All of the determined parameters with statistics are available in Table 1.

Table 1. Parameters of oxidative stress and antioxidant status determined in blood plasma and RBCs of neurotypical (i.e., control) children and children with ASD.

PLASMA	Control (n = 14–17)	ASD (n = 18–28)	Statistics (p Value)
AOPP (μmol/g prot)	0.664 ± 0.3	0.998 ± 0.73	0.11
FRUC (mmol/g prot)	0.017 ± 0.003	0.022 ± 0.003	**0.015**
TBARS (μmol/L)	1.55 ± 0.3	1.62 ± 0.34	0.56
FRAP (μmol/L)	570 ± 81	536 ± 100	0.24
GSH/GSSG	3.34 ± 0.29	3.01 ± 0.45	0.06
CAT (U/mL)	0.636 ± 0.35	1.21 ± 0.39	**<0.0001**
SOD (inhibition rate %)	2415 ± 232	2392 ± 234	0.77
GPx (U/L)	139 ± 21	155 ± 25	**0.046**
ERYTHROCYTES	**Control (n = 15–17)**	**ASD (n = 29–31)**	**Statistics (p Value)**
AOPP (μmol/g Hb)	7.25 ± 0.9	7.88 ± 0.8	**0.015**
FRUC (mmol/g Hb)	3.49 ± 0.67	3.73 ± 0.94	0.37
TBARS (μmol/L)	63.4 ± 22.3	93.7 ± 38.6	**0.007**
FRAP (μmol/L)	44,937 ± 4636	45,771 ± 5595	0.61
GSH/GSSG	0.68 ± 0.07	0.71 ± 0.15	0.43
CAT (U/mg Hb)	5.09 ± 0.67	5.52 ± 0.77	0.06
SOD (inhibition rate %)	830 (590; 2660)	710 (500; 880)	0.15
GPx (U/g Hb)	155 ± 46	160 ± 47	0.73

Abbreviations: ASD—autism spectrum disorder, AOPP—advanced oxidation protein products, FRUC—fructosamine, TBARS—thiobarbituric acid reactive substances, FRAP—ferric reducing antioxidant power, GSH/GSSG—the ratio of reduced to oxidized glutathione, CAT—catalase, SOD—superoxide dismutase, GPx—glutathione peroxidase, Hb—hemoglobin. Data are presented as means ± standard deviations or as median with interquartile range where appropriate. Statistical significance between groups was assessed using the unpaired t-test when data were normally distributed or the Mann–Whitney test when the distribution was non-Gaussian. The p-values in bold are statistically significant.

3.2. *Erythrocyte Parameters*

Regarding the alteration in the size of a single RBC expressed by MCV and the variability in the size of RBCs-RDW, there were no differences between neurotypical children and children with ASD (for MCV: 79.9 ± 3.1 fL in neurotypical vs. 79.5 ± 4.4 fL in ASD, p = 0.76; for RDW-SD: 37.3 ± 2.4 fL in neurotypical vs. 38.4 ± 3.2 fL in ASD, p = 0.74). Both parameters are a part of the standard complete blood count, while their deviations are related to nutritional deficiencies, certain inflammatory diseases as well as problems with the RBC production in general [31]. The lack of changes in the MCV and RDW parameters indicates the absence of significant problems with the production of RBCs, so we can exclude a more obvious nutrient deficiency that interferes with erythropoiesis in our ASD children. However, RBC deformability was lower (p = 0.017, Figure 1a) and the osmotic resistance of RBCs was higher (p = 0.026, Figure 1b) in children with ASD compared with neurotypical ones. The nitric oxide concentration (p = 0.04) as well as the level of reactive oxygen species (p = 0.03) were also higher in RBCs from children with ASD compared with neurotypical children (Figure 1c,d). There was no significant difference in RBC Ca^{2+} content between the groups (in arbitrary units: 2491 ± 607 in neurotypical vs. 2309 ± 660 in ASD, p = 0.38).

The RBC morphology was significantly altered in children with ASD, as type I and II echinocytes were more frequent compared with neurotypical controls (p < 0.0001). The percentages of individual shapes of RBCs in peripheral blood smears are presented in Table 2, while representative photographs of the RBC morphology are shown in Figure 2.

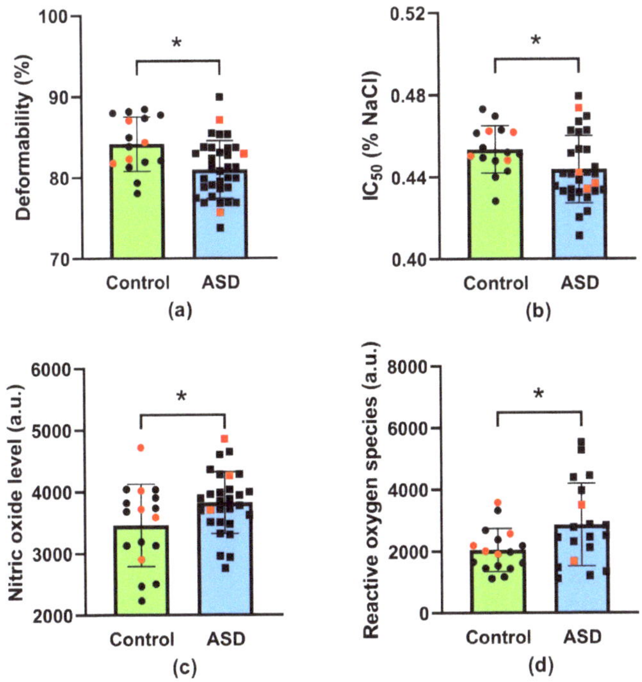

Figure 1. Erythrocyte deformability (**a**), erythrocyte osmotic resistance, presented as NaCl concentration at which 50% of hemolysis occurs (IC_{50}) (**b**), nitric oxide level (**c**) and the level of reactive oxygen species (**d**) in erythrocytes of neurotypical (i.e., control) children (n = 16–17) and children with autism spectrum disorder (ASD, n = 19–34). * $p < 0.05$ versus control group. Data are presented as means ± standard deviations. Statistical significance between groups was assessed using the unpaired t-test. Red dots indicate female study participants.

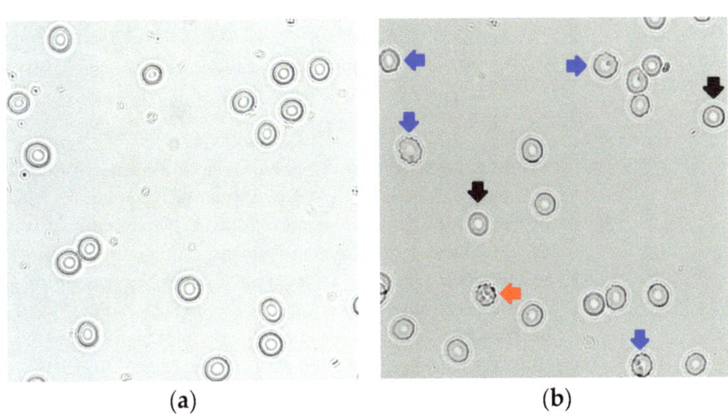

Figure 2. Representative photographs of erythrocytes in neurotypical children (**a**) and children with autism spectrum disorder (**b**). Black arrows: normal erythrocytes; blue arrows: echinocyte I (irregularly contoured erythrocytes); red arrow: echinocyte II (erythrocyte with spicules).

Table 2. Erythrocyte morphology in children with ASD and neurotypical controls.

Erythrocyte Shape	Control (*n* = 10)	ASD (*n* = 7)
Normal	95.1%	81.6%
Echinocyte I	4.4%	15.7%
Echinocyte II	0.5%	2.7%

ASD—autism spectrum disorder. Data are presented as the percentage of normal erythrocytes and echinocytes (echinocyte I—irregularly contoured erythrocytes; echinocyte II—erythrocytes with spicules) from the total erythrocyte count, while an average of 750 erythrocytes were analyzed from each individual.

3.3. *Na,K-ATPase Enzyme Kinetics in Erythrocyte Membranes*

When activating the Na,K-ATPase with an increasing concentration of sodium ions, the enzyme activity was higher in samples from ASD children as compared with control children. In the presence of the lowest applied concentration of NaCl (2 mmol/L), the activation of the enzyme was higher by 13% in the ASD group. At a higher presence of NaCl, the difference gradually increased, and in the presence of 100 mmol/L NaCl, the increase in enzyme activity amounted to 22% in the ASD group (Figure 3a).

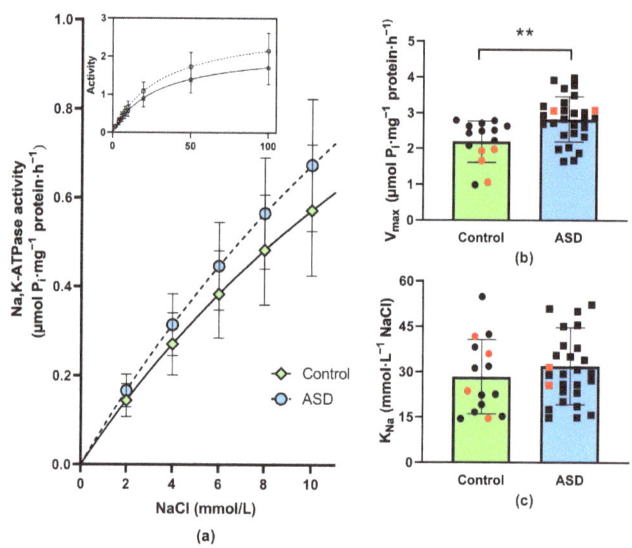

Figure 3. Na,K-ATPase enzyme activity as a function of NaCl concentrations ranging from 2 to 10 mmol/L. Inset—Na,K-ATPase enzyme activation in the whole investigated concentration range of NaCl (**a**). Kinetic parameters of Na,K-ATPase enzyme: maximal velocity of the enzyme reaction (V_{max}) (**b**) and concentration of Na^+ required to achieve the half-maximal velocity of the enzyme reaction (K_{Na}) (**c**) in erythrocyte membranes of neurotypical (i.e., control) children (*n* = 15) and children with autism spectrum disorder (ASD, *n* = 28). ** $p < 0.001$ versus the control group. Data are presented as means ± standard deviations. Statistical significance between groups was assessed using the unpaired *t*-test. Red dots indicate female study participants.

The evaluation of the obtained data according to Michaelis–Menten equation resulted in a statistically significant increase in the V_{max} value by 25% in the ASD group (Figure 3b). The slight increase in K_{Na} (+11%) did not reach the level of statistical significance (Figure 3c).

4. Discussion

In this study, we aimed to assess the oxidative and antioxidative status in the blood plasma as well as in the RBCs of ASD individuals and neurotypical controls. Noteworthy, the same set of parameters reflecting possible oxidative stress was measured in both types of samples in order to effectively match and compare potential changes in the ASD condition.

4.1. Changes in Oxidative Stress Markers in Children with ASD

Protein oxidation, quantified by determining the AOPP concentration, has previously been shown to be higher in the plasma of ASD individuals, but also in siblings of children with ASD, compared with controls with no ASD relationship [32]. In the ASD children involved in our study, an AOPP increase was not significant in the plasma, while it was statistically significant in the RBCs, suggesting greater oxidation of RBC proteins than those proteins occurring in the blood plasma. A possible discrepancy in AOPP plasma concentration in children with ASD between the previously published study (where AOPP is reported in µmol per mL of plasma volume) [32] and our data may be due to the use of different units, as we adjusted the AOPP concentration to the plasma protein level.

The more intense RBC oxidative damage in ASD children proposed here was also demonstrated when focusing on lipid peroxidation. TBARS concentration was significantly higher in the RBCs but not in the plasma of our ASD children. These data support the idea that in ASD, the level of oxidative stress is higher in RBCs than at the systemic level, i.e., than that determined in blood plasma. No differences in AOPP and TBARS plasma concentrations between children with ASD and controls were described previously [33,34], while an increase in TBARS levels in RBCs in children with ASD was documented in [19], in line with our observations. Despite the statistically significant differences in the RBC concentrations of AOPP and TBARS, the assessment of antioxidant power by determining FRAP, as well as the general marker of oxidative stress, the GSH/GSSG ratio, showed that the antioxidant status is not completely disturbed in the RBCs of ASD children. The observed trend toward a lower GSG/GSSG ratio in the ASD group compared with the control group can be considered consistent with other studies that determined this ratio in the blood plasma of ASD individuals, where it was lower compared with controls [35–38].

Focusing on antioxidant enzymes, more data have already been published, but are often inconsistent. CAT activity was shown to be lower [34,39], higher [40], as well as unchanged [19] in the RBCs of ASD children in comparison with the controls. Our data showed that an increase in CAT activity did not reach the level of statistical significance in RBCs, but appeared highly significant in the blood plasma of ASD children when compared with neurotypical ones. This finding is in agreement with previously published enhanced circulating CAT activities in ASD patients [33], although a decrease in the activity of circulating SOD was reported in these patients as well, consistently with recently published data [41]. Although we did not notice modifications in SOD activity in ASD patients, either in the blood plasma or in the RBCs, our experiments support other observations also reporting no changes in SOD activity, either in the RBCs or in the plasma of ASD individuals [19,42,43]. In contrast, SOD activity in the RBCs of ASD patients has been found to be higher [34,39,40,44] as well as lower [45] compared with neurotypical individuals. In addition, another study reported age-related changes in RBC SOD activity. Before the age of 6 years, SOD activity was lower in ASD children, while it was not changed in older children [46]. This observation could at least partially explain the inconsistent findings published by different research teams, although we did not observe any correlation between the antioxidant enzyme activities and the age of participants. In the RBCs of ASD patients, we observed unchanged GPx activity, consistent with a previously published paper [47], although a decrease in this enzyme activity in RBCs has also been reported [34,45]. Conflicting results regarding GPx activity have been obtained by analyzing plasma samples as well—it has been found to be higher [42,44] and lower [45] in ASD individuals compared with controls, while our data support its increase in children with ASD. Since ASD represents a somewhat heterogenous group of neurodevelopmental disorders, it can be assumed that oxidative stress is not equally responsible for behavioral disturbances in different individuals.

4.2. Are Erythrocyte Properties Altered in Children with ASD?

To the best of our knowledge, there is no study reporting no differences between ASD and neurotypical individuals when focusing on RBCs. Oxidative stress was suggested to be at least partially responsible for the multiple morphological alterations to RBCs taken from ASD individuals, as incubation with an antioxidant significantly restored the RBC morphology [21]. Abnormal RBC morphology in ASD was also observed in another study [20]. Scanning electron microscopy was applied in both studies [20,21]. Our study revealed observable abnormalities in RBC membrane morphology using light microscopy, with evidently limited resolution compared to scanning electron microscopy. The formation of echinocytes, which was more frequent in our ASD children, may have been caused not only by intrinsic but also by extrinsic factors, e.g., due to contact with the glass surface [48], which should be taken into account. Nevertheless, since the handling of all blood samples was identical, we can hypothesize increased echinocyte formation due to intrinsic factors occurring in ASD. In line with these findings, the triad consisting of the observed alterations to RBC shape and cytoskeletal β-actin, together with oxidative modifications of the RBC membrane, was suggested to be a potential measurable marker for ASD [20]. The transition from a discoid shape to echinocytes was also observed during the storage of RBCs, simultaneously with an increase in their stiffness [49]. It was reported that the RBC rigidity increases progressively with the degree of echinocytosis [50]. This is consistent with the findings of our study showing an increase in osmotic resistance concomitant with a decrease in RBC deformability, indicating an increase in RBC membrane rigidity.

In the search for mechanisms that could be responsible for such differences in RBC mechanical properties between ASD and neurotypical children, the finding of increased oxidative damage to RBCs (discussed in the previous subsection) supports our observation of significantly higher levels of reactive oxygen species in RBCs from ASD children (visualized using the redox-sensitive fluorescent probe DCF). In addition, higher NO level in the RBCs of ASD children might also be responsible for RBC membrane alterations leading to a decrease in RBC deformability, as an adequate NO concentration was shown to be a prerequisite for optimal RBC deformability [51]. An enhancement in NO production by RBCs in ASD was also suggested previously, but only via the measurement of nitrate and nitrite concentrations [42].

Despite the observed statistically significant differences in RBC deformability, osmotic resistance, NO and reactive oxygen species, some RBC parameters were revealed to be equal in both the control and ASD groups. An increase in calcium ion concentration triggers the eryptosis, i.e., suicidal RBC death [52], which is not applicable in our participants with ASD. Similarly, the unchanged MCV and RDW parameters in ASD children do not support the theoretical idea that basic hematological analysis could help in the diagnosis of ASD.

4.3. Na,K-ATPase Enzyme Kinetics in Erythrocyte Membranes and ASD

Study of enzymes maintaining the proper intracellular ion concentration showed that changes in Na,K-ATPase activity are probably linked with ASD. Post-mortem studies of brain samples from children with autism (13 ± 3.7 years old) showed that this enzyme, with a crucial role in intercellular signal transmission via maintaining the optimal homeostasis of sodium and potassium ions, had higher activity in the frontal cortex and in the cerebellum compared with samples from age-matched (12.5 ± 3.5 years old) control subjects [53]. Another study of metabolic biomarkers in blood plasma indicated an impairment of energy metabolism in the brains of children with autism (3 to 15 years old), as indicated by a 40% increase in the lactate level compared with typically developing control children (unspecified age). In those individuals with ASD, the activity of Na,K-ATPase in RBCs was higher by more than 70% as compared to the control subjects [54], while these alterations observed in autistic children were accompanied by signs of oxidative stress [44]. In contrast, another study documented oxidative stress in the blood plasma of autistic children (aged 5 to 12 years) that showed significantly lower activity (by 66%) of the Na,K-ATPase enzyme in the RBCs of ASD compared with the RBCs of control children [19].

In all of the abovementioned studies focused on Na,K-ATPase, the enzyme activity was measured in the presence of certain specific conditions of ionic and substrate concentrations. In the present study, by varying the concentration of sodium within a wide range corresponding to physiologically relevant intracellular but also extracellular levels, we aimed to provide more detailed information concerning the alteration of Na,K-ATPase levels in the RBCs of ASD patients. Based on the results presented in our study, we can assume that the elevated enzyme activity in the RBCs of ASD children in all of the investigated concentrations of sodium was caused by the higher presence of active Na,K-ATPase molecules, as suggested by the increased V_{max} value. Our kinetic measurements also provide information concerning the ability of the enzyme to bind sodium. The similarities of the K_{Na} values in both groups of children exclude the possibility of larger structural changes in the sodium-binding region of the enzyme in the autistic as well as the control group.

4.4. Sex-Specific Differences

It is widely accepted that sex differences occur in the prevalence and manifestation of ASD, while reported male-to-female ratios range from 2:1 to 5:1 in favor of males [55–57]. In the available databases, studies are focused on the possible causes of this phenomenon [58–61], but so far without definitive conclusions. The low number of girls included in our study, as well as the unequal number of participating girls and boys, represent a significant limitation to present possible sex differences. However, it is worth mentioning that we did not observe clear sex-specific differences in the parameters that our study focused on. The only exception might be in the case of the number of active Na,K-ATPase molecules represented by the V_{max} value in the control group, with the tendency being for lower values in girls than in boys (Figure 3b).

Although we were unable to observe any signs of differences in oxidative stress markers between girls and boys, it was observed that the male brain is more susceptible to suspected neurotoxic substances and the female brain is better protected against oxidative damage [61]. Via the use of experimental animals, it was proposed that possible female protective factors could include a more efficient glutathione regeneration cycle and greater antioxidant capacity of the brain tissue [62]. However, this is an important area that should be explored further.

5. Conclusions

The obtained data indicate greater oxidative damage to proteins and lipids in the RBCs than to those in the blood plasma of ASD subjects. In addition, antioxidant enzyme activities (CAT and GPx) were increased in plasma samples of children with ASD compared with controls, while not in the RBCs. Changes in oxidative status could at least partially contribute to the deterioration of RBC deformability observed in children with ASD. Moreover, significantly higher NO levels and Na,K-ATPase enzyme activity were detected in the RBCs of ASD individuals in comparison with neurotypical children. Abnormalities in the antioxidant status and RBC properties may be involved in the pathomechanisms of ASD and contribute to its clinical manifestations.

Author Contributions: Conceptualization, J.R.; methodology, T.J., N.V. and J.R.; software, N.V.; validation, T.J., N.V. and J.R.; formal analysis, T.J., N.V. and J.R.; investigation, T.J., D.R., K.J., D.S., N.V. and J.R.; resources, M.K. and A.T.; data curation, T.J., N.V. and J.R.; writing—original draft preparation, N.V. and J.R.; writing—review and editing, T.J., D.R., K.J., A.T., D.S., N.V. and J.R.; visualization, T.J.; supervision, K.J., N.V. and J.R.; project administration, J.R.; funding acquisition, A.T. and J.R. All authors have read and agreed to the published version of the manuscript.

Funding: This research was funded by the Scientific Grant Agency of the Ministry of Education, Science, Research and Sport of the Slovak Republic, grant numbers: VEGA 1/0193/21 and VEGA 1/0062/21, then by Comenius University in Bratislava grant number: UK/376/2023, as well as by The Slovak Research and Development Agency, grant number: APVV-20-0139.

Institutional Review Board Statement: This study was conducted in accordance with the Declaration of Helsinki, and was approved by the Ethics Committee of Faculty of Medicine, Comenius University and University Hospital in Bratislava, Slovak Republic (protocol code: 78/2021, date of approval: 2 August 2021).

Informed Consent Statement: Informed consent was obtained from all caregivers of the children involved in this study.

Data Availability Statement: The data supporting the findings of this study are available in this article.

Acknowledgments: The authors thank Dagmar Ciganekova and Jana Meryova for their excellent technical assistance.

Conflicts of Interest: The authors declare no conflict of interest.

References

1. Radosinska, J.; Vrbjar, N. Erythrocyte Deformability and Na,K-ATPase Activity in Various Pathophysiological Situations and Their Protection by Selected Nutritional Antioxidants in Humans. *IJMS* **2021**, *22*, 11924. [CrossRef] [PubMed]
2. Radosinska, J.; Vrbjar, N. The Role of Red Blood Cell Deformability and Na,K-ATPase Function in Selected Risk Factors of Cardiovascular Diseases in Humans: Focus on Hypertension, Diabetes Mellitus and Hypercholesterolemia. *Physiol. Res.* **2016**, *65* (Suppl. S1), S43–S54. [CrossRef]
3. Lord, C.; Elsabbagh, M.; Baird, G.; Veenstra-Vanderweele, J. Autism Spectrum Disorder. *Lancet* **2018**, *392*, 508–520. [CrossRef] [PubMed]
4. Marí-Bauset, S.; Zazpe, I.; Mari-Sanchis, A.; Llopis-González, A.; Morales-Suárez-Varela, M. Food Selectivity in Autism Spectrum Disorders: A Systematic Review. *J. Child Neurol.* **2014**, *29*, 1554–1561. [CrossRef]
5. Al-Beltagi, M.; Saeed, N.K.; Bediwy, A.S.; Elbeltagi, R.; Alhawamdeh, R. Role of Gastrointestinal Health in Managing Children with Autism Spectrum Disorder. *World J. Clin. Pediatr.* **2023**, *12*, 171–196. [CrossRef] [PubMed]
6. Stigler, K.A.; Sweeten, T.L.; Posey, D.J.; McDougle, C.J. Autism and Immune Factors: A Comprehensive Review. *Res. Autism Spectr. Disord.* **2009**, *3*, 840–860. [CrossRef]
7. Bartakovicova, K.; Kemenyova, P.; Belica, I.; Janik Szapuova, Z.; Stebelova, K.; Waczulikova, I.; Ostatnikova, D.; Babinska, K. Sleep Problems and 6-Sulfatoxymelatonin as Possible Predictors of Symptom Severity, Adaptive and Maladaptive Behavior in Children with Autism Spectrum Disorder. *IJERPH* **2022**, *19*, 7594. [CrossRef]
8. Muskens, J.B.; Ester, W.A.; Klip, H.; Zinkstok, J.; Van Dongen-Boomsma, M.; Staal, W.G. Novel Insights into Somatic Comorbidities in Children and Adolescents Across Psychiatric Diagnoses: An Explorative Study. *Child Psychiatry Hum. Dev.* **2023**. [CrossRef]
9. Micai, M.; Fatta, L.M.; Gila, L.; Caruso, A.; Salvitti, T.; Fulceri, F.; Ciaramella, A.; D'Amico, R.; Del Giovane, C.; Bertelli, M.; et al. Prevalence of Co-Occurring Conditions in Children and Adults with Autism Spectrum Disorder: A Systematic Review and Meta-Analysis. *Neurosci. Biobehav. Rev.* **2023**, *155*, 105436. [CrossRef]
10. Bjørklund, G.; Meguid, N.A.; El-Bana, M.A.; Tinkov, A.A.; Saad, K.; Dadar, M.; Hemimi, M.; Skalny, A.V.; Hosnedlová, B.; Kizek, R.; et al. Oxidative Stress in Autism Spectrum Disorder. *Mol. Neurobiol.* **2020**, *57*, 2314–2332. [CrossRef]
11. Chauhan, A.; Chauhan, V.; Brown, W.T.; Cohen, I. Oxidative Stress in Autism: Increased Lipid Peroxidation and Reduced Serum Levels of Ceruloplasmin and Transferrin—The Antioxidant Proteins. *Life Sci.* **2004**, *75*, 2539–2549. [CrossRef] [PubMed]
12. Anwar, A.; Marini, M.; Abruzzo, P.M.; Bolotta, A.; Ghezzo, A.; Visconti, P.; Thornalley, P.J.; Rabbani, N. Quantitation of Plasma Thiamine, Related Metabolites and Plasma Protein Oxidative Damage Markers in Children with Autism Spectrum Disorder and Healthy Controls. *Free Radic. Res.* **2016**, *50*, S85–S90. [CrossRef] [PubMed]
13. Raymond, L.J.; Deth, R.C.; Ralston, N.V.C. Potential Role of Selenoenzymes and Antioxidant Metabolism in Relation to Autism Etiology and Pathology. *Autism Res. Treat.* **2014**, *2014*, 164938. [CrossRef] [PubMed]
14. Rose, S.; Bennuri, S.C.; Wynne, R.; Melnyk, S.; James, S.J.; Frye, R.E. Mitochondrial and Redox Abnormalities in Autism Lymphoblastoid Cells: A Sibling Control Study. *FASEB J.* **2017**, *31*, 904–909. [CrossRef]
15. Frye, R.E.; Lionnard, L.; Singh, I.; Karim, M.A.; Chajra, H.; Frechet, M.; Kissa, K.; Racine, V.; Ammanamanchi, A.; McCarty, P.J.; et al. Mitochondrial Morphology Is Associated with Respiratory Chain Uncoupling in Autism Spectrum Disorder. *Transl. Psychiatry* **2021**, *11*, 527. [CrossRef]
16. Richardson, K.J.; Kuck, L.; Simmonds, M.J. Beyond Oxygen Transport: Active Role of Erythrocytes in the Regulation of Blood Flow. *Am. J. Physiol.-Heart Circ. Physiol.* **2020**, *319*, H866–H872. [CrossRef]
17. Alexy, T.; Detterich, J.; Connes, P.; Toth, K.; Nader, E.; Kenyeres, P.; Arriola-Montenegro, J.; Ulker, P.; Simmonds, M.J. Physical Properties of Blood and Their Relationship to Clinical Conditions. *Front. Physiol.* **2022**, *13*, 906768. [CrossRef]
18. Jasenovec, T.; Radosinska, D.; Celusakova, H.; Filcikova, D.; Babinska, K.; Ostatnikova, D.; Radosinska, J. Erythrocyte Deformability in Children with Autism Spectrum Disorder: Correlation with Clinical Features. *Physiol. Res.* **2019**, *68*, S307–S313. [CrossRef]
19. Ghezzo, A.; Visconti, P.; Abruzzo, P.M.; Bolotta, A.; Ferreri, C.; Gobbi, G.; Malisardi, G.; Manfredini, S.; Marini, M.; Nanetti, L.; et al. Oxidative Stress and Erythrocyte Membrane Alterations in Children with Autism: Correlation with Clinical Features. *PLoS ONE* **2013**, *8*, e66418. [CrossRef]

20. Ciccoli, L.; De Felice, C.; Paccagnini, E.; Leoncini, S.; Pecorelli, A.; Signorini, C.; Belmonte, G.; Guerranti, R.; Cortelazzo, A.; Gentile, M.; et al. Erythrocyte Shape Abnormalities, Membrane Oxidative Damage, and β-Actin Alterations: An Unrecognized Triad in Classical Autism. *Mediat. Inflamm.* **2013**, *2013*, 432616. [CrossRef]
21. Bolotta, A.; Battistelli, M.; Falcieri, E.; Ghezzo, A.; Manara, M.C.; Manfredini, S.; Marini, M.; Posar, A.; Visconti, P.; Abruzzo, P.M. Oxidative Stress in Autistic Children Alters Erythrocyte Shape in the Absence of Quantitative Protein Alterations and of Loss of Membrane Phospholipid Asymmetry. *Oxidative Med. Cell. Longev.* **2018**, *2018*, 6430601. [CrossRef] [PubMed]
22. Witko-Sarsat, V.; Gausson, V.; Descamps-Latscha, B. Are Advanced Oxidation Protein Products Potential Uremic Toxins? *Kidney Int.* **2003**, *63*, S11–S14. [CrossRef]
23. Benzie, I.F.F.; Strain, J.J. The Ferric Reducing Ability of Plasma (FRAP) as a Measure of "Antioxidant Power": The FRAP Assay. *Anal. Biochem.* **1996**, *239*, 70–76. [CrossRef] [PubMed]
24. San-Gil, F.; Schier, G.M.; Moses, R.G.; Gan, I.E. Improved Estimation of Fructosamine, as a Measure of Glycated Serum Protein, with the Technicon RA-1000 Analyzer. *Clin. Chem.* **1985**, *31*, 2005–2006. [CrossRef]
25. Aebi, H. Catalase in Vitro. *Methods Enzym.* **1984**, *105*, 121–126. [CrossRef]
26. Radosinska, J.; Kollarova, M.; Jasenovec, T.; Radosinska, D.; Vrbjar, N.; Balis, P.; Puzserova, A. Aging in Normotensive and Spontaneously Hypertensive Rats: Focus on Erythrocyte Properties. *Biology* **2023**, *12*, 1030. [CrossRef] [PubMed]
27. Jasenovec, T.; Radosinska, D.; Kollarova, M.; Balis, P.; Zorad, S.; Vrbjar, N.; Bernatova, I.; Cacanyiova, S.; Tothova, L.; Radosinska, J. Effects of Taxifolin in Spontaneously Hypertensive Rats with a Focus on Erythrocyte Quality. *Life* **2022**, *12*, 2045. [CrossRef]
28. Lang, E.; Pozdeev, V.I.; Gatidis, S.; Qadri, S.M.; Häussinger, D.; Kubitz, R.; Herebian, D.; Mayatepek, E.; Lang, F.; Lang, K.S.; et al. Bile Acid-Induced Suicidal Erythrocyte Death. *Cell Physiol. Biochem.* **2016**, *38*, 1500–1509. [CrossRef]
29. Jasenovec, T.; Radosinska, D.; Kollarova, M.; Balis, P.; Ferenczyova, K.; Kalocayova, B.; Bartekova, M.; Tothova, L.; Radosinska, J. Beneficial Effect of Quercetin on Erythrocyte Properties in Type 2 Diabetic Rats. *Molecules* **2021**, *26*, 4868. [CrossRef]
30. Rabbani, N.; Thornalley, P.J. Protein Glycation—Biomarkers of Metabolic Dysfunction and Early-Stage Decline in Health in the Era of Precision Medicine. *Redox Biol.* **2021**, *42*, 101920. [CrossRef]
31. Brzeźniakiewicz-Janus, K.; Rupa-Matysek, J.; Tukiendorf, A.; Janus, T.; Franków, M.; Lancé, M.D.; Gil, L. Red Blood Cells Mean Corpuscular Volume (MCV) and Red Blood Distribution Width (RDW) Parameters as Potential Indicators of Regenerative Potential in Older Patients and Predictors of Acute Mortality—Preliminary Report. *Stem. Cell Rev. Rep.* **2020**, *16*, 711–717. [CrossRef] [PubMed]
32. Nasrallah, O.; Alzeer, S. Measuring Some Oxidative Stress Biomarkers in Autistic Syrian Children and Their Siblings: A Case-Control Study. *Biomark. Insights* **2022**, *17*, 117727192211239. [CrossRef]
33. Yenkoyan, K.; Harutyunyan, H.; Harutyunyan, A. A Certain Role of SOD/CAT Imbalance in Pathogenesis of Autism Spectrum Disorders. *Free Radic. Biol. Med.* **2018**, *123*, 85–95. [CrossRef]
34. László, A.; Novák, Z.; Szőllősi-Varga, I.; Hai, D.Q.; Vetró, Á.; Kovács, A. Blood Lipid Peroxidation, Antioxidant Enzyme Activities and Hemorheological Changes in Autistic Children. *Ideggyogy. Szle.* **2013**, *66*, 23–28. [PubMed]
35. James, S.J.; Cutler, P.; Melnyk, S.; Jernigan, S.; Janak, L.; Gaylor, D.W.; Neubrander, J.A. Metabolic Biomarkers of Increased Oxidative Stress and Impaired Methylation Capacity in Children with Autism. *Am. J. Clin. Nutr.* **2004**, *80*, 1611–1617. [CrossRef] [PubMed]
36. Melnyk, S.; Fuchs, G.J.; Schulz, E.; Lopez, M.; Kahler, S.G.; Fussell, J.J.; Bellando, J.; Pavliv, O.; Rose, S.; Seidel, L.; et al. Metabolic Imbalance Associated with Methylation Dysregulation and Oxidative Damage in Children with Autism. *J. Autism Dev. Disord.* **2012**, *42*, 367–377. [CrossRef] [PubMed]
37. Al-Yafee, Y.A.; Al- Ayadhi, L.Y.; Haq, S.H.; El-Ansary, A.K. Novel Metabolic Biomarkers Related to Sulfur-Dependent Detoxification Pathways in Autistic Patients of Saudi Arabia. *BMC Neurol.* **2011**, *11*, 139. [CrossRef]
38. Bjørklund, G.; Tinkov, A.A.; Hosnedlová, B.; Kizek, R.; Ajsuvakova, O.P.; Chirumbolo, S.; Skalnaya, M.G.; Peana, M.; Dadar, M.; El-Ansary, A.; et al. The Role of Glutathione Redox Imbalance in Autism Spectrum Disorder: A Review. *Free Radic. Biol. Med.* **2020**, *160*, 149–162. [CrossRef]
39. Zoroglu, S.S.; Armutcu, F.; Ozen, S.; Gurel, A.; Sivasli, E.; Yetkin, O.; Meram, I. Increased Oxidative Stress and Altered Activities of Erythrocyte Free Radical Scavenging Enzymes in Autism. *Eur. Arch. Psychiatry Clin. Neurosci.* **2004**, *254*, 143–147. [CrossRef]
40. Altun, H.; Şahin, N.; Kurutaş, E.B.; Karaaslan, U.; Sevgen, F.H.; Fındıklı, E. Assessment of Malondialdehyde Levels, Superoxide Dismutase, and Catalase Activity in Children with Autism Spectrum Disorders. *Psychiatry Clin. Psychopharmacol.* **2018**, *28*, 408–415. [CrossRef]
41. Vellingiri, B.; Venkatesan, D.; Iyer, M.; Mohan, G.; Krishnan, P.; Sai Krishna, K.; Narayanasamy, A.; Gopalakrishnan, A.V.; Kumar, N.S.; Subramaniam, M.D. Concurrent Assessment of Oxidative Stress and MT-ATP6 Gene Profiling to Facilitate Diagnosis of Autism Spectrum Disorder (ASD) in Tamil Nadu Population. *J. Mol. Neurosci.* **2023**, *73*, 214–224. [CrossRef] [PubMed]
42. Söğüt, S.; Zoroğlu, S.S.; Özyurt, H.; Ramazan Yılmaz, H.; Özuğurlu, F.; Sivaslı, E.; Yetkin, Ö.; Yanık, M.; Tutkun, H.; Savaş, H.A.; et al. Changes in Nitric Oxide Levels and Antioxidant Enzyme Activities May Have a Role in the Pathophysiological Mechanisms Involved in Autism. *Clin. Chim. Acta* **2003**, *331*, 111–117. [CrossRef] [PubMed]
43. Imataka, G.; Yui, K.; Shiko, Y.; Kawasaki, Y.; Sasaki, H.; Shiroki, R.; Yoshihara, S. Urinary and Plasma Antioxidants in Behavioral Symptoms of Individuals with Autism Spectrum Disorder. *Front. Psychiatry* **2021**, *12*, 684445. [CrossRef] [PubMed]
44. Al-Gadani, Y.; El-Ansary, A.; Attas, O.; Al-Ayadhi, L. Metabolic Biomarkers Related to Oxidative Stress and Antioxidant Status in Saudi Autistic Children. *Clin. Biochem.* **2009**, *42*, 1032–1040. [CrossRef]

45. Yorbik, O.; Sayal, A.; Akay, C.; Akbiyik, D.I.; Sohmen, T. Investigation of Antioxidant Enzymes in Children with Autistic Disorder. *Prostaglandins Leukot. Essent. Fat. Acids* **2002**, *67*, 341–343. [CrossRef]
46. Meguid, N.A.; Dardir, A.A.; Abdel-Raouf, E.R.; Hashish, A. Evaluation of Oxidative Stress in Autism: Defective Antioxidant Enzymes and Increased Lipid Peroxidation. *Biol. Trace Elem. Res.* **2011**, *143*, 58–65. [CrossRef]
47. Paşca, S.P.; Nemeş, B.; Vlase, L.; Gagyi, C.E.; Dronca, E.; Miu, A.C.; Dronca, M. High Levels of Homocysteine and Low Serum Paraoxonase 1 Arylesterase Activity in Children with Autism. *Life Sci.* **2006**, *78*, 2244–2248. [CrossRef]
48. Bessis, M. Red Cell Shapes. An Illustrated Classification and Its Rationale. *Nouv. Rev. Fr. Hematol.* **1972**, *12*, 721–745.
49. Xu, Z.; Zheng, Y.; Wang, X.; Shehata, N.; Wang, C.; Sun, Y. Stiffness Increase of Red Blood Cells during Storage. *Microsyst. Nanoeng.* **2018**, *4*, 17103. [CrossRef]
50. Chabanel, A.; Schachter, D.; Chien, S. Increased Rigidity of Red Blood Cell Membrane in Young Spontaneously Hypertensive Rats. *Hypertension* **1987**, *10*, 603–607. [CrossRef]
51. Starzyk, D.; Korbut, R.; Gryglewski, R.J. The Role of Nitric Oxide in Regulation of Deformability of Red Blood Cells in Acute Phase of Endotoxaemia in Rats. *J. Physiol. Pharmacol.* **1997**, *48*, 731–735. [PubMed]
52. Lang, E.; Qadri, S.M.; Lang, F. Killing Me Softly—Suicidal Erythrocyte Death. *Int. J. Biochem. Cell Biol.* **2012**, *44*, 1236–1243. [CrossRef] [PubMed]
53. Ji, L.; Chauhan, A.; Brown, W.T.; Chauhan, V. Increased Activities of Na+/K+-ATPase and Ca^{2+}/Mg^{2+}-ATPase in the Frontal Cortex and Cerebellum of Autistic Individuals. *Life Sci.* **2009**, *85*, 788–793. [CrossRef]
54. Al-Mosalem, O.A.; El-Ansary, A.; Attas, O.; Al-Ayadhi, L. Metabolic Biomarkers Related to Energy Metabolism in Saudi Autistic Children. *Clin. Biochem.* **2009**, *42*, 949–957. [CrossRef] [PubMed]
55. Napolitano, A.; Schiavi, S.; La Rosa, P.; Rossi-Espagnet, M.C.; Petrillo, S.; Bottino, F.; Tagliente, E.; Longo, D.; Lupi, E.; Casula, L.; et al. Sex Differences in Autism Spectrum Disorder: Diagnostic, Neurobiological, and Behavioral Features. *Front. Psychiatry* **2022**, *13*, 889636. [CrossRef] [PubMed]
56. Maenner, M.J.; Shaw, K.A.; Bakian, A.V.; Bilder, D.A.; Durkin, M.S.; Esler, A.; Furnier, S.M.; Hallas, L.; Hall-Lande, J.; Hudson, A.; et al. Prevalence and Characteristics of Autism Spectrum Disorder Among Children Aged 8 Years—Autism and Developmental Disabilities Monitoring Network, 11 Sites, United States, 2018. *MMWR Surveill. Summ.* **2021**, *70*, 1–16. [CrossRef] [PubMed]
57. Edwards, H.; Wright, S.; Sargeant, C.; Cortese, S.; Wood-Downie, H. Research Review: A Systematic Review and META-ANALYSIS of Sex Differences in Narrow Constructs of Restricted and Repetitive Behaviours and Interests in Autistic Children, Adolescents, and Adults. *Child Psychol. Psychiatry* **2023**, jcpp.13855. [CrossRef]
58. Gunes, S.; Ekinci, O.; Celik, T. Iron Deficiency Parameters in Autism Spectrum Disorder: Clinical Correlates and Associated Factors. *Ital. J. Pediatr.* **2017**, *43*, 86. [CrossRef]
59. Gockley, J.; Willsey, A.J.; Dong, S.; Dougherty, J.D.; Constantino, J.N.; Sanders, S.J. The Female Protective Effect in Autism Spectrum Disorder Is Not Mediated by a Single Genetic Locus. *Mol. Autism* **2015**, *6*, 25. [CrossRef]
60. Werling, D.M.; Geschwind, D.H. Sex Differences in Autism Spectrum Disorders. *Curr. Opin. Neurol.* **2013**, *26*, 146–153. [CrossRef]
61. Kern, J.K.; Geier, D.A.; Homme, K.G.; King, P.G.; Bjørklund, G.; Chirumbolo, S.; Geier, M.R. Developmental Neurotoxicants and the Vulnerable Male Brain: A Systematic Review of Suspected Neurotoxicants That Disproportionally Affect Males. *Acta Neurobiol. Exp.* **2017**, *77*, 269–296. [CrossRef]
62. Dukhande, V.V.; Isaac, A.O.; Chatterji, T.; Lai, J.C.K. Reduced Glutathione Regenerating Enzymes Undergo Developmental Decline and Sexual Dimorphism in the Rat Cerebral Cortex. *Brain Res.* **2009**, *1286*, 19–24. [CrossRef] [PubMed]

antioxidants

Article

Red Blood Cells from Individuals with Lesch–Nyhan Syndrome: Multi-Omics Insights into a Novel S162N Mutation Causing Hypoxanthine-Guanine Phosphoribosyltransferase Deficiency

Julie A. Reisz [1], Monika Dzieciatkowska [1], Daniel Stephenson [1], Fabia Gamboni [1], D. Holmes Morton [2] and Angelo D'Alessandro [1,*]

[1] Department of Biochemistry and Molecular Genetics, University of Colorado Anschutz Medical Campus, Aurora, CO 80045, USA; julie.haines@cuanschutz.edu (J.A.R.); monika.dzieciatkowska@cuanschutz.edu (M.D.); daniel.stephenson@cuanschutz.edu (D.S.); fabia.gamboni@cuanschutz.edu (F.G.)

[2] Central Pennsylvania Clinic, A Medical Home for Special Children and Adults, Belleville, PA 17004, USA; dholmesmorton@gmail.com

* Correspondence: angelo.dalessandro@cuanschutz.edu; Tel.: +1-303-7240096

Abstract: Lesch–Nyhan syndrome (LN) is an is an X-linked recessive inborn error of metabolism that arises from a deficiency of purine salvage enzyme hypoxanthine-guanine phosphoribosyltransferase (HPRT). The disease manifests severely, causing intellectual deficits and other neural abnormalities, hypercoagulability, uncontrolled self-injury, and gout. While allopurinol is used to alleviate gout, other symptoms are less understood, impeding treatment. Herein, we present a high-throughput multi-omics analysis of red blood cells (RBCs) from three pediatric siblings carrying a novel S162N HPRT1 mutation. RBCs from both parents—the mother, a heterozygous carrier, and the father, a clinically healthy control—were also analyzed. Global metabolite analysis of LN RBCs shows accumulation of glycolytic intermediates upstream of pyruvate kinase, unsaturated fatty acids, and long chain acylcarnitines. Similarly, highly unsaturated phosphatidylcholines are also elevated in LN RBCs, while free choline is decreased. Intracellular iron, zinc, selenium, and potassium are also decreased in LN RBCs. Global proteomics documented changes in RBC membrane proteins, hemoglobin, redox homeostasis proteins, and the enrichment of coagulation proteins. These changes were accompanied by elevation in protein glutamine deamidation and methylation in the LN children and carrier mother. Treatment with allopurinol incompletely reversed the observed phenotypes in the two older siblings currently on this treatment. This unique data set provides novel opportunities for investigations aimed at potential therapies for LN-associated sequelae.

Keywords: Lesch–Nyhan; purine salvage; hypoxanthine-guanine phosphoribosyltransferase deficiency; metabolomics; redox proteomics; lipidomics

Citation: Reisz, J.A.; Dzieciatkowska, M.; Stephenson, D.; Gamboni, F.; Morton, D.H.; D'Alessandro, A. Red Blood Cells from Individuals with Lesch–Nyhan Syndrome: Multi-Omics Insights into a Novel S162N Mutation Causing Hypoxanthine-Guanine Phosphoribosyltransferase Deficiency. *Antioxidants* **2023**, *12*, 1699. https://doi.org/10.3390/antiox12091699

Academic Editor: Marco G. Alves

Received: 12 July 2023
Revised: 14 August 2023
Accepted: 25 August 2023
Published: 31 August 2023

1. Introduction

Lesch–Nyhan (LN) syndrome is a disease that affects approximately one in every three-hundred-and-eighty-thousand births. Over 2000 known genetic mutations contribute to the etiology of LN, a rare X-linked recessive disorder [1,2]. A wealth of mutations have been described within the intronic or exonic regions 1–9 on chromosome Xq26-27, genetic abnormalities that manifest biochemically with the deficient activity of hypoxanthine-guanine phosphoribosyltransferase (HPRT). HPRT is a purine salvage enzyme responsible for the conversion of hypoxanthine to IMP and guanine to GMP. As such, deficient HPRT activity results in the accumulation of hypoxanthine, feeding further oxidation by xanthine oxidase to urate, with concomitant excess generation of hydrogen peroxide, ultimately leading to the development of gout [3].

Clinical manifestations of the disease are severe and heterogenous; symptoms can include intellectual deficits, reduced gray and white matter in the brain, self-injurious behavior, hypercoagulability, macrocytic anemia, and gout [4,5]. With X-linked recessive genetic etiology, only males develop LN disease, though females may be heterozygous and, in some cases, display mild symptoms. While advances in the understanding of the mechanisms leading to gout have informed interventions with xanthine oxidase-targeting molecules, like allopurinol, further advancements in the understanding of the molecular manifestations of this disease are necessary to advance more successful interventions beyond the treatment of gout [6].

The breakdown and deamination of adenosine triphosphate to hypoxanthine is a hallmark of hypoxia-induced mitochondrial dysfunction [7,8], under either environmental or pathological conditions (e.g., ischemic or hemorrhagic hypoxia) [9]. In vivo, hypoxia is counteracted by red blood cells (RBCs), the dominant cell type in the blood and the most numerous cell type in the human body (~84% of total cells in an adult individual) [10]. Circulating RBCs are tasked with oxygen carrying and delivery, a process they have evolved to fulfill by losing organelles (including nuclei and mitochondria) to maximize hemoglobin content. These molecular adaptations are accompanied by the incapacity of RBCs to synthesize new proteins, for example, to replace oxidatively damaged components in the face of ongoing Fenton and Haber–Weiss chemistry as a result of the excess load of oxygen (up to 1 billion molecules per cell) and oxygen-coordinating iron in a mature erythrocyte [10]. In this sense, RBCs are extremely sensitive to perturbations in metabolic and redox homeostasis: For example, lacking mitochondria, RBC exposure to oxidant stress in vivo or ex vivo promotes purine deamination via AMP deaminase 3, which in turn favors the accumulation of inosine monophosphate, a precursor to hypoxanthine [11]. Ultimately, elevated RBC hypoxanthine is associated with altered morphology and increased extravascular hemolysis via splenic sequestration and erythrophagocytosis [12].

The goal of the Lesch–Nyhan study described herein is to provide metabolic insights that could inform effective treatments through dietary and/or metabolic interventions. A comprehensive molecular profiling of RBCs from LN patients has not previously been described and is the focus of the present study. Herein, we enrolled a family of five with the goal of leveraging a multiomics approach to describe in detail, for the first time, the molecular derangements in RBCs from three male LN patients as compared to the carrier mother and the clinically healthy father.

2. Materials and Methods

2.1. Subject Recruitment and Sample Collection

Blood samples were collected through venipuncture from a family of volunteers, including father (clinically healthy), mother (carrier), and three male children diagnosed with Lesch–Nyhan syndrome at the Central Pennsylvania Clinic and at Boston Children's and Lancaster General Hospitals under institutionally reviewed protocols (No. 2014-12) and upon the signing of informed consent. The children were each homozygous for the HPRT1 c.485 G>A;p.Ser162Asn variant and presented with severe disability from generalized dystonia, self-injury, and increased serum uric acid. At the time of blood collection, their ages were 4 years, 2.5 years, and 4 months old. Owing to the very young age, only the two older siblings were being treated with allopurinol (mean dose 6.44 mg/Kg of weight per day) [13]. RBCs were separated from whole blood through centrifugation for 10 min at 4 °C and 2000× *g*.

2.2. High Throughput Metabolomics

Metabolomics analyses were performed as previously described [14]. Red blood cell samples were thawed on ice, then a 10 μL aliquot was treated with 90 μL of ice cold 5:3:2 MeOH:MeCN:water (*v*/*v*/*v*) then vortexed for 30 min at 4 °C. Supernatants were clarified by centrifugation (10 min, 12,000× *g*, 4 °C). The resulting metabolite extracts were analyzed (10 μL per injection) using ultra-high-pressure liquid chromatography coupled to mass spectrometry (UHPLC-MS—Vanquish and Q Exactive, ThermoFisher, Bremen, Germany). Metabolites were resolved on a Phenomenex Kinetex C18 column

(2.1 × 150 mm, 1.7 µm) at 45 °C using a 5-min gradient method in positive and negative ion modes (separate runs) over the scan range 65–975 *m*/*z* exactly as previously described [14]. Oxylipins were resolved on a Waters ACQUITY UPLC BEH C18 column (2.1 × 100 mm, 1.7 µm) at 60 °C using mobile phase (A) of 20:80:0.02 MeCN:water:formic acid (FA) and a mobile phase (B) of 20:80:0.02 MeCN:isopropanol:FA. For negative mode analysis, the chromatographic gradient was as follows: 0.35 mL/min flowrate, 0% B 0–0.5 min, 25% B at 1 min, 40% B at 2.5 min, 55% B at 2.6 min, 70% B at 4.5 min, 100% B at 4.6–6 min, and 0% B at 6.1–7 min. The Q Exactive MS was operated in negative ion mode, scanning in full MS mode (2 µscans) from 150 to 1500 *m/z* at 70,000 resolution, with 4 kV spray voltage, 45 sheath gas, and 15 auxiliary gas. Following data acquisition, .raw files were converted to .mzXML using RawConverter version 1.2.0.1, then metabolites were assigned and peaks were integrated using Maven (Princeton University, Princeton, NJ, USA) in conjunction with the KEGG database and an in-house standard library. Quality control was assessed as using technical replicates run at the beginning, end, and middle of each sequence as previously described [14].

2.3. Transition Metal Analysis

Transition metals ^{23}Na, ^{24}Mg, ^{39}K, ^{44}Ca, ^{57}Fe, ^{63}Cu, ^{77}Se, and ^{66}Zn were measured using inductively coupled plasma–mass spectrometry on a Thermo iCAP RQ ICP-MS coupled to an ESI SC-4DX FAST autosampler system using 10 µL of RBCs exactly as previously described [15]. Final dilutions of 1:250 and 1:3750 were then analyzed via ICP-MS. Different dilutions were used to ensure that all analytes fell within the calibration curves (selenium reads fell just below calibrant 1; however, all intensities were approximately 10× above the blanks, so values were extrapolated). All chemicals and materials used for ICP-MS analysis were obtained from ThermoFisher, Waltham, MA, USA; ICP-MS calibrants and solutions were obtained from SPEX CertiPrep. Instrument performance was assessed by infusing an internal standard mix via a peristaltic pump and monitoring signal throughout the run. Additionally, separate quality controls of a known concentration (75 ppb) of each analyte were injected at the beginning, throughout the run between samples, and at the end of the run. The acceptance criterion for all QCs was ±25% of the known concentration. Thermo Scientific Qtegra software version 2.10.3324.131 was used for all data acquisition and analysis.

2.4. Untargeted Lipidomics

Total lipids were extracted as previously described [16]: 10 µL of RBCs were mixed with 90 µL of cold methanol. Samples were then briefly vortexed and incubated at −20 °C for 30 min. Following incubation, samples were centrifuged at 12,700 RPM for 10 min at 4 °C and 80 µL of supernatant was transferred to a new tube for analysis. Lipid extracts were analyzed (10 µL per injection) on a Thermo Vanquish UHPLC/Q Exactive MS system using a 5-min lipidomics gradient and a Kinetex C18 column (30 × 2.1 mm, 1.7 µm, Phenomenex) held at 50 °C. Mobile phase A: 25:75 MeCN:water with 5 mM ammonium acetate; Mobile phase B: 90:10 isopropanol:MeCN with 5 mM ammonium acetate. The gradient and flow rate were as follows: 0.3 mL/min of 10% B at 0 min, 0.3 mL/min of 95% B at 3 min, 0.3 mL/min of 95% B at 4.2 min, 0.45 mL/min 10% B at 4.3 min, 0.4 mL/min of 10% B at 4.9 min, and 0.3 mL/min of 10% B at 5 min. Samples were run in positive and negative ion modes (both ESI, separate runs) at 125 to 1500 *m*/*z* and 70,000 resolution, 4 kV spray voltage, 45 sheath gas, and 25 auxiliary gas. The MS was run in data-dependent acquisition mode (ddMS^2) with top10 fragmentation. Raw MS data files were searched using LipidSearch v 5.0 (ThermoFisher).

2.5. Global Proteomics

Proteomics analyses were performed as described [17]. A volume of 10 µL of RBCs were lysed in 90 µL of distilled water. Then, 5 µL of lysed RBCs were mixed with 45 µL of 5% SDS and then vortexed. Samples were reduced with 10 mM DTT at 55 °C for 30 min, cooled to room temperature, and then alkylated with 25 mM iodoacetamide in the dark for 30 min. Next, a final concentration of 1.2% phosphoric acid and then six volumes of binding buffer (90% methanol; 100 mM triethylammonium bicarbonate, TEAB; pH 7.1) were added to each sample. After gentle mixing, the protein solution was loaded to an S-Trap 96-well plate, spun at 1500× *g* for 2 min, and the flow-through collected and reloaded onto the 96-well plate. This step was repeated three times and then the 96-well plate was washed with 200 µL of binding buffer three times. Finally, 1 µg of sequencing-grade trypsin (Promega, Madison, WI, USA) and 125 µL of digestion buffer (50 mM TEAB) were added onto the filter and were digested at 37 °C for 6 h. To elute peptides, three stepwise buffers were applied, with 100 µL of each with one more repeat, including 50 mM TEAB, 0.2% formic acid (FA), 50% acetonitrile, and 0.2% FA. The peptide solutions were pooled, lyophilized, and resuspended in 500 µL of 0.1 % FA.

Each sample was loaded onto individual Evotips for desalting and was then washed with 200 µL 0.1% FA, followed by the addition of 100 µL storage solvent (0.1% FA) to keep the Evotips wet until analysis. The Evosep One system (Evosep, Odense, Denmark) was used to separate peptides on a Pepsep column (150 µm inter diameter, 15 cm) packed with ReproSil C18 1.9 µm, 120A resin. The system was coupled to a timsTOF Pro mass spectrometer (Bruker Daltonics, Bremen, Germany) via a nano-electrospray ion source (Captive Spray, Bruker Daltonics). The mass spectrometer was operated in PASEF mode. The ramp time was set to 100 ms, and 10 PASEF MS/MS scans per topN acquisition cycle were acquired. MS and MS/MS spectra were recorded from m/z 100 to 1700. The ion mobility was scanned from 0.7 to 1.50 Vs/cm^2. Precursors for data-dependent acquisition were isolated within ±1 Th and were fragmented with an ion mobility-dependent collision energy, which was linearly increased from 20 to 59 eV in positive mode. Low-abundance precursor ions with an intensity above a threshold of 500 counts but below a target value of 20,000 counts were repeatedly scheduled and otherwise dynamically excluded for 0.4 min.

2.6. Database Searching and Protein Identification

MS/MS spectra were extracted from raw data files and converted into .mgf files using MS Convert (ProteoWizard, version 3.0). Peptide spectral matching was performed with Mascot version 2.5 against the Uniprot human database. Mass tolerances were ±15 ppm for parent ions and ± 0.4 Da for fragment ions. Trypsin specificity was used, allowing for 1 missed cleavage. Met oxidation, Cys dioxidation, Cys thiol conversion to dehydroalanine, protein N-terminal acetylation, isopeptide bond formation with loss of ammonia (K), and peptide N-terminal pyroglutamic acid formation were set as variable modifications with Cys carbamidomethylation set as a fixed modification.

Scaffold (v 4.8, Proteome Software, Portland, OR, USA) was used to validate MS/MS-based peptide and protein identifications. Peptide identifications were accepted if they could be established at greater than 95.0% probability as specified by the Peptide Prophet algorithm. Protein identifications were accepted if they could be established at greater than 99.0% probability, and contained at least two identified unique peptides.

Oxidized cysteine content was determined by summing the spectral counts for peptides containing Cys sulfinic acid (Cys-SO2H, i.e., deoxidation) and Cys conversion to dehydroalanine, a stable end-product of oxidized cysteine intermediates. The irreversibly oxidized Cys content was normalized to total peptide spectral counts for the corresponding protein. Asparagine deamidation and glutamate/aspartate methylation were determined using database searches with these variable modifications. Results were normalized to total peptide spectral counts for the corresponding protein.

2.7. Statistics and Visualization

Statistical analysis was conducted using MetaboAnalyst v 5.0 with the following data normalizations: metabolomics data were sum-normalized, log transformed, and autoscaled; lipidomics data were sum-normalized and autoscaled; proteomics data were sum-normalized and autoscaled; Violin plots, bar graphs, and their corresponding statistics (one-way ANOVA with Kruskal–Wallis multiple comparisons test) were prepared in GraphPad Prism v 9.5.1. Network analysis and pathway analysis were performed in OmicsNet v 2.0, using as input the significantly altered metabolites and proteins ($p < 0.05$), the STRING database for protein–protein interactions, and the KEGG database for protein–metabolite interactions. The pathway analysis was performed in the Function Explorer node against the KEGG database.

3. Results

3.1. Clinical Presentation and Hematological Parameters

Two young Amish brothers presented at the Central Pennsylvania Clinic with severe disability from generalized dystonia with delays of gross motor milestones, poor head and trunk control, and uncontrolled movement of the extremities. The disability of the first boy was previously misdiagnosed as cerebral palsy attributed to an injury at birth. Genetic testing began when the second affected boy presented with similar symptoms. Recognized causes of Amish cerebral palsy were ruled-out by biochemical and targeted PCR-based mutation tests and a microarray-based genotype of parents for 1200-pathogenic variants in Lancaster County Amish population. This panel includes testing for many genetic causes of dystonia, spasticity, and ataxia, as well as by a 256-gene-panel of neurological disorders at InVitae, including glutaric aciduria type 1 (GA1), propionic acidemia, cobalamin-C and methylenetetrahydrofolate reductase (MTHFR) deficiencies, and congenital Parkinsonism. In retrospect, these screening panels did not include HPRT1. An HPRT1-variant of unknown significance was eventually found by whole-exome sequencing of the parents and two boys by two independent sequencing labs—Regeneron-Clinic for Special Children and NCGM by Neuberg—Central Pennsylvania Clinic. The affected male children inherited the single X-chromosome HPRT1 c.485 G>A;p.Ser162Asn variant, which their mother carries along with a normal allele.

After the HPRT1 variant was reported, serum uric acid levels were measured to help confirm the significance of the variant. A third affected infant in the family was diagnosed by targeted mutation testing of cord blood. Dystonia in the youngest child became apparent at 4–6 months of age; increased uric acid in the neonate was similar to that in the older siblings. At the time of initial work-up, none of the children displayed the self-injurious behavior characteristic of Lesch–Nyhan (LN). The middle boy began lip, tongue, and finger biting between 18 and 24 months of age. His older and younger siblings do not injure themselves.

The children's mother is of Lancaster County Amish descent. She carries the X-linked variant in HPRT1 c.485 G>A;p.Ser162Asn, which appears to be a novel variant of unknown significance that structurally neighbors the binding site residues E134-T142 (Figure 1A,B) [2,18]. This variant is not seen in a database of more than 10,000 Amish whole exomes—a University of Maryland and Regeneron database—or in gnomAD (Broad Institute), which contains 250,000 exomes and whole genomes. No other cases of Lesch–Nyhan disease have been found in children of the mother's sisters or her own mother's generation. Inquiries by one of the authors (DHM) at Clinics for Amish Children with Genetic Disorders in Ohio, Kentucky, Indiana, Wisconsin, and Ontario, Canada did not uncover additional cases of the HPRT1 c.485 G>A;p.Ser162Asn mutation. We thus conclude that it is likely that this mutation is de novo in the boys' mother. Whole-exome sequencing and dense microarray karyotype now allow the recognition of single family and single case de novo X-linked and autosomal-dominant mutations as well as rare autosomal recessive mutations and compound heterozygous cases that could not previously be found by homozygosity or genome-wide-SNP mapping.

This novel HPRT1-variant can now be classified as pathogenic and severe because of the neonatal onset of generalized dystonia and the emergence of self-injury in infancy. Increased uric acid is a biochemical feature of the disorder. Allopurinol is given to reduce the risk of uric acid kidney and bladder stones but the suppression of uric acid to normal levels does not improve the movement disorder or self-injury.

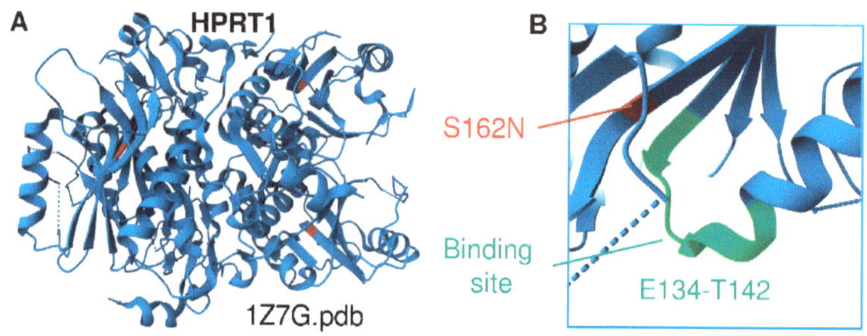

Figure 1. (**A**) Structural representation of HPRT1 (PDB ID 1Z7G). (**B**) Location of the Ser162Asn mutation identified in this cohort, a previously unreported HPRT1 mutation that is adjacent to the substrate binding site (E134-T142 region).

None of the three affected children has presented with overt anemia as judged by low hemoglobin or hematocrit (Table 1). One child has increased mean corpuscular volume (MCV) and red cell distribution width (RDW). The MCVs of the other two children are high normal but remain within the normal range. All three children had high serum uric acid concentrations at the time of diagnosis −5 to 6.5 mg/dL (normal level <2–4.5 mg/dL). Allopurinol suppressed serum uric acid to the normal range and reduced their risk of uric acid kidney stones; however, the normalization of serum uric acid did not improve their movement disorder.

Table 1. Complete blood count (CBC) parameters for the three enrolled patients with c.485 G>A;p.Ser162Asn HPRT mutation. Abbreviations: WBC = white blood cell × 10^9/L, RBC = red blood cell in 10^6/mL, Hgb = hemoglobin beta in g/dL, Hct = hematocrit in %, MCV = mean corpuscular volume in fL, MCH = mean corpuscular hemoglobin in pg, MCHC = mean corpuscular hemoglobin concentration in g/dL, platelet count × 10^9/L, RDW = red cell distribution width in fL.

	Patient 1		**Patient 2**			**Patient 3**	
Age at draw (yrs)	*4.25*	*5.00*	*2.75*	*3.25*	*3.67*	*0.75*	*1.50*
WBC count	7.3	5.9	7.8	8.6	6.0	6.2	5.6
RBC count	4.03	4.05	4.23	4.18	3.89	4.58	4.29
Hgb	12.5	12.2	12.4	12.7	12.9	12.3	12.5
Hct	37.6	38.6	37.6	40.0	39.3	37.1	39.1
MCV	93.3	95.3	88.9	95.7	101.0	81	91.1
MCH	31.0	30.1	29.3	30.4	33.2	26.9	29.1
MCHC	33.2	31.6	33.0	31.8	32.8	33.2	32.0
Platelet count	308	337	300	298	318	310	310
RDW	14.4	15.9	15.2	14.3	14.5	15.4	17.8

Data in red and blue indicate values that are above or below normal range, respectively.

3.2. Multiomics Signatures of LN in Patients with the HPRT1 c.485 G>A;p.Ser162Asn Mutation

In addition to the three affected children, their two parents voluntarily enrolled in this study: the father with normal HPRT and the mother who is heterozygous for HPRT deficiency (herein referred to as LN carrier). The family are members of the Pennsylvania Amish community, an underrepresented population in scientific research [19]. RBC samples were processed via mass spectrometry-based metabolomics, lipidomics, proteomics, and trace metal analysis (Figure 2A). Given the rarity of these specimens, two blood draws were obtained from each subject on the same day, then each sample was processed in technical triplicate (i.e., six measurements per subject). Comparing the LN RBCs (i.e., offspring) to the merged LN carrier and control (i.e., parents), we observed changes in particular to the metabolite and protein compartments (Figure 2B). Specifically, the two biomarkers of LN

RBCs with the highest significance are the HPRT protein and its product GMP, both sharply decreased in LN.

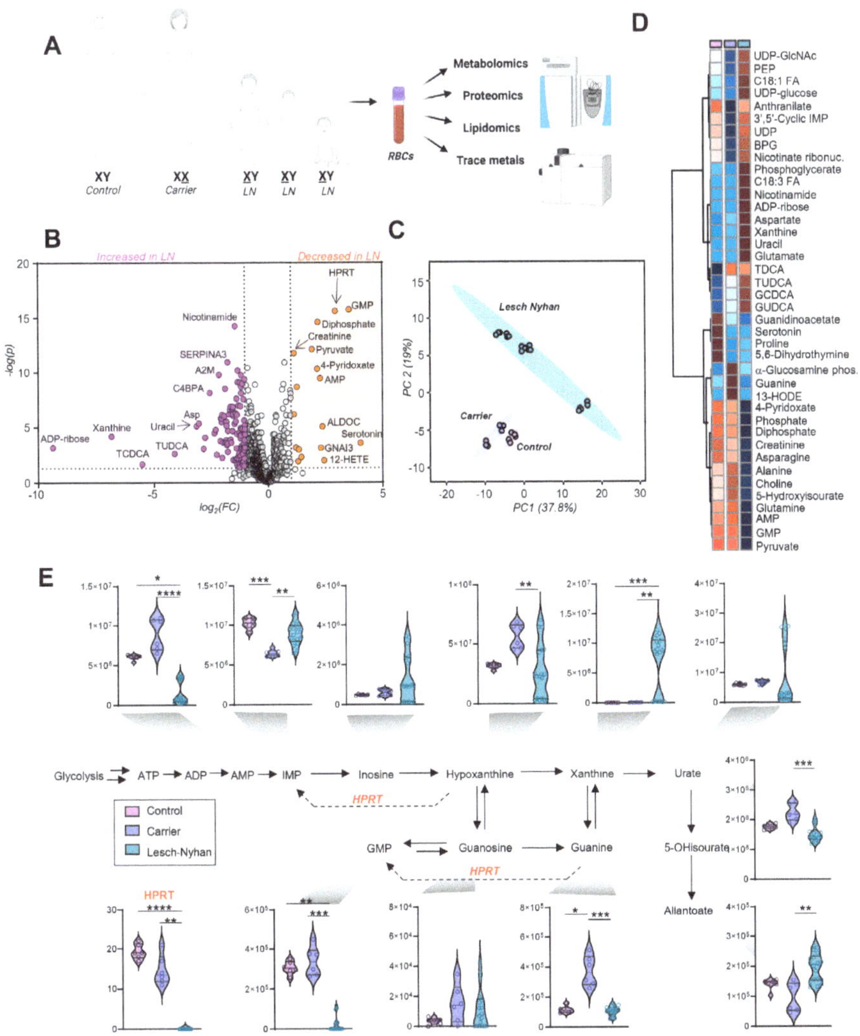

Figure 2. Multiomics analysis of red blood cells (RBCs) from 3 individuals with Lesch–Nyhan disease and their parents—a heterozygous carrier mother and clinically healthy control father. (**A**) Omics study design; (**B**) Volcano plot of all omics results, LN vs. nonLN. Orange indicates molecules significantly elevated in LN relative to carrier and control, purple indicates significantly decreased molecules in LN. (**C**) Principal component analysis (PCA) of metabolomics results; (**D**) Hierarchical clustering analysis of the top 40 metabolites by ANOVA. Group means are shown. (**E**) Purine metabolism and salvage. Data are peak areas (au), statistics are one-way ANOVA with Kruskal–Wallis multiple comparisons test. * $p < 0.05$, ** $p < 0.01$, *** $p < 0.001$, **** $p < 0.0001$.

High throughput metabolomics with a targeted approach to compounds in energy and redox metabolism yielded measurements of 192 metabolites (Table S1). Principal component analysis (Figure 2C) suggests a differential metabolic phenotype of the LN RBCs in comparison to both the carrier and control RBCs, two groups whose profiles are more similar. Global changes in metabolite profiles are represented via heat map with hierarchical clustering of the top 40 metabolites by one-way ANOVA (Figure 2D). The metabolites affected in LN RBCs implicate numerous pathways related to nitrogen metabolism and illustrate the far-reaching effects of HPRT deficiency.

3.3. Dysregulation of Purine Metabolism Is a Hallmark of LN Patient RBCs, Only in Part Recapitulated in the Mutation-Carrying Mother's RBCs

An overview of purine metabolism (Figure 2E) illustrates the effects of LN syndrome on RBCs and the degree to which allopurinol treatment normalizes purine levels in the two older siblings on the treatment. Expectedly, LN RBCs are deficient in HPRT at the protein level along with HPRT product GMP and upstream substrate guanine, though they have normal levels of IMP, as a result of the increased deamination of AMP in the three siblings but not in the mother carrying the mutation in heterozygosity without clinically relevant manifestations. In this view, it is interesting to note that LN carrier RBCs differ from control RBCs for several metabolites including HPRT product IMP (reduced in carrier) and substrate guanine (increased in carrier).

Other metabolic derangements include the elevation of hypoxanthine precursor and IMP dephosphorylation product—inosine. Importantly, allopurinol and its bioactive form oxopurinol are structural isomers of hypoxanthine and xanthine, respectively, and each pair of isobaric compounds coelutes due to the partners' high degree of structural similarity (i.e., polarity). Though exact levels of hypoxanthine and xanthine cannot be discerned from the aforementioned isobaric drug and its catabolite, we observe that—in the two siblings on treatment—allopurinol normalizes RBC levels of urate in LN subjects. However, products downstream of urate, including 5-hydroxyisourate (5-OHisourate) and allantoate (non-enzymatic oxidation product in uricase-deficient human RBCs [20]), are less affected by the treatment with allopurinol.

3.4. Altered Glycolysis in LN RBCs Manifests with Dysfunctional Flux through Pyruvate Kinase

We next moved to an in-depth assessment of steady state levels of RBC energy metabolites. Though levels of glucose are not changed in LN RBCs, the depletion of hexose phosphate compounds and the accumulation of glycolytic intermediates downstream to glyceraldehyde 3-phosphate dehydrogenase (GAPDH) are observed in LN patients (Figure 3A). This signature becomes apparent beginning with bisphosphoglycerate (BPG, Figure 3A) through phosphoenolpyruvate (PEP). Allopurinol treatment appears to partially normalize the levels of these intermediates. There exists a sharp change in trend from PEP to pyruvate, whereby LN RBCs that possess abundant PEP also have significantly lower levels of pyruvate relative to the control and LN carrier RBCs. The conversion of PEP to pyruvate is catalyzed by the redox-sensitive enzyme [21] pyruvate kinase (PK), a rate-limiting and irreversible final step of glycolysis that catalyzes the second pay-off (ATP-generating) reaction. Though the PEP-to-pyruvate ratio is suggestive of a defect in PK expression or activity, changes in PK expression were not observed in the global proteomics analysis (Figure 3A). While deficiency in PK activity is not uncommon in the Amish Mennonite community [22], these patients do not carry any of the known PK mutations, which is suggestive of an oxidation-dependent inhibitory effect on this enzyme. However, post-translational modifications to active site cysteines that could explain a depressed PK catalytic activity—as is known in cancer and immune cells for various PK isoforms [21,23]—were not observed in this study. Glucose metabolism via the pentose phosphate pathway (PPP—Figure 3A), which is regulated by another enzyme encoded by a chromosome X-linked gene—glucose 6-phosphate dehydrogenase—was not broadly altered by HPRT status, though it is interesting to note that the RBCs of each of the three LN offspring had varying levels of the PPP end-product ribose phosphate. Interestingly, compared to the mother carrying the mutation, the LN siblings showed increases in the levels of 6-phosphogluconate, the entry step metabolite of the oxidative phase, NADPH-generating arm of the PPP—suggestive of moderate increases in the activation of antioxidant pathways in the LN group (Figure 3A).

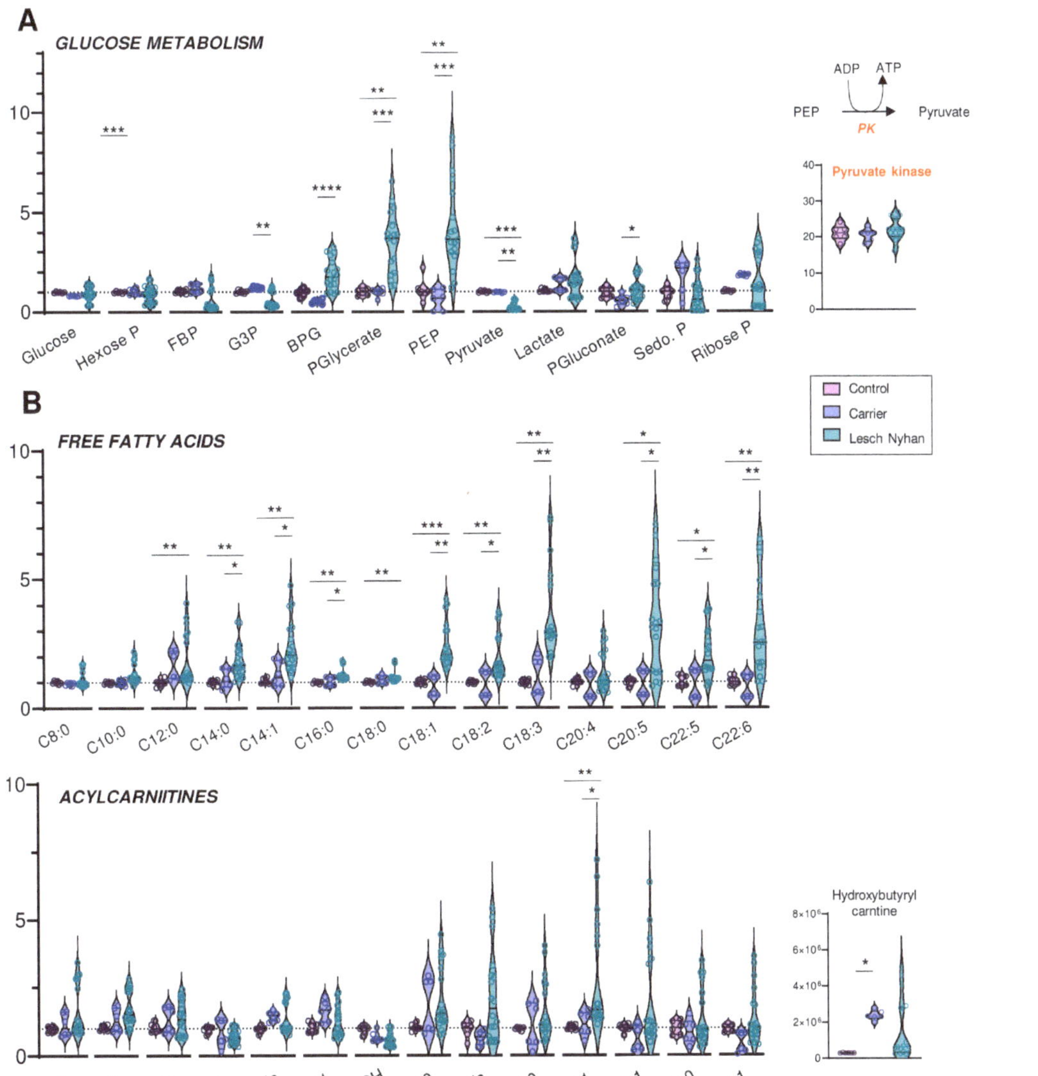

Figure 3. (**A**) Overview of glycolysis and the pentose phosphate pathway. (**B**) Levels of free fatty acids (**top**) and acylcarnitines (**bottom**) by increasing acyl chain length. Data are presented normalized to control samples. Statistics are one-way ANOVA with Kruskal–Wallis multiple comparisons test. * $p < 0.05$, ** $p < 0.01$, *** $p < 0.001$, **** $p < 0.0001$.

3.5. RBCs from LN Patients Have Increased Free Fatty Acids and Acylcarnitines, a Hallmark of Altered Membrane Integrity and Deformability

Steady-state levels of many fatty acids are elevated in LN RBCs relative to carrier and control (Figure 3B, upper panel). In particular, the trend appears for fatty acyl chain lengths of C12 and longer, though C20:4 (i.e., arachidonic acid) is not changed and no significant increases in hydroxyeicosatetraenoic peroxidation products were observed in this group. The enrichment in fatty acids is more evident in the long chain poly- and highly unsaturated fatty acids (C18 and higher, 2 to 6 double bonds), and similar to glycolytic intermediates, only partially normalized by allopurinol. Acylcarnitines also trend toward elevation in the LN RBCs, though allopurinol more completely normalizes the levels of carnitines than RBCs, so few acylcarnitines have significant changes (Figure 3B, lower panel). Hydroxybutyrylcarnitine (C4-OH) had the greatest fold change difference for both LN and carrier vs. control though these values were also normalized with allopurinol treatment, suggestive of cross-talk between purine oxidation and membrane lipid remodeling.

3.6. LN Significantly Impacts Glutaminolysis, Conjugated Bile Acids and Tryptophan-Derived Inflammatory and Neurotransmitter Metabolites: Metabolic Signatures of Hepatic and Neural Dysfunction and Microbiome Dysbiosis

We next looked deeper into other metabolic pathways enriched in the top 40 heat map (Figure 2D). Figure 4A illustrates transamination substrates and products; we report in LN RBCs lower ratios of amide-containing amino acids glutamine and asparagine relative to their ester-containing counterparts glutamate and aspartate, respectively. In these subjects, there was not a measurable difference between LN carrier and control RBCs. Glutathione pools, including reduced (GSH) and oxidized (GSSG) forms, were elevated in RBCs from LN individuals (Figure S1A). Urea cycle metabolites and polyamines were largely unchanged, except for metabolites downstream of arginine: citrulline and creatine are increased in the carrier and LN while guanidinoacetate and creatine are both decreased in the carrier and LN RBCs relative to control patient RBCs (Figure S1B).

We observed low levels of methylation products, including guanidinoacetate, creatine, creatinine, choline, and glucosamine, as well as important precursors for purine nucleotides—asparagine, aspartate, glutamine, phosphate—and the depletion of the end products of purine synthesis—guanine, guanosine, GMP, GTP, and AMP. Substrates required to support purine de novo synthesis, including glycine and upstream metabolite serine, are not altered in LN RBCs though we note that samples from the LN individual not receiving allopurinol yielded six of the nine lowest glycine readings (Figure S1C). Other requisite substrates, such as phosphoribosyl pyrophosphate (PRPP), were not detected.

Several bile acids (Figure 4B) are increased in LN RBCs, especially cholate- and taurine-conjugated bile acids, metabolites that are deconjugated by the gut microbiome and accumulate in the bloodstream as a result of inflammatory conditions triggering dysbiosis [24]. Given the neural manifestations of LN, it is interesting to note that RBCs harbor several of the enzymes that can synthesize neurotransmitters [25]—making the measurement of these compounds in blood cells a viable strategy to infer the dysregulation of such metabolic pathways in the central nervous system [10]. Here, we observed that adrenaline and dopamine are increased in LN RBCs—as opposed to results in murine models reporting decreases [26]—without changes to upstream essential amino acid tyrosine (Figure 4C). Similarly, while no changes were observed in the levels of the precursor tryptophan, serotonin levels were sharply decreased in both the carrier and LN RBCs, an observation that warrants additional investigation to the extent that dysfunctional serotonin synthesis and reuptake is etiologically linked to depression [27], and obsessive-compulsive disorders like facial grimacing are common manifestation of LN, including in the patients enrolled in this study. Of note, kynurenine levels were elevated in the sibling not on allopurinol, which is relevant in that this tryptophan metabolite is elevated in response to cGAS-STING-interferon activation secondary to infections (e.g., SARS-CoV-2). Kynurenine catabolism has been linked to adverse neurological manifestations in other populations with genetic anomalies, like subjects suffering from Trisomy 21 [28] or Aicardi–Goutieres syndrome [29].

Finally, to complete the small molecule analysis, we quantified levels of transition metals in RBCs from these subjects and report decreased levels of Fe, Zn, Se, and K in LN RBCs relative to both carrier and control (Figure 4D; see Table S1 for the full metallomics data set), which is directly relevant to erythropoiesis [4,30], redox biochemistry (including the synthesis of selenoproteins that counteract lipid peroxidation [31]), and neurological conditions [32].

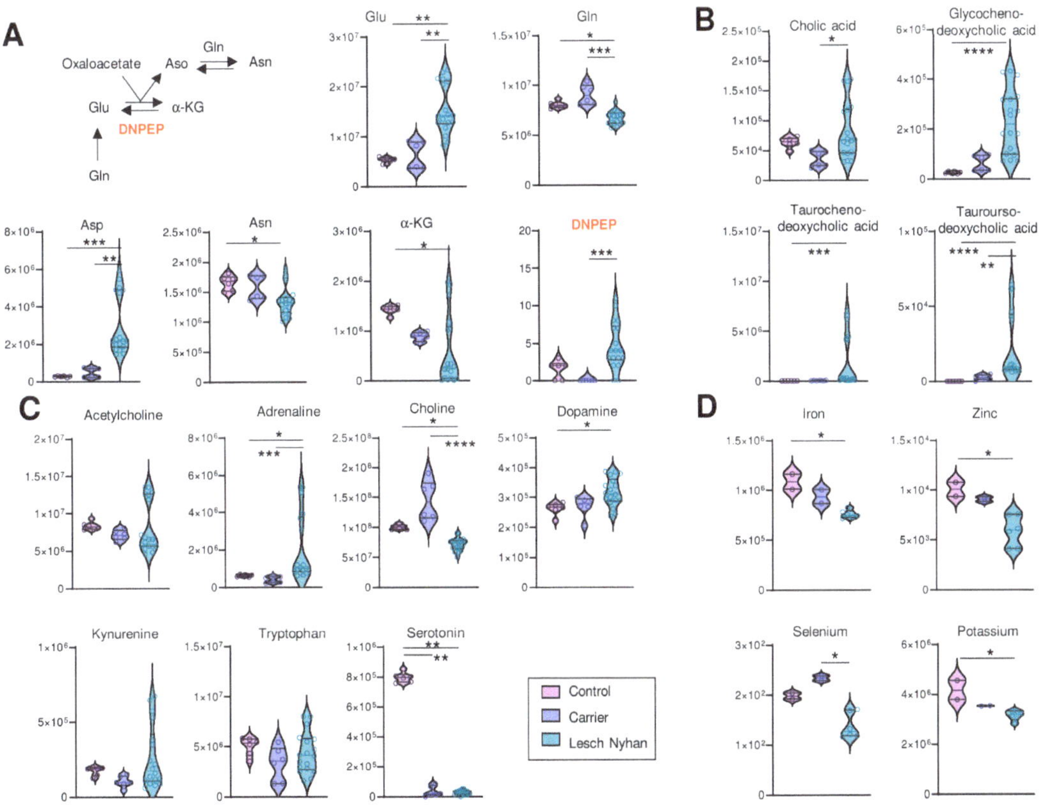

Figure 4. (**A**) Overview of transamination reactions reveals decreased amido-containing amino acids Gln and Asn in LN RBCs and increased carboxylate amino acids Glu and Asp. (**B**) RBC bile acids are elevated in LN subjects' RBCs. (**C**) Levels of neurometabolites. (**A–C**) Data are peak areas (au). (**D**) Trace metal analysis (ppb). Statistics are one-way ANOVA with Kruskal–Wallis multiple comparisons test. * $p < 0.05$, ** $p < 0.01$, *** $p < 0.001$, **** $p < 0.0001$.

3.7. The RBC Lipidome in LN Is Characterized by Elevation in Phosphatidylcholines and Depression in Sphingomyelins, Ceramides and Phosphatidylethanolamines

Untargeted lipidomics followed by database matching of MS^2 data and identification yielded 257 named lipids, 103 of which were differentially abundant with $p < 0.05$ by one-way ANOVA. A PLS-DA of named lipids illustrates, similar to the metabolomics data, that the LN carrier RBCs have an intermediate lipid phenotype relative to the control and LN RBCs (Figure 5A). In fact, allopurinol treatment in two of the three LN subjects partially, yet incompletely, restores the lipid phenotype to levels compared to controls in the treated siblings relative to the youngest LN patient who is not receiving the drug. A heat map with hierarchical clustering of the top 50 lipids (by p value) reveals snapshot regions of the RBC lipidome that appear to be impacted most by HPRT heterozygosity and/or HPRT deficiency (Figure 5B). In particular, phosphatidylcholines (PCs) with highly unsaturated acyl chains are increased in LN RBCs (Figure 5C). Sphingomyelins, among others, are increased in the LN carrier but not in LN RBCs and thus warrant additional investigation to disentangle the effect of sex. These same SMs are lower in the two LN patients receiving allopurinol than in their untreated sibling. A view into the effects of allopurinol treatment on LN RBCs is presented via a volcano plot (Figure 5D) illustrating an allopurinol-associated increase in PCs and PEs and a corresponding decrease in SMs along with acylcarnitines (Figure 5D,E, here validated with unsupervised approaches on top of the targeted measurements described above in Figure 3B). Based on this analysis, we observed that the LN siblings presented an elevation in the RBC levels of phosphatidylcholines (PCs), accompanied by decreases in sphingomyelins, ceramides, and phosphatidylethanolamines (Figure 5E).

3.8. The RBC Proteome in LN Is Characterized by Depletion of Structural Proteins, Elevation in the Levels of Acute Phase Response and Complement Proteins, and Elevated Protein Cysteine Oxidation

A global proteomics analysis of tryptic digests reveals 147 RBC proteins that were significantly altered, using one-way ANOVA ($p < 0.05$). RBCs from individuals with LN are characterized by a phenotypic shift evident on PLS-DA whereby RBCs from the LN carrier display an intermediate phenotype between the LN subjects and the control RBCs (Figure 6A,B). A hierarchical clustering analysis of the top 50 altered proteins (by p value) illustrates the protein expression changes that prominently characterize the LN RBC phenotype (Figure 6C). Specifically, HPRT peptides are undetectable in LN RBCs, in accordance with the complete lack of activity known in LN erythrocytes [33], and are higher in the carrier mother than in the control father (1.3-fold increase, $p = 0.021$), as anticipated for a chromosome X-encoded protein product. In addition, LN RBCs have decreased levels of hemoglobins, consistent with LN-associated anemia. It is crucial to note that RBCs in LN patients are typically macrocytic, thus hemoglobin copies per cell are increased; here, equal aliquots of RBCs reveal less hemoglobin in LN specimens compared to the carrier and control. In addition to hemoglobins, LN RBCs display decreased membrane and cytoskeletal proteins—dematin (DMTN), aquaporin 1 (AQP1), protein 4.1 (EPB41), Ras-related protein (RAB1B), and carbonic anhydrase 3 (CA3). We also observe a decrease in the LN RBCs of proteins involved in redox homeostasis, including 3-mercaptopyruvate sulfurtransferase (MPST)—an enzyme involved in the synthesis of cysteine—and H_2S, aldehyde dehydrogenase (ALDH1A1)—which scavenges aldehydes produced by lipid peroxidation—and catalase (CAT), a key H_2O_2 detoxifying heme enzyme.

Conversely, a separate subset of the RBC proteome is enriched in LN subjects. Affected proteins include several related to heme and iron dynamics—uroporphyrinogen decarboxylase (UROD), ceruloplasmin (CP), hemopexin (HPX), and serotransferrin (TF)—and membrane structure and function—centractin (ACTR1A) and fibronectin (FN1). Additionally, purine salvage enzyme adenine phosphoribosyltransferase (APRT), which produces AMP, is elevated in LN subjects perhaps as a compensation for dysregulation induced by the absence of functional HPRT. Finally, we observed elevated levels of complement system proteins (C4BPA, CFH, CP, CFB) along with increases in proteins involved in coagulation and hemostasis, such as the serpins, fibrinogens, kininogen 1 (KNG1), and prothrombin (F2). These acute phase response proteins have been previously associated with the results of inflammatory events that promote deposition on the RBC membrane and prime the RBC for increased susceptibility to intra- and extra-vascular hemolysis [34], especially in response to interferon-triggering infections like SARS-CoV-2 [17].

As anticipated above for PK, RBC metabolism is modulated in part by disruptions to redox homeostasis. A particular challenge in RBCs is the inability to synthesize proteins de novo (lack of organelles), thus RBCs are uniquely susceptible to oxidative damage and reliant on repair/degradation pathways. We assessed levels of modified proteins by quantifying (a) the oxidation of protein cysteines to form sulfinic acid (P-SO_2H) and the elimination of sulfur to form protein dehydroalanine (DHA); (b) carboxylate side chain methylation (Glu, Asp) along with asparagine deamidation followed by methylation of the L-isoaspartyl groups resulting from this reaction (Figure 6D).

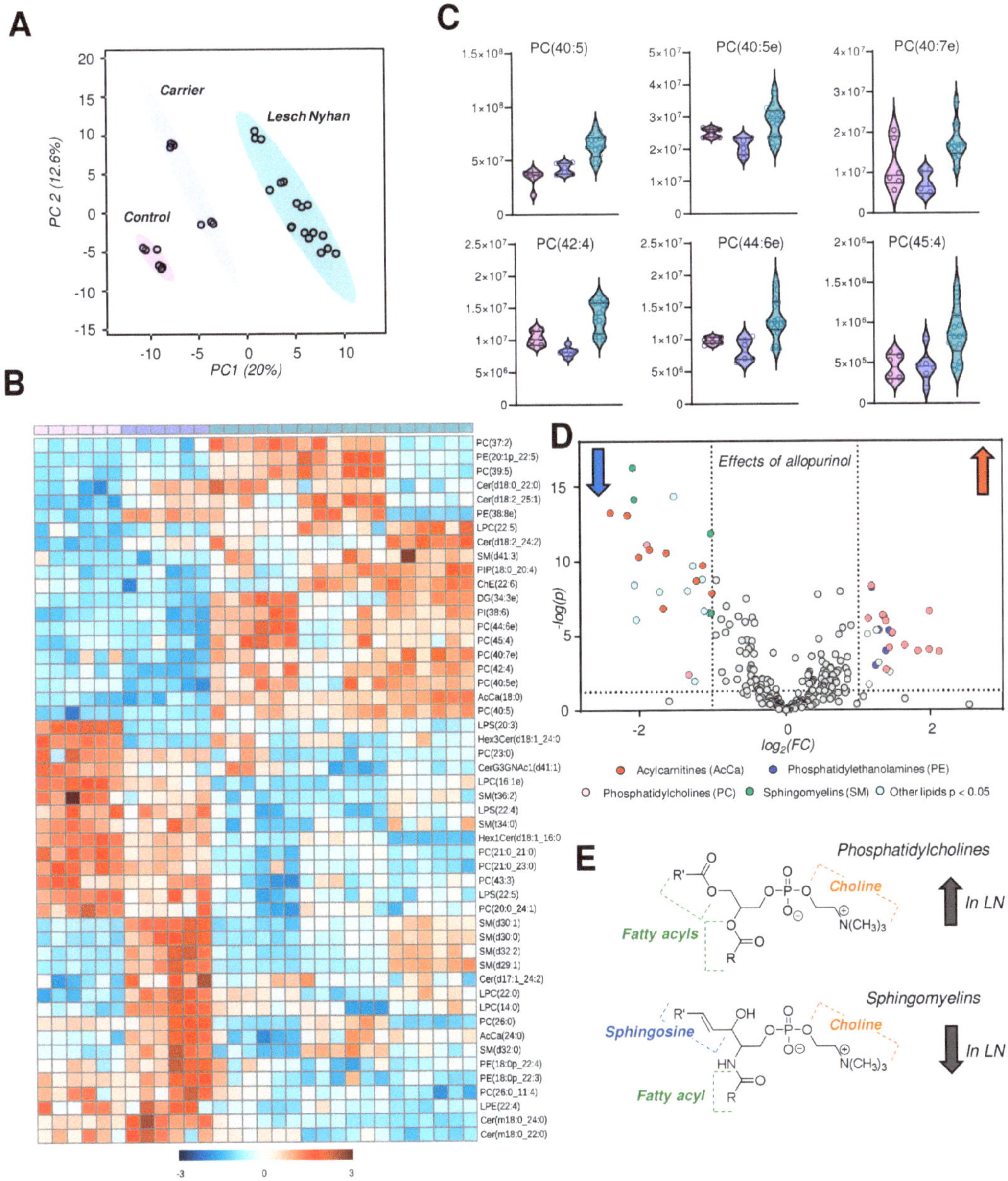

Figure 5. Lipidomics profiles of RBC samples from Lesch–Nyhan subjects versus carrier and control. (**A**) Partial least squares-discriminant analysis (PLS-DA) reveals a clustering of samples along component 1 (x-axis) illustrating that LN carrier RBCs possess an intermediate phenotype between LN and control RBCs. (**B**) Heat map with hierarchical clustering of the top 50 metabolites by one-way ANOVA. (**C**) Violin plots of highly unsaturated phosphatidylcholines. (**D**) Volcano plot of the effects of allopurinol treatment (2 LN subjects) versus untreated (1 LN subject) on the RBC lipidome. The red shaded region indicates lipids that are elevated in allopurinol-exposed RBCs; the blue shaded region indicates lipids that are decreased in allopurinol-exposed RBCs. (**E**) General structures of phosphatidylcholines (**top**) and sphingomyelins (**bottom**).

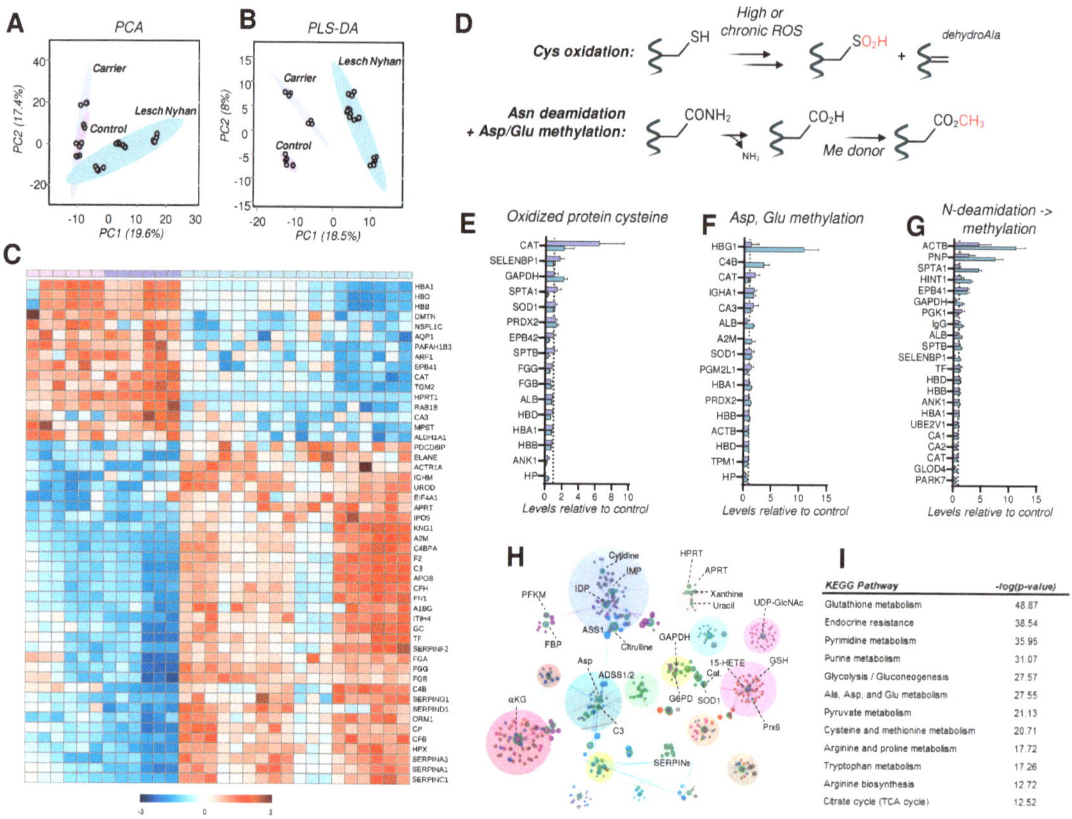

Figure 6. Proteomic profiling of RBCs from Lesch–Nyhan subjects. (**A**) Unsupervised principal component analysis demonstrates overlapping phenotypes in carrier and control RBCs. (**B**) Partial least squares-discriminant analysis illustrates the clustering of biological groups and the intermediate phenotype of the carrier RBCs. (**C**) Global proteomics results as visualized via heat map with hierarchical clustering of the top 50 proteins by *p*-value (one-way ANOVA). (**D**) Schematic of post-translational modifications (PTMs) investigated as readouts for proteome oxidation, including cysteine oxidation, asparagine deamidiation, and methylation of glutamate and aspartate. (**E**) Irreversibly oxidized protein cysteine (sulfinic acid and dehydroalanine) content as a function of total protein levels. Levels are presented relative to control RBCs (dashed line at x = 1). (**F**) Methylation of protein glutamate and aspartate as a function of total protein levels. Levels are presented relative to control RBCs (dashed line at x = 1). (**G**) Asparagine deamidation with subsequent aspartate methylation catalyzed by PIMT1 and other methylating enzymes. Levels are presented relative to control RBCs (dashed line at x = 1). (**H**) Network analysis of protein-protein and protein-metabolite interactions. Labeled note color denotes metabolic pathway: green—nucleotides, blue—amino acids, purple—energy metabolism, red—redox metabolism. (**I**) Pathway analysis of integrated proteomics and metabolomics data. Abbreviations found in Supplementary Table S1.

With respect to oxidized cysteine, we identified 16 proteins in the global proteomics analysis with a measurable formation of these oxidized cysteine products (Figure 6E). Spectral counts of peptide sulfinic acids and peptide dehydroalanine were summed then subsequently normalized to total peptide levels (i.e., spectral counts) for the corresponding protein. As displayed in Figure 6E, the majority of these proteins are unexpectedly less oxidized (at cysteine) in the LN RBCs than in carrier or control (i.e., levels < 1). The few notable exceptions are catalase (CAT), the key heme-containing H_2O_2 detoxifying enzyme, methanethiol oxidase (SELENBP1), glycolytic enzyme and known ROS target glyceraldehyde 3-phosphate dehydrogenase (GAPDH) [35], and spectrin alpha (SPTA1). Of these, there is more oxidized GAPDH in LN RBCs relative to carrier and control; for the remainder, LN RBCs have lower levels of oxidized cysteine-containing peptides.

Oxidant stress in RBCs leads to damage of asparagine and glutamine residues by deamidation followed by methylation of the resulting isoaspartate and isoglutamate residues [36]. The methylation of (iso)aspartate and glutamate side chains was observed in hemoglobin gamma (HBG1), complement C4-B (C4B), albumin (ALB), and alpha-2-macroglobulin (A2M) at the highest levels in LN RBCs versus carrier and control (Figure 6F). In contrast, only catalase (CAT) has the highest methylated Asp and Glu content in the carrier. Here, we report elevated N-deamidation/methylation occurring in LN RBCs on β-actin (ACTB), purine nucleoside phosphorylase (PNP), SPTA1, adenosine 5′-monophosphoramidase (HINT1), and protein 4.1 (EPB41)—a marker of RBC aging in vivo—relative to both carrier and control (Figure 6G). Levels of this modification in ACTB, PNP, and EPB41 were lowest in the control. These PTM observations warrant additional experimentation, particularly if the RBC lifespan is different in LN subjects versus the general population. A shortened RBC lifespan would be expected to result in a greater proportion of young RBCs in circulation, which would possess fewer age-associated oxidation events.

Finally, metabolomics and proteomics data sets were integrated into an interaction network (Figure 6H), revealing links among: (a) nucleotides and HPRT (green dots); (b) redox metabolism (red dots); and (c) energy metabolism (blue and purple dots). A pathway analysis against the KEGG database supports the alterations of these pathways in LN RBCs (Figure 6I).

4. Discussion

Herein, we report the first integrated multi-omics investigation of how Lesch–Nyhan disease caused by an HPRT1 c.485 G>A;p.Ser162Asn variant impacts the molecular composition of RBCs. Without means to synthesize ATP de novo, RBCs are reliant on the purine salvage pathway to restore ATP pools. Though hypoxanthine is not a direct substrate for purine salvage production of ATP, enzymopathies in the pathway disrupt metabolic homeostasis; in the case of Lesch–Nyhan disease, HPRT deficiency results in the underutilization of hypoxanthine and xanthine, leading to the accumulation of urate and purine monophosphates. Disruptions to the purine salvage pathway in RBCs have been noted in the context of iatrogenic or environmental stimuli that trigger oxidant stress to the erythrocyte, such as blood storage [11], irradiation, exercise [37,38], cryostimulation [39], propionic acidemia, exposure to hypoxia ex vivo and in vivo, and Down syndrome (i.e., trisomy 21) [40].

Here, we leverage a unique and rare sample set voluntarily donated by a family carrying a novel S162N HPRT mutation to perform a comprehensive multiomics analysis. Global metabolomics profiles suggest that the HPRT deficiency drives the RBC phenotypic changes; far fewer differences are observed when comparing the control (HPRT normal) to the carrier RBCs, which is consistent with the lack of clinical manifestations in the mother carrying the mutation. The proteomics characterization also abides by this theme in that control and carrier are similar to each other and LN RBCs offer a shifted phenotype. In contrast, global lipidomic profiles reveal marked differences between the control (father) and carrier (mother) RBCs (Figure 4B, bottom half). With low n, it is not possible to disentangle the effect of sex alone on the observed changes, though it is important to note that the two subjects are similar in age.

Though with few subjects, given the rarity of this mutation, the treatment of two of the three LN individuals with allopurinol allows for the preliminary insight into how such treatment impacts RBCs and where dysregulated metabolic phenotypes are restored, even if partially. For example, RBC free fatty acids and acylcarnitine levels are partially normalized in LN patients receiving allopurinol, confirming prior reports about a crosstalk between purine oxidation and activation of the Lands cycle [41–43]. The proteomics profiles of RBCs from allopurinol-treated subjects suggest a decrease in disease severity, though the effect is relatively modest (Figure 5C). Lipidomics profiles are unique in that LN RBCs display significant differences that appear to be imparted by allopurinol treatment, in particular among the phosphatidylcholines (PCs) and sphingomyelins. Interestingly, PC-derived choline could contribute methyl groups to fuel protein isoaspartyl methylation upon deamidation of asparagine residues, a protein damage repair pathway that is regulated by PIMT in mature RBCs [36]. Interestingly, genetic ablation of PIMT is lethal in mice and—like LN—manifests with neurological sequelae (seizures at 6–8 weeks of age when in homozygosity [44], accelerated aging of the brain when in heterozygosity [45]) as well as an RBC oxidant stress phenotype [36]. In this view, it is interesting to speculate that deamidation of N162 in the novel mutant HPRT investigated in this study would restore the negative charge of the serine residue in the dominant canonical HPRT, provided that isoaspartyl formation does not rearrange the HPRT protein backbone and further compromise its enzymatic activity. This observation is relevant in that it suggests that the supplementation of methyl group donors (e.g., choline, betaine, vitamin B5, *S*-adenosylmethionine) may partially restore the phenotype observed in this patient population as a co-adjuvant to allopurinol treatment.

Though only observed in one of the three children from the small patient cohort in this study, the macrocytic phenotype common in individuals with LN results from dysfunctional one-carbon metabolism—critical to hematopoiesis [40], a common trait between LN and another genetic condition accompanied by macrocytic anemia—Down syndrome [46]. Interestingly, elevated kynurenine in Trisomy 21 is associated with the accumulation of its neurotoxicant catabolites [28], another phenotypic overlap between the LN patients investigated here and Down syndrome. In this view, it is worth noting that the activation of the interferon-indole 2,3-dioxygenase (IDO)-kynurenine axis is common among LN, Down syndrome, and Aicardi–Goutieres syndrome, also associated with clinical neurological manifestations [29,47]. Similarly, the elevated synthesis of neurotoxic kynurenine metabolites is a feature of glutaric aciduria type 1 (GA1). Clinical parallels exist between Lesch–Nyhan and glutaric aciduria 1 (GA1), a pathology caused by the incomplete catabolism of lysine and tryptophan. For example, in both disease states infants appear normal at birth and the movement disorder emerges over the first 1–2 years of life. Over the 35-year period of 1988–2023, GA1 research provided insights that have led to increasingly effective, preventative treatments. The outcomes of affected infants have improved from 95% disabled to 95% with good neurological outcomes [48]. Metabolic intervention in GA1 consists of a low lysine diet with supplementation of a lysine-free, tryptophan-reduced, amino acid mixture, and oral supplementation of L-carnitine. Similar metabolic interventions could be tested to mitigate the neurological manifestations of LN.

Since the kynurenine pathway is downstream of the cGAS-STING viral/pathogen sensing pathway, this observation could contribute to explaining why disease severity worsens in LN patients upon infection [49–52]. Finally, it is interesting to speculate that increased reticulocytosis or mitochondrial-containing mature erythrocytes—as it is observed for example in sickle cell patients [53] or lupus erythematosus [54]—results in an increased basal activation of humoral immune responses downstream of interferon signaling . Whether this phenomenon occurs in LN is not clear, though our data clearly point at the elevation of carboxylic acids and mitochondrial proteins in circulating RBCs.

Severe depletion of serotonin in both LN patients and the mutation-carrying mother is consistent with elevated tryptophan catabolism towards the kynurenine pathway at the expense of the synthesis of this important metabolite, an additional factor that contributes to explaining the neurological disorder in the LN population.

We are not aware of other studies that have investigated the impact of allopurinol on the RBC lipidome, which could be translationally relevant for patients with gout; such characterizations are warranted, as dynamic changes to the lipidome could alter RBC membrane structure and therefore cellular function. Despite benefits of purine catabolism, energy metabolism—especially the payoff steps of glycolysis downstream to pyruvate kinase—appeared depressed in the LN patients, suggestive of a potential benefit to the use of PK activators (e.g., mitapivat [55]) as a co-adjuvant in the treatment of LN.

5. Conclusions

In conclusion, this multiomics investigation provides a wealth of heretofore unknown molecular level information about how HPRT deficiency impacts the RBC and allows for the contextualization of the rare Lesch–Nyhan disease alongside other diseases and physiological changes (e.g., exercise) that impact purine metabolism. We recognize as a limitation of this study the limited size of the patient cohort. Further, we acknowledge that, though various HPRT mutations are known to result in LN disease, the molecular profiles described here are derived from RBCs specifically carrying the HPRT1 c.485 G>A;p.Ser162Asn mutation. The data herein could inform future complementary treatment to present regimens like allopurinol, for example, via supplementation of substrates to fuel compensatory regulatory pathways, here observed to be active in LN patients and the healthy carrier of this mutation, such as protein isoaspartyl methylation.

Supplementary Materials: The following supporting information can be downloaded at: https://www.mdpi.com/article/10.3390/antiox12091699/s1, Omics data generated in this study is included in Supplementary Table S1. An overview of redox and nitrogen metabolism is included in Figure S1.

Author Contributions: Conceptualization, D.H.M. and A.D.; Methodology, A.D., J.A.R., M.D. and D.S.; Investigation, J.A.R., M.D., D.S. and F.G.; Data Curation, J.A.R., M.D., D.S. and F.G.; Writing—Original Draft Preparation, J.A.R.; Writing—Review & Editing, all authors; Supervision, A.D. and D.H.M.; Project Administration, A.D. and D.H.M.; Funding Acquisition, A.D. All authors have read and agreed to the published version of the manuscript.

Funding: A.D. was supported by funds by the National Heart, Lung, and Blood Institute (R01HL146442, R01HL149714, R01HL148151, R01HL161004).

Institutional Review Board Statement: The study was conducted in accordance with the Declaration of Helsinki; specimens were collected at the Central Pennsylvania Clinic, Boston Children's Hospital, and Lancaster General Hospitals under institutionally reviewed Protocol No. 2014-12 and upon signing of informed consent.

Informed Consent Statement: Informed consent was obtained from all subjects involved in the study.

Data Availability Statement: Data generated in this study are available in Supplementary Table S1 and/or from the authors upon reasonable request.

Acknowledgments: The authors would like to acknowledge Lacrisha Diven, RN, Research Coordinator, Central PA Clinic, who maintains data about population genetics, carrier & disease rates for Kish Valley. The authors also kindly thank the family who volunteered for this study.

Conflicts of Interest: The authors declare that A.D. is a founder of Omix Technologies Inc. and Altis Biosciences LLC. A.D. is also a consultant for Hemanext Inc. and Macopharma Inc. The other authors also declare no conflict of interest.

References

1. Mak, B.S.; Chi, C.-S.; Tsai, C.-R.; Lee, W.-J.; Lin, H.-Y. New mutations of the HPRT gene in Lesch-Nyhan syndrome. *Pediatr. Neurol.* **2000**, *23*, 332–335. [CrossRef] [PubMed]
2. Fu, R.; Ceballos-Picot, I.; Torres, R.J.; Larovere, L.E.; Yamada, Y.; Nguyen, K.V.; Hegde, M.; Visser, J.E.; Schretlen, D.J.; Nyhan, W.L.; et al. Genotype–phenotype correlations in neurogenetics: Lesch-Nyhan disease as a model disorder. *Brain* **2013**, *137*, 1282–1303. [CrossRef] [PubMed]
3. Bell, S.; Kolobova, I.; Crapper, L.; Ernst, C. Lesch-Nyhan Syndrome: Models, Theories, and Therapies. *Mol. Syndromol.* **2016**, *7*, 302–311. [CrossRef]
4. Cakmakli, H.F.; Torres, R.J.; Menendez, A.; Yalcin-Cakmakli, G.; Porter, C.C.; Puig, J.G.; Jinnah, H.A. Macrocytic anemia in Lesch-Nyhan disease and its variants. *Genet. Med.* **2019**, *21*, 353–360. [CrossRef]
5. Nyhan, W.L. Clinical Features of the Lesch-Nyhan Syndrome. *Arch. Intern. Med.* **1972**, *130*, 186–192. [CrossRef]

6. Kranen, S.; Keough, D.; Gordon, R.B.; Emmerson, B.T. Xanthine-containing calculi during allopurinol therapy. *J. Urol.* **1985**, *133*, 658–659. [CrossRef]
7. Saugstad, O.D. Hypoxanthine as a measurement of hypoxia. *Pediatr. Res.* **1975**, *9*, 158–161. [CrossRef]
8. Saugstad, O.D. Hypoxanthine as an indicator of hypoxia: Its role in health and disease through free radical production. *Pediatr. Res.* **1988**, *23*, 143–150. [CrossRef]
9. Chouchani, E.T.; Pell, V.R.; Gaude, E.; Aksentijevic, D.; Sundier, S.Y.; Robb, E.L.; Logan, A.; Nadtochiy, S.M.; Ord, E.N.J.; Smith, A.C.; et al. Ischaemic accumulation of succinate controls reperfusion injury through mitochondrial ROS. *Nature* **2014**, *515*, 431–435. [CrossRef]
10. Nemkov, T.; Reisz, J.A.; Xia, Y.; Zimring, J.C.; D'Alessandro, A. Red blood cells as an organ? How deep omics characterization of the most abundant cell in the human body highlights other systemic metabolic functions beyond oxygen transport. *Expert Rev. Proteom.* **2018**, *15*, 855–864. [CrossRef]
11. Nemkov, T.; Sun, K.; Reisz, J.A.; Song, A.; Yoshida, T.; Dunham, A.; Wither, M.J.; Francis, R.O.; Roach, R.C.; Dzieciatkowska, M.; et al. Hypoxia modulates the purine salvage pathway and decreases red blood cell and supernatant levels of hypoxanthine during refrigerated storage. *Haematologica* **2018**, *103*, 361–372. [CrossRef] [PubMed]
12. Roussel, C.; Morel, A.; Dussiot, M.; Marin, M.; Colard, M.; Fricot-Monsinjon, A.; Martinez, A.; Chambrion, C.; Henry, B.; Casimir, M.; et al. Rapid clearance of storage-induced microerythrocytes alters transfusion recovery. *Blood* **2021**, *137*, 2285–2298. [CrossRef] [PubMed]
13. Torres, R.J.; Prior, C.; Puig, J.G. Efficacy and safety of allopurinol in patients with the Lesch-Nyhan syndrome and partial hypoxanthine- phosphoribosyltransferase deficiency: A follow-up study of 18 Spanish patients. *Nucleosides Nucleotides Nucleic Acids* **2006**, *25*, 1077–1082. [CrossRef]
14. Nemkov, T.; Reisz, J.A.; Gehrke, S.; Hansen, K.C.; D'Alessandro, A. High-Throughput Metabolomics: Isocratic and Gradient Mass Spectrometry-Based Methods. *Methods Mol. Biol.* **2019**, *1978*, 13–26. [CrossRef] [PubMed]
15. Stephenson, D.; Nemkov, T.; Qadri, S.M.; Sheffield, W.P.; D'Alessandro, A. Inductively-Coupled Plasma Mass Spectrometry-Novel Insights From an Old Technology Into Stressed Red Blood Cell Physiology. *Front. Physiol.* **2022**, *13*, 828087. [CrossRef]
16. Reisz, J.A.; Zheng, C.; D'Alessandro, A.; Nemkov, T. Untargeted and Semi-targeted Lipid Analysis of Biological Samples Using Mass Spectrometry-Based Metabolomics. *Methods Mol. Biol.* **2019**, *1978*, 121–135. [CrossRef]
17. Thomas, T.; Stefanoni, D.; Dzieciatkowska, M.; Issaian, A.; Nemkov, T.; Hill, R.C.; Francis, R.O.; Hudson, K.E.; Buehler, P.W.; Zimring, J.C.; et al. Evidence of Structural Protein Damage and Membrane Lipid Remodeling in Red Blood Cells from COVID-19 Patients. *J. Proteome Res.* **2020**, *19*, 4455–4469. [CrossRef]
18. Nguyen, K.V.; Naviaux, R.K.; Nyhan, W.L. Novel mutation in the human HPRT1 gene and the Lesch-Nyhan disease. *Nucleosides Nucleotides Nucleic Acids* **2017**, *36*, 704–711. [CrossRef]
19. Conlin, W.E. Ethical Considerations for Treating the Old Order Amish. *Ethics Behav.* **2021**, *31*, 419–432. [CrossRef]
20. Tian, R.; Yang, C.; Chai, S.M.; Guo, H.; Seim, I.; Yang, G. Evolutionary impacts of purine metabolism genes on mammalian oxidative stress adaptation. *Zool. Res.* **2022**, *43*, 241–254. [CrossRef]
21. van de Wetering, C.; Manuel, A.M.; Sharafi, M.; Aboushousha, R.; Qian, X.; Erickson, C.; MacPherson, M.; Chan, G.; Adcock, I.M.; ZounematKermani, N.; et al. Glutathione-S-transferase P promotes glycolysis in asthma in association with oxidation of pyruvate kinase M2. *Redox Biol.* **2021**, *47*, 102160. [CrossRef]
22. Roy, M.K.; Cendali, F.; Ooyama, G.; Gamboni, F.; Morton, H.; D'Alessandro, A. Red Blood Cell Metabolism in Pyruvate Kinase Deficient Patients. *Front. Physiol.* **2021**, *12*, 735543. [CrossRef] [PubMed]
23. Prakasam, G.; Iqbal, M.A.; Bamezai, R.N.K.; Mazurek, S. Posttranslational Modifications of Pyruvate Kinase M2: Tweaks that Benefit Cancer. *Front. Oncol.* **2018**, *8*, 22. [CrossRef] [PubMed]
24. Ridlon, J.M.; Kang, D.J.; Hylemon, P.B.; Bajaj, J.S. Bile acids and the gut microbiome. *Curr. Opin. Gastroenterol.* **2014**, *30*, 332–338. [CrossRef] [PubMed]
25. D'Alessandro, A.; Dzieciatkowska, M.; Nemkov, T.; Hansen, K.C. Red blood cell proteomics update: Is there more to discover? *Blood Transfus.* **2017**, *15*, 182–187. [CrossRef]
26. Meek, S.; Thomson, A.J.; Sutherland, L.; Sharp, M.G.F.; Thomson, J.; Bishop, V.; Meddle, S.L.; Gloaguen, Y.; Weidt, S.; Singh-Dolt, K.; et al. Reduced levels of dopamine and altered metabolism in brains of HPRT knock-out rats: A new rodent model of Lesch-Nyhan Disease. *Sci. Rep.* **2016**, *6*, 25592. [CrossRef]
27. Meltzer, H.Y. Role of serotonin in depression. *Ann. N. Y. Acad. Sci.* **1990**, *600*, 486–500. [CrossRef]
28. Powers, R.K.; Culp-Hill, R.; Ludwig, M.P.; Smith, K.P.; Waugh, K.A.; Minter, R.; Tuttle, K.D.; Lewis, H.C.; Rachubinski, A.L.; Granrath, R.E.; et al. Trisomy 21 activates the kynurenine pathway via increased dosage of interferon receptors. *Nat. Commun.* **2019**, *10*, 4766. [CrossRef]
29. Unali, G.; Crivicich, G.; Pagani, I.; Abou-Alezz, M.; Folchini, F.; Valeri, E.; Matafora, V.; Reisz, J.A.; Giordano, A.M.S.; Cuccovillo, I.; et al. Interferon-inducible phospholipids govern IFITM3-dependent endosomal antiviral immunity. *EMBO J.* **2023**, *42*, e112234. [CrossRef]
30. Camaschella, C.; Pagani, A.; Nai, A.; Silvestri, L. The mutual control of iron and erythropoiesis. *Int. J. Lab. Hematol.* **2016**, *38* (Suppl. S1), 20–26. [CrossRef]

31. Stolwijk, J.M.; Stefely, J.A.; Veling, M.T.; van 't Erve, T.J.; Wagner, B.A.; Raife, T.J.; Buettner, G.R. Red blood cells contain enzymatically active GPx4 whose abundance anticorrelates with hemolysis during blood bank storage. *Redox Biol.* **2021**, *46*, 102073. [CrossRef] [PubMed]
32. Dominiak, A.; Wilkaniec, A.; Wroczyński, P.; Adamczyk, A. Selenium in the Therapy of Neurological Diseases. Where is it Going? *Curr. Neuropharmacol.* **2016**, *14*, 282–299. [CrossRef] [PubMed]
33. Nyhan, W.L. Lesch-Nyhan Disease. *J. Hist. Neurosci.* **2005**, *14*, 1–10. [CrossRef]
34. Luo, S.; Hu, D.; Wang, M.; Zipfel, P.F.; Hu, Y. Complement in Hemolysis- and Thrombosis-Related Diseases. *Front. Immunol.* **2020**, *11*, 1212. [CrossRef]
35. Reisz, J.A.; Wither, M.J.; Dzieciatkowska, M.; Nemkov, T.; Issaian, A.; Yoshida, T.; Dunham, A.J.; Hill, R.C.; Hansen, K.C.; D'Alessandro, A. Oxidative modifications of glyceraldehyde 3-phosphate dehydrogenase regulate metabolic reprogramming of stored red blood cells. *Blood* **2016**, *128*, e32–e42. [CrossRef] [PubMed]
36. D'Alessandro, A.; Hay, A.; Dzieciatkowska, M.; Brown, B.C.; Morrison, E.J.; Hansen, K.C.; Zimring, J.C. Protein-L-isoaspartate O-methyltransferase is required for in vivo control of oxidative damage in red blood cells. *Haematologica* **2021**, *106*, 2726–2739. [CrossRef]
37. Dudzinska, W.; Suska, M.; Lubkowska, A.; Jakubowska, K.; Olszewska, M.; Safranow, K.; Chlubek, D. Comparison of human erythrocyte purine nucleotide metabolism and blood purine and pyrimidine degradation product concentrations before and after acute exercise in trained and sedentary subjects. *J. Physiol. Sci.* **2018**, *68*, 293–305. [CrossRef]
38. Zielinski, J.; Slominska, E.M.; Krol-Zielinska, M.; Krasinski, Z.; Kusy, K. Purine metabolism in sprint- vs endurance-trained athletes aged 20-90 years. *Sci. Rep.* **2019**, *9*, 12075. [CrossRef]
39. Dudzinska, W.; Lubkowska, A. Changes in the Concentration of Purine and Pyridine as a Response to Single Whole-Body Cryostimulation. *Front. Physiol.* **2021**, *12*, 634816. [CrossRef]
40. D'Alessandro, A.; Anastasiadi, A.T.; Tzounakas, V.L.; Nemkov, T.; Reisz, J.A.; Kriebardis, A.G.; Zimring, J.C.; Spitalnik, S.L.; Busch, M.P. Red Blood Cell Metabolism In Vivo and In Vitro. *Metabolites* **2023**, *13*, 793. [CrossRef]
41. Wu, H.; Bogdanov, M.; Zhang, Y.; Sun, K.; Zhao, S.; Song, A.; Luo, R.; Parchim, N.F.; Liu, H.; Huang, A.; et al. Hypoxia-mediated impaired erythrocyte Lands' Cycle is pathogenic for sickle cell disease. *Sci. Rep.* **2016**, *6*, 29637. [CrossRef] [PubMed]
42. Arduini, A.; Mancinelli, G.; Radatti, G.L.; Dottori, S.; Molajoni, F.; Ramsay, R.R. Role of carnitine and carnitine palmitoyltransferase as integral components of the pathway for membrane phospholipid fatty acid turnover in intact human erythrocytes. *J. Biol. Chem.* **1992**, *267*, 12673–12681. [CrossRef]
43. Nemkov, T.; Skinner, S.C.; Nader, E.; Stefanoni, D.; Robert, M.; Cendali, F.; Stauffer, E.; Cibiel, A.; Boisson, C.; Connes, P.; et al. Acute Cycling Exercise Induces Changes in Red Blood Cell Deformability and Membrane Lipid Remodeling. *Int. J. Mol. Sci.* **2021**, *22*, 896. [CrossRef] [PubMed]
44. Kim, E.; Lowenson, J.D.; MacLaren, D.C.; Clarke, S.; Young, S.G. Deficiency of a protein-repair enzyme results in the accumulation of altered proteins, retardation of growth, and fatal seizures in mice. *Proc. Natl. Acad. Sci. USA* **1997**, *94*, 6132–6137. [CrossRef]
45. Qin, Z.; Dimitrijevic, A.; Aswad, D.W. Accelerated protein damage in brains of PIMT+/- mice; a possible model for the variability of cognitive decline in human aging. *Neurobiol. Aging* **2015**, *36*, 1029–1036. [CrossRef]
46. Culp-Hill, R.; Zheng, C.; Reisz, J.A.; Smith, K.; Rachubinski, A.; Nemkov, T.; Butcher, E.; Granrath, R.; Hansen, K.C.; Espinosa, J.M.; et al. Red blood cell metabolism in Down syndrome: Hints on metabolic derangements in aging. *Blood Adv* **2017**, *1*, 2776–2780. [CrossRef] [PubMed]
47. Giordano, A.M.S.; Luciani, M.; Gatto, F.; Abou Alezz, M.; Beghe, C.; Della Volpe, L.; Migliara, A.; Valsoni, S.; Genua, M.; Dzieciatkowska, M.; et al. DNA damage contributes to neurotoxic inflammation in Aicardi-Goutieres syndrome astrocytes. *J. Exp. Med.* **2022**, *219*, e20211121. [CrossRef]
48. Strauss, K.A.; Williams, K.B.; Carson, V.J.; Poskitt, L.; Bowser, L.E.; Young, M.; Robinson, D.L.; Hendrickson, C.; Beiler, K.; Taylor, C.M.; et al. Glutaric acidemia type 1: Treatment and outcome of 168 patients over three decades. *Mol. Genet. Metab.* **2020**, *131*, 325–340. [CrossRef]
49. Cameron, J.S.; Moro, F.; Simmonds, H.A. Gout, uric acid and purine metabolism in paediatric nephrology. *Pediatr. Nephrol.* **1993**, *7*, 105–118. [CrossRef]
50. Lynch, B.J.; Noetzel, M.J. Recurrent coma and Lesch-Nyhan syndrome. *Pediatr. Neurol.* **1991**, *7*, 389–391. [CrossRef]
51. Oka, T.; Utsunomiya, M.; Ichikawa, Y.; Koide, T.; Takaha, M.; Mimaki, T.; Sonoda, T. Xanthine calculi in the patient with the Lesch-Nyhan syndrome associated with urinary tract infection. *Urol. Int.* **1985**, *40*, 138–140. [CrossRef]
52. Bassermann, R.; Gutensohn, W.; Jahn, H.; Springmann, J.S. Pathological and immunological observations in a case of Lesch-Nyhan-syndrome. *Eur. J. Pediatr.* **1979**, *132*, 93–98. [CrossRef]
53. Gallivan, A.; Alejandro, M.; Kanu, A.; Zekaryas, N.; Horneman, H.; Hong, L.K.; Vinchinsky, E.; Lavelle, D.; Diamond, A.M.; Molokie, R.E.; et al. Reticulocyte mitochondrial retention increases reactive oxygen species and oxygen consumption in mouse models of sickle cell disease and phlebotomy-induced anemia. *Exp. Hematol.* **2023**, *122*, 55–62. [CrossRef] [PubMed]

54. Kim, J.; Gupta, R.; Blanco, L.P.; Yang, S.; Shteinfer-Kuzmine, A.; Wang, K.; Zhu, J.; Yoon, H.E.; Wang, X.; Kerkhofs, M.; et al. VDAC oligomers form mitochondrial pores to release mtDNA fragments and promote lupus-like disease. *Science* **2019**, *366*, 1531–1536. [CrossRef] [PubMed]
55. Al-Samkari, H.; Galacteros, F.; Glenthoj, A.; Rothman, J.A.; Andres, O.; Grace, R.F.; Morado-Arias, M.; Layton, D.M.; Onodera, K.; Verhovsek, M.; et al. Mitapivat versus Placebo for Pyruvate Kinase Deficiency. *N. Engl. J. Med.* **2022**, *386*, 1432–1442. [CrossRef] [PubMed]

antioxidants

MDPI

Article

Erythrocyte Oxidative Status in People with Obesity: Relation to Tissue Losses, Glucose Levels, and Weight Reduction

Beata Szlachta [1], Anna Birková [2,*], Beáta Čižmárová [2], Anna Głogowska-Gruszka [1], Paulina Zalejska-Fiolka [1], Maria Dydoń [1] and Jolanta Zalejska-Fiolka [1]

[1] Department of Biochemistry, Faculty of Medical Science, Zabrze Medical University of Silesia, 40-055 Katowice, Poland; b.szlachta@uk-brandenburg.de (B.S.); aglogowska@sum.edu.pl (A.G.-G.); s91883@365.sum.edu.pl (P.Z.-F.); m.dydon@nzozkruszyna.pl (M.D.); jzalejskafiolka@sum.edu.pl (J.Z.-F.)

[2] Department of Medical and Clinical Biochemistry, Pavol Jozef Šafárik University, 040 11 Košice, Slovakia; beata.cizmarova@upjs.sk

* Correspondence: anna.birkova@upjs.sk

Abstract: Background: This study aimed to investigate the impact of reductions in various body mass components on the erythrocyte oxidative status and glycemic state of people with obesity (PWO). Methods: A total of 53 PWO followed a six-month individualized low-calorie diet with exercise, during which anthropometric, biochemical, and oxidative parameters were measured. The participants were divided into groups based on weight (W), visceral fat area (VFA), total body water (TBW), and skeletal muscle mass (SMM) losses, as well as normoglycemia (NG) and hyperglycemia (HG). Results: Weight reduction normalized glycemia and influenced erythrocyte enzyme activity. Regardless of the tissue type lost (VFA, TBW, or SMM), glutathione peroxidase activity decreased in all groups, accompanied by an increase in glutathione reductase activity. Lipofuscin (LPS) and malondialdehyde (MDA) concentrations decreased regardless of the type of tissue lost. The α-/γ-tocopherol ratio increased in those losing >10% body weight, >15% VFA, and >5% TBW. In the NG group, compared to the HG group, there was a decrease in glutathione peroxidase and an increase in glutathione reductase, with these changes being stronger in the HG group. The LPS and MDA concentrations decreased in both groups. Significant correlations were observed between glucose reduction and changes in catalase, retinol, and α-tocopherol, as well as between VFA reduction and changes in vitamin E, L-LPS, and the activities of L-GR and L-GST. Conclusions: This analysis highlights the complex interactions between glucose metabolism, oxidative state, and erythrocyte membrane integrity, crucial for understanding diabetes and its management. This study shows the significant metabolic adaptability of erythrocytes in response to systemic changes induced by obesity and hyperglycemia, suggesting potential therapeutic targets to improve metabolic health in obese individuals.

Keywords: bioimpedance; weight loss; visceral fat loss; total body water loss; skeletal muscle mass loss; erythrocyte antioxidant enzymes; lipofuscin; malondialdehyde; glycemia

Citation: Szlachta, B.; Birková, A.; Čižmárová, B.; Głogowska-Gruszka, A.; Zalejska-Fiolka, P.; Dydoń, M.; Zalejska-Fiolka, J. Erythrocyte Oxidative Status in People with Obesity: Relation to Tissue Losses, Glucose Levels, and Weight Reduction. *Antioxidants* **2024**, *13*, 960. https://doi.org/10.3390/antiox13080960

Academic Editors: Angelo D'Alessandro and Alkmini T. Anastasiadi

Received: 5 July 2024
Revised: 2 August 2024
Accepted: 5 August 2024
Published: 7 August 2024

1. Introduction

Obesity is a serious, chronic, complex disease that affects both children and adults. It is considered as a global health problem because its incidence is increasing worldwide. The prevalence of obesity has nearly tripled since 1975. Globally, in 2016, 39% of adults (aged 18 and over) were overweight and 13% were obese, while in 2020, 39 million children under the age of 5 were overweight or obese [1]. Eurostat reports that obesity rates, as well as problems with weight, are rapidly rising in most EU member states. In 2019, it was estimated that 52.7% of the adult population in the EU was overweight [2]. Obesity is associated with many serious conditions, such as cardiovascular diseases, diabetes mellitus, musculoskeletal diseases, certain types of cancer, liver dysfunction, dementia, sleep disorders, asthma, and infertility, while it is also associated with a reduction in life expectancy [3–6].

There is increasing evidence suggesting that oxidative stress plays an important role in the pathogenesis of obesity and its associated diseases. Also, a significantly positive correlation between BMI and biomarkers of oxidative stress has been observed [6]. In obese people, increased values of oxidative stress markers in plasma, serum, and urine have been observed, namely malondialdehyde, 8-epi-prostaglandin F2α, F-2 isoprostanes, thiobarbituric acid reactive species, and protein carbonyls [6,7].

In blood, erythrocytes (red blood cells, RBCs) are a promising model for the study of oxidative stress in the organism, as they are exposed to shear stress induced by blood flow, as well as oxidative stress in the circulation of blood. RBCs have an unparalleled organelle-free structure and unique biochemical and biophysical properties that define their functionality. Their lipid and protein composition changes during their lifetime, and specifically, these changes can be observed at the level of an RBC's plasma membrane, consisting of 19.5% (*w*/*w*) water, 39.5% proteins, 35.1% lipids, and 5.8% carbohydrates [8]. Thanks to the simple isolation of erythrocytes and their membranes, as well as -omics approaches, there is a wealth of exact knowledge about the composition of molecules in erythrocytes. Detailed lipidomic analyses of RBC membranes point to a relation between membrane lipid composition and the rheological behavior of RBCs in the microvascular network [9]. Recently, Bryk et al. conducted a proteome analysis of RBCs to better elucidate their specific structure. In this study, the authors identified around 2500 proteins within RBCs. Among the 20 most abundant proteins in whole RBCs are chaperones and proteins mainly involved in the transport of gasses, cell antioxidant defense, membrane transport, and the major metabolic pathways in RBCs, which influence the glycolysis and motility of the cells [10].

In many pathological conditions (e.g., diabetes, cardiovascular diseases, and hemoglobinopathies) and aging, there are changes in RBCs' biochemical and biophysical properties, and such changes can be used as markers of pathology in various diseases [11]. Oxidative stress, formed by excess reactive oxygen species (ROS) and eliminated by the action of enzymatic and non-enzymatic antioxidants in the human body, can also influence the structure and function of RBCs. RBCs contain antioxidant systems that fundamentally contribute to the function and integrity of these cells and maintain redox regulation in the body [12,13]. The antioxidant system of RBCs includes antioxidant enzymes, namely superoxide dismutase (SOD; RBCs contain Cu/Zn SOD), glutathione peroxidase (GPX), glutathione reductase (GR), catalase (KAT), peroxidases (Prx1, Prx2, and Prx6), and thioredoxin reductase (TR), and tripeptide glutathione (GSH) is considered as the main intracellular low-molecular-weight antioxidant. Other non-enzymatic antioxidants found in RBCs include vitamin C and vitamin E [14]. In obese individuals, there is evidence of reduced levels of antioxidant enzymes in RBCs (SOD, CAT, and GPX). Decreased levels of vitamin A, vitamin C, vitamin E, and β-carotene have also been observed in people with obesity (PWO), thought to be related to greater oxidative damage. In a previous study, it was shown that supplementation with antioxidants led to a reduction in oxidative stress [6,15].

In our previous publication [16] on the effect of reductions in various body mass components on the serum oxidative, glycemic, and lipid parameters of PWO, we found different patterns of activity changes in SOD isoenzymes. Lower mitochondrial SOD activity correlated with greater muscle and water loss, contrasting with cytosolic isoenzyme activity, which was lower in those retaining muscle and water. We also found that glucose level influenced weight reduction and redox state. Hyperglycemia is associated with higher total SOD levels, while improved glycemia increases mitochondrial SOD activity. Overall, we concluded that weight reduction reduces cardiovascular disease risk, yet active antioxidant molecules are more abundant in hyperglycemic cases, increasing further after glycemic improvement and weight loss. This prior study suggested that body weight alone inadequately predicts patients' glycemic and redox statuses.

The present study discusses erythrocyte oxidative status in PWO, including individuals with and without hyperglycemia, who underwent a weight-reduction program (WRP).

We also examined oxidative state depending on the tissue reduced during weight loss (muscle, fat, and water).

2. Materials and Methods

2.1. Ethical Permission

The study was approved by the Bioethical Committee of the Medical University of Silesia (SUM), with the reference number KNW/0022/KB1/19/I/16 (10 May 2016). The study was performed in accordance with the standards of the Declaration of Helsinki. The patients were informed about the procedure and potential risks before they were asked to participate. Written informed consent to participate in the study and publish the research results was obtained from all patients before the study started. Patients' data were protected by a RODO statement (General Data Protection Regulation). To avoid bias, all samples were anonymized and numbered.

2.2. Study Population

This study was based on reports indicating that body mass reduction can ameliorate metabolic impairment. It was a prospective, single-center study.

The study inclusion criteria included the following: body mass index (BMI) > 30 kg/m^2, no treatment affecting lipid and glucose metabolism, and informed consent to participate in the study. The study exclusion criteria included the following: lack of consent to participate in the study; disturbances of consciousness; severe hepatic, renal, respiratory, or circulatory insufficiency; chronic alcohol abuse; pregnancy; treatment-resistant depression; history of a serious nervous system injury; presence of an implanted cardiac pacemaker; and/or treatment affecting lipid and glucose metabolism.

A total of 53 people with obesity (PWO) (37 women and 16 men) were enrolled into a 6-month weight-reduction program (WRP), which consisted of a balanced, individualized, low-calorie diet, moderate exercise, and health education. They were monitored monthly during the follow-up visits, performed by a doctor and a dietitian. The patients were weighed, motivated to continue participating in the program, and received support in the form of training on how to calculate calorie intake and assess the quality of food products. Weight reduction was monitored until the achievement of a healthy body weight or a reduction in initial weight of 5–15%.

2.3. Dietary and Physical Activity Intervention

As stated, the patients were enrolled into a weight-reduction program (WRP), which was based on reports showing that weight loss improves lipid and glucose levels. Prior to the treatment program, all patients followed a poor-quality diet that was high in saturated fats, salt, simple sugars, refined grains, and preservatives, with a low intake of antioxidant-rich fruits and vegetables and minimal physical activity. During the first visit, after a medical examination and the completion of a questionnaire (which included history of obesity, health problems, food preferences, and physical activity), we determined each patient's basal metabolic rate (BMR), comprehensive metabolic rate (CPM), and daily energy deficit (DDE) in kcal per day. We recommended a slight calorie restriction, calculated as 15% of energy based on CPM. Each patient received personalized instructions based on recommendations from the Polish National Food and Nutrition Institute, including the provision of resources such as a seven-day menu and a list of products, along with their glycemic index (IG) values, that could ensure sufficient water and vegetable/fruit intake, as outlined in Table 1 below.

The patients were advised to increase their vegetable consumption to 500 g per day and their fruit consumption to 200–300 g per day (only fruit until 4–5 p.m.) and to have five meals per day that were appropriate in terms of meal composition and caloric value. In addition to following a proper diet, the patients were encouraged to engage in moderate physical exercise for at least 40 min three times a week. The patients were also advised to only consume snacks with low or medium glycemic index (IG) values, if needed.

Table 1. Composition of patients' diet, designed according to the healthy nutrition recommendations of the Polish National Food and Nutrition Institute.

Nutrient	% of Total Energy Intake and Limitations	
Carbohydrates	45–65%	Limiting the intake of free sugars to less than 10% of total energy intake.
Fats	25–35%	Limiting the intake of saturated fatty acids to less than 10% of total energy intake and intake of 3–6% mono- and polyunsaturated fatty acids in the form of fish and vegetable oils.
Proteins	10–15%	Animal and vegetable sources.

During the follow-up visits, which occurred every month, the patients were weighed, and their waist and hip circumferences were measured. Weight-loss results, as well as the quantity and composition of their diet, were recorded continuously. The patients were motivated to keep partaking in the WRP and were provided with support in the form of training on how to calculate calorie intake and assess the quality of food products.

Weight reduction was monitored until patients reached a healthy body weight or achieved a 5–15% reduction in their initial weight, which typically took six months.

2.4. Dividing Patients into Groups

No significant differences in glucose levels, BMI, VFA, or BFM (body fat mass) were observed between the male and female participants. Due to the similarity in responses, data from both sexes were analyzed collectively. The participants were divided into groups based on weight loss (WL), visceral fat area (VFA) loss, total body water (TBW) loss, and skeletal muscle mass (SMM) loss, with cut-off lines for each category as follows:

Group	Subgroup designation	
Weight loss (WL)	WL < 10%	WL > 10%
Visceral fat area loss (VFAL)	VFA < 15%	VFA > 15%
Total body water loss (TBWL)	TBW < 5%	TBW > 5%
Skeletal muscle mass loss (SMML)	SMM < 5%	SMM > 5%

Weight reduction was monitored until patients reached a healthy body weight or achieved a 5–15% reduction in their initial weight, which typically took six months. Hence, we used a breakdown for PWO based on whether they achieved a weight reduction of less or more than 15% [17]. The ranges applied for VFA, SMM, and TW were based on the obtained research results and statistical analyses.

The participants were also divided into normo- (NG) and hyperglycemic (HG) groups based on the initial fasting glucose concentration.

2.5. Biochemical Assessment

Fasting blood samples were collected at the Metabolic Clinic in Miasteczko Śląskie, Poland, from the cubital vein using clotting tubes to obtain serum (9 mL) and clot activator tubes (8 mL) to obtain plasma and erythrocytes. After centrifugation (10 min, 3000 rpm, 4 °C), the plasma was aliquoted, and 10% hemolysate was prepared in re-distilled water. All samples were stored at −80 °C at the Department of Biochemistry, Faculty of Medical Science, Medical University of Silesia, Katowice, Poland. To avoid bias, all samples were anonymized and numbered. Following good laboratory practices, biochemical markers were assessed during the first and last visits.

Glucose and insulin concentrations were determined in serum. Glucose (Glc) was measured using a BS-200E biochemical analyzer (Mindray, Shenzhen, China) and Alpha Diagnostics reagents (San Antonio, TX, USA). Insulin was determined using INS-IRMA kits (KIP1251-KIP1254, DIA Source Immuno Assays S.A., Louvain, Belgium). The repeatability values (within-run precision coefficients of variation) for glucose and insulin were 0.6%

and 4.1%, respectively. The reproducibility values (between-run precision coefficients of variation) for the above-mentioned parameters were 1.6% and 7.7%, respectively. The HOMA-IR coefficient was calculated as follows:

$$[\text{fasting insulin concentration } [\mu\text{IU/mL}] \times \text{fasting glucose concentration } [\text{mg/dL}])]/405.$$

2.6. Oxidative Status Parameters

Analysis of oxidative status parameters was performed in 10% erythrocyte hemolysate. Hemoglobin concentration was determined using the Drabkin method [18]. Analyses of the activity of L-SOD, L-KAT, L-GPX, L-GR, L-GST, L-LPS, and L-MDA were conducted using a PerkinElmer automated analyzer (PerkinElmer, Waltham, MA, USA). The activity of superoxide dismutase (L-SOD) was assessed using the Oyanagui method [19] and expressed in nitric units NU/mgHb. Catalase (L-KAT) activity was measured using the method of Johansson and Borg [20] and expressed as kIU/g Hb. Glutathione peroxidase (L-GPX) activity was measured using the kinetic method of Paglia and Valentine [21] and expressed as IU/g Hb. The activity of glutathione reductase (L-GR) was measured according to Richterich [22] and expressed as IU/g Hb. The activity of glutathione S-transferase (L-GST) was measured according to the kinetic method of Habig and Jakoby [23] and expressed as IU/g Hb.

Lipofuscin (L-LPS) concentration was determined according to Jain [24] and is expressed as RF (relative fluorescence lipid extract)/gHb. Malondialdehyde (L-MDA) concentration was measured fluorometrically according to the methodology of Ohkawa, Ohishi, and Yagi [25] and expressed in μmol/gHb.

Assessments of α-tocopherol, γ-tocopherol, and retinol concentrations were conducted in plasma using the HPLC (KNAUER HPLC equipment) assay outlined by Lim C.K. [26]. HPLC conditions: KNAUER HPLC equipment, HPLC pump 64, EUROCHROM 2000 for data analysis. Column: PS Spheribond 80-5 ODS2, 125 × 4.6 mm with pre-column. Mobile phase: 4% (*v*/*v*) water in methanol (POCH Gliwice, Poland, no 621991154). Flow rate: 1 [mL/min]; analysis time: 20 min; temp.: 40 °C. Fluorescence detector: Shimadzu RF-10Axl, wavelength absorbance for α-tocopherol ext. 325 nm em. 470 nm; for retinol ext. 292 nm em. 330 nm.

2.7. Body Mass Parameters

Body mass parameters were measured using the InBody S10 body composition analyzer (Biospace, Cerritos, CA, USA), which utilizes bioelectrical impedance spectroscopy. This method was selected for its efficiency and accuracy [27]. The InBody S10 has received ISO 9001:2015 and ISO 13485:2016 certification, as well as EN60601-1, EN60601-1-2, and CE MDD (Directive 93/42/EEC) medical certificates [28–32].

A standardized measurement protocol was implemented to obtain accurate bioimpedance results. This included recording age, sex, height, and body mass and conducting measurements at consistent time slots. Specific guidelines were followed, such as fasting for 3 h, abstaining from alcohol for 12 h, and avoiding medications that affect water balance. Subjects removed accessories, heavy clothing, and jewelry, and measurements were taken after standing upright for at least 5 min. Repeatability was assessed with 12 measurements on the same individual, resulting in coefficients of variation below 3% for all parameters.

2.8. Statistical Analysis

The results presented herein represent results from before and after the 6-month weight-reduction program. The calculated differences between these values are expressed as delta (Δ). All results are shown as the mean ± SD. Statistical analysis was performed with SPSS Statistics 22 (IBM, Armonk, NY, USA). The Kolmogorov–Smirnov test was used for normality testing. A paired *t*-test was used to determine differences in clinical parameters between values within groups before and after the dietary intervention. An unpaired *t*-test was used to determine differences in clinical parameters between groups, and assumptions of equal variances were tested using Levene's test for equality of variances. The Pearson correlation coefficient was used to express the strength and direction of linear relationships between two parameters. Statistical significance was assumed at a p-value of <0.05.

3. Results

Table 2 shows a comparison of the determined parameters for the entire cohort of patients (without dividing them into groups) before and after the WRP. In the whole group of PWO (Table 2), weight reduction did not lead to significant changes in the activity of L-SOD and L-KAT or in the concentrations of retinol, alpha-tocopherol, and gamma-tocopherol. At the same time, increases in L-GST and L-GR activity and the α/γ-tocopherol ratio were observed. In contrast, the concentrations of glucose, MDA, and LPS decreased, and the insulin resistance index (HOMA-IR) and the activity of L-GPX were significantly reduced. In terms of body mass composition, a significant reduction was observed in all analyzed parameters (weight, VFA, TBW, and SMM) during the experiment.

Table 2. General description of the whole group of probands. Data are expressed as the mean $\pm$ SD. A t-test was used for a comparison of the parameters before and after weight-reduction program (WRP). The difference is significant if $p < 0.05$.

Parameter	Before WRP	After WRP	p
Weight	101.0 ± 21.0	90.7 ± 19.0	**2.1*10^{-13}**
VFA	142.2 ± 34.3	115.5 ± 33.5	**2.1*10^{-15}**
TBW	44.2 ± 9.5	41.2 ± 8.7	**6.4*10^{-5}**
SMM	33.8 ± 7.7	31.2 ± 6.9	**0.0002**
Glc	117.7 ± 38.7	90.6 ± 16.9	**8.2*10^{-6}**
HOMA IR	3.70 ± 3.38	2.53 ± 2.16	**0.005**
L-SOD	179.4 ± 15.4	176.9.0 ± 18.2	0.49
L-GPX	61.91 ± 10.6	52.6 ± 15.7	**4.9*10^{-9}**
L-GST	0.22 ± 0.11	0.26 ± 0.13	**0.015**
L-KAT	472.4 ± 75.3	483.6 ± 69.5	0.23
L-GR	8.41 ± 2.1	9.13 ± 1.9	**0.00011**
L-LPS	1100.9 ± 235.1	913.9 ± 234.4	**8.2*10^{-6}**
L-MDA	0.39 ± 0.09	0.33 ± 0.05	**5.1*10^{-5}**
Retinol	0.62 ± 0.16	0.62 ± 0.16	0.70
α-tocopherol	26.24 ± 7.55	26.92 ± 8.53	0.38
γ-tocopherol	1.62 ± 0.59	1.54 ± 0.57	0.30
Ratio α/γ tocopherol	17.40 ± 5.55	18.88 ± 7.44	**0.042**

Legend: WRP—weight-reduction program; weight [kg]; VFA—visceral fat area [cm^2]; TBW—total body water [L]; SMM—skeletal muscle mass [kg]; HOMA-IR—homeostatic model assessment for insulin resistance; L-SOD—hemolysate superoxide dismutase [NU/mgHb]; L-GPX—hemolysate glutathione peroxidase [IU/gHb]; L-GST—hemolysate glutathione transferase [IU/gHb]; L-GR—hemolysate glutathione reductase [IU/gHb]; L-KAT—hemolysate catalase [kIU/gHb]; L-LPS—hemolysate lipofuscin [RF/gHb]; L-MDA—hemolysate malondialdehyde [μmol/gHb]; retinol [μM/L]; α-tocopherol [μM/L]; and γ-tocopherol [μM/L].

3.1. Oxidative Stress Parameter Levels Related to Specific Compartment Loss

The results regarding the selected oxidative stress parameters for all the study groups are presented in Table 3.

Table 3. Statistical analysis of selected anti- and pro-oxidative parameters in individual groups of people with obesity (PWO). Data are expressed as mean ± SD. For within-group (p (X_1 vs. X_2)) and between-group (#p) comparisons of the differences between measurements before and after the WRP a *t*-test was used.

Parameter	X	WEIGHT LOSS		VISCERAL FAT AREA LOSS		TOTAL BODY WATER LOSS		SKELETAL MUSCE MASS LOSS	
		WL < 10% N = 34	**WL > 10% #p N = 19**	**VFA < 15% N = 21**	**VFA > 15% #p N = 31**	**TBW < 5% N = 27**	**TBW > 5% #p N = 20**	**SMM < 5% N = 24**	**SMM > 5% #p N = 24**
L-SOD	1	177.7 ± 15.03	182.5 ± 15.4 $^{\#0.29}$	178.2 ± 14.8	180.8 ± 15.9 $^{\#0.56}$	179.4 ± 13.45	181.3 ± 17.8 $^{\#0.68}$	180.91 ± 13.82	179.84 ± 16.68 $^{\#0.81}$
	2	180.1 ± 19.8	171.2 ± 14.0 $^{\#0.09}$	175.0 ± 20.1	176.8 ± 15.7 $^{\#0.72}$	181.4 ± 16.68	171.8 ± 16.0 $^{\#0.06}$	183.57 ± 17.37	171.07 ± 13.65 $^{\#0.009}$
p (X_1 vs. X_2)		0.29	**0.041**	0.58	0.36	0.63	0.12	0.54	0.10
L-GPX	1	61.0 ± 11.1	63.6 ± 9.8 $^{\#0.41}$	64.84 ± 10.44	59.76 ± 10.6 $^{\#0.10}$	58.3 ± 10.22	64.04 ± 10.4 $^{\#0.07}$	56.39 ± 9.34	65.67 ± 9.74 $^{\#0.002}$
	2	51.7 ± 15.5	54.1 ± 16.3 $^{\#0.60}$	57.47 ± 15.79	49.43 ± 15.2 $^{\#0.07}$	49.92 ± 15.3	54.2 ± 17.8 $^{\#0.38}$	47.92 ± 15.33	56.13 ± 16.36 $^{\#0.08}$
p (X_1 vs. X_2)		**$8.2*10^{-8}$**	**0.0066**	**0.0007**	**$6*10^{-6}$**	**0.0002**	**0.0003**	**0.0003**	**0.0002**
L-GST	1	0.22 ± 0.11	0.21 ± 0.09 $^{\#0.82}$	0.21 ± 0.11	0.22 ± 0.10 $^{\#0.78}$	0.23 ± 0.12	0.21 ± 0.1 $^{\#0.45}$	0.25 ± 0.12	0.19 ± 0.09 $^{\#0.08}$
	2	0.25 ± 0.13	0.28 ± 0.14 $^{\#0.49}$	0.21 ± 0.12	0.29 ± 0.13 $^{\#0.046}$	0.28 ± 0.14	0.24 ± 0.14 $^{\#0.40}$	0.29 ± 0.14	0.23 ± 1.13 $^{\#0.12}$
p		0.10	0.08	0.97	**0.013**	**0.038**	0.40	**0.052**	0.26
L-KAT	1	495.5 ± 74.0	428.6 ± 57.2 $^{\#0.002}$	469.8 ± 66.6	471.7 ± 82.0 $^{\#0.93}$	469.2 ± 85.6	479.3 ± 69.6 $^{\#0.67}$	484.1 ± 80.4	462.1 ± 75.2 $^{\#0.34}$
	2	496.3 ± 69.8	461.5 ± 64.9 $^{\#0.08}$	487.8 ± 63.8	476.2 ± 70.7 $^{\#0.55}$	481.2 ± 65.8	484.8 ± 77.1 $^{\#0.86}$	482.1 ± 61.9	481.0 ± 78.1 $^{\#0.96}$
p (X_1 vs. X_2)		0.93	**0.047**	0.12	0.84	0.32	0.87	0.87	0.22
L-GR	1	8.77 ± 2.34	7.74 ± 1.35 $^{\#0.05}$	8.24 ± 1.98	8.51 ± 2.24 $^{\#0.67}$	8.71 ± 2.37	8.28 ± 1.67 $^{\#0.50}$	8.80 ± 2.45	8.15 ± 1.65 $^{\#0.29}$
	2	9.15 ± 2.21	9.11 ± 1.04 $^{\#0.93}$	8.88 ± 2.04	9.28 ± 1.76 $^{\#0.47}$	9.56 ± 1.94	8.89 ± 1.76 $^{\#0.24}$	9.60 ± 2.03	8.82 ± 1.71 $^{\#0.16}$
p (X_1 vs. X_2)		**0.048**	**0.0003**	**0.046**	**0.001**	**0.007**	**0.004**	**0.015**	**0.002**
L-LPS	1	1114.7 ± 257.9	1074.8 ± 188.7 $^{\#0.57}$	1082.5 ± 301.8	1117.9 ± 182.5 $^{\#0.60}$	1121.0 ± 241.5	1104.8 ± 195.5 $^{\#0.81}$	1150.8 ± 234.1	1058.4 ± 214.3 $^{\#0.16}$
	2	914.5 ± 249.4	913.0 ± 212.4 $^{\#0.98}$	811.6 ± 216.4	980.1 ± 226.5 $^{\#0.01}$	987.5 ± 219.7	865.5 ± 235.8 $^{\#0.08}$	1026.8 ± 223.6	829.0 ± 204.8 $^{\#0.003}$
p (X_1 vs. X_2)		**0.0005**	**0.0051**	**0.0007**	**0.002**	**0.02**	**0.0003**	**0.06**	**$1.85*10^{-5}$**
L-MDA	1	0.39 ± 0.11	0.38 ± 0.05 $^{\#0.44}$	0.37 ± 0.09	0.39 ± 0.10 $^{\#0.53}$	0.38 ± 0.09	0.41 ± 0.10 $^{\#0.32}$	0.39 ± 0.09	0.39 ± 0.10 $^{\#0.79}$
	2	0.32 ± 0.05	0.34 ± 0.04 $^{\#0.12}$	0.31 ± 0.05	0.34 ± 0.04 $^{\#0.015}$	0.34 ± 0.04	0.33 ± 0.04 $^{\#0.43}$	0.34 ± 0.04	0.32 ± 0.04 $^{\#0.22}$
p (X_1 vs. X_2)		**0.0006**	**0.027**	**0.003**	**0.010**	**0.019**	**0.0096**	**0.023**	**0.008**
Retinol	1	0.62 ± 0.17	0.63 ± 0.15 $^{\#0.78}$	0.67 ± 0.18	0.59 ± 0.15 $^{\#0.10}$	0.60 ± 0.19	0.64 ± 0.15 $^{\#0.40}$	0.58 ± 0.19	0.66 ± 0.13 $^{\#0.13}$
	2	0.63 ± 0.15	0.60 ± 0.17 $^{\#0.56}$	0.65 ± 0.17	0.59 ± 0.15 $^{\#0.24}$	0.62 ± 0.16	0.60 ± 0.18 $^{\#0.66}$	0.62 ± 0.17	0.60 ± 0.16 $^{\#0.73}$
p (X_1 vs. X_2)		0.71	0.26	0.52	0.83	0.42	0.10	0.16	**0.030**
α-tocopherol	1	26.44 ± 8.53	26.59 ± 6.13 $^{\#0.94}$	27.36 ± 8.40	24.63 ± 5.38 $^{\#0.20}$	25.53 ± 7.63	26.75 ± 6.44 $^{\#0.58}$	24.75 ± 7.11	27.12 ± 7.07 $^{\#0.26}$
	2	27.33 ± 9.70	26.03 ± 6.28 $^{\#0.57}$	27.52 ± 8.55	25.27 ± 6.19 $^{\#0.29}$	27.17 ± 7.90	25.51 ± 7.15 $^{\#0.48}$	27.33 ± 8.36	25.52 ± 6.48 $^{\#0.42}$
p (X_1 vs. X_2)		0.23	0.65	0.90	0.54	0.16	0.28	**0.028**	0.16

Table 3. *Cont.*

Parameter	X	WEIGHT LOSS		VISCERAL FAT AREA LOSS		TOTAL BODY WATER LOSS		SKELETAL MUSCE MASS LOSS	
		WL < 10% N = 34	**WL > 10% #p N = 19**	**VFA < 15% N = 21**	**VFA > 15% #p N = 31**	**TBW < 5% N = 27**	**TBW > 5% #p N = 20**	**SMM < 5% N = 24**	**SMM > 5% #p N = 24**
γ-tocopherol	1	1.65 ± 0.64	1.55 ± 0.50 #0.53	1.64 ± 0.68	1.55 ± 0.50 #0.61	1.55 ± 0.60	1.73 ± 0.57 #0.33	1.53 ± 0.57	1.72 ± 0.59 #0.29
	2	1.68 ± 0.65	1.28 ± 0.26 #0.003	1.62 ± 0.55	1.43 ± 0.54 #0.21	1.62 ± 0.63	1.39 ± 0.37 #0.13	1.64 ± 0.67	1.44 ± 0.38 #0.21
p (X_1 vs. X_2)		0.78	**0.026**	0.90	0.26	0.55	**0.0016**	0.34	**0.019**
Ratio α-/γ-tocopherol	1	16.96 ± 5.26	18.35 ± 6.22 #0.43	17.97 ± 5.45	17.06 ± 5.86 #0.58	17.38 ± 4.91	16.98 ± 6.86 #0.82	17.06 ± 4.71	17.19 ± 6.69 #0.94
	2	17.23 ± 4.75	21.87 ± 10.43 #0.09	17.57 ± 4.22	19.74 ± 9.27 #0.27	18.01 ± 5.89	19.84 ± 10.2 #0.50	18.03 ± 6.20	19.17 ± 9.43 #0.63
p (X_1 vs. X_2)		0.72	**0.029**	0.66	**0.015**	0.56	**0.027**	0.39	0.10

Legend: X_1—before WRP; X_2—after WRP; weight [kg]; VFA—visceral fat area [cm^2]; TBW—total body water [L]; SMM—skeletal muscle mass [kg]; L-SOD—hemolysate superoxide dismutase [NU/mgHb]; L-GPX—hemolysate glutathione peroxidase [IU/gHb]; L-GST—hemolysate glutathione transferase [IU/gHb]; L-GR—hemolysate glutathione reductase [IU/gHb]; L-KAT—hemolysate catalase [kIU/gHb]; L-LPS—hemolysate lipofuscin [RF/gHb]; and L-MDA—hemolysate malondialdehyde [μmol/gHb]; Retinol [μmol/L]; α-tocopherol [μmol/L]; γ-tocopherol [μmol/L].

The activity of L-SOD decreased significantly in group WL > 10%. A trend showing a decrease in L-SOD was also visible in groups SMM > 5% and TBW > 5%. The activity of L-GPX and the concentration of L-LPS and L-MDA decreased significantly in all study groups. The activity of L-GST increased significantly only in VFA > 15%, TBW < 5%, and SMM < 5%. The activity of L-GR increased significantly in all study groups. There were no changes in L-KAT activity in all studied groups, with the exception of WL > 10%. In this group, the L-KAT activity increased significantly after the WRP. The concentration of α-tocopherol increased significantly in the SMM < 5% group, while the concentration of γ-tocopherol decreased significantly in the WL > 10%, TBW > 5%, and SMM > 5% groups. The ratio of α-tocopherol to γ-tocopherol increased in WL > 10%, VFA > 15%, and TBW > 5%.

Despite the higher L-SOD activity before body mass reduction in the TBW < 5% group, the activity of this parameter was significantly lower after weight reduction compared to the TBW > 5% group. Also, in the SMM > 5% group, L-SOD activity was significantly lower compared to the SMM < 5% group after weight reduction. L-GPX activity was statistically significantly higher in the TBW > 5% and SMM > 5% groups before weight reduction compared to the TBW < 5% and SMM < 5% groups, respectively. The activity of L-GPX was also higher in the SMM > 5% group after reduction compared to the SMM < 5% group. L-GST activity was significantly lower in the SMM > 5% group before weight reduction and in the VFA > 15% group compared to the VFA < 15% group after weight reduction. However, L-LPS decreased in all groups, with the lowest values being found for patients with greater SMM loss and lower VFA loss. The concentration of γ-tocopherol was lower in WL > 10% than WL < 10% after the WRP, but the ratio of α-/γ-tocopherol was higher in WL > 10% than WL < 10% after the body mass reduction program.

3.2. Effects of Glucose Adjustment and VFA Reduction on Antioxidant Properties of Erythrocytes

Correlation analysis revealed few relations between the changes in the measured parameters. The results are shown as delta:

$$\Delta = \text{parameter before study (WRP) (X\_1)} - \text{parameter at the end of the study (WRP) (X\_2)}$$

The most significant (Figure 1) correlations were between glucose reduction vs. catalase; between retinol and α-tocopherol changes; and between VFA reduction vs. vitamin E, L-LPS, L-GR, and L-GST activity changes. With a higher glucose decrease after the WRP, there was a higher increase in L-KAT activity (r = −0.31; p = 0.028). The opposite relation was observed between glucose decrease and retinol and α-tocopherol changes, as the glucose decrease correlated with retinol (r = 0.39; p = 0.0056) and α-tocopherol decreases (r = 0.42; p = 0.0025). The WRP-related decrease in VFA correlates negatively with the decreases in LPS (r = −0.28; p = 0.048*) and γ-tocopherol (r = 0.33; p = 0.018*) and with the increase in the α/γ-tocopherol ratio (r = −0.44; p = 0.0013). Also, the higher the VFA reduction, the higher the increases in L-GR (r^2 = −0.28; p = 0.049) and L-GST activity (r = −0.30; p = 0.038).

3.3. Results of Studied Parameters in Normo- and Hyperglycemia Patients

Considering the correlations between glucose and some anti- and pro-oxidant parameters, we decided to show the changes in oxidative status in the normo- and hyperglycemic groups (Table 4). There were no differences in loss of weight, VFA, TBW, or SMM, or in L-KAT activity and retinol concentration between the normo- (NG) and hyperglycemic (HG) groups. There was no change in L-SOD activity in either group, but L-SOD activity was lower in the HG group after the WRP in comparison to the NG group. The L-GPX activity was lower in the NG group before the WRP and decreased in both groups after the WRP, with no differences observed between the groups. The L-GST activity decreased in the NG group after the WRP, and the activity was lower in the HG group after the program. There were no differences between the groups in L-GR activity, and the activity of L-GR decreased in both groups after the WRP. The L-LPS concentration was lower in the HG group both before and after the WRP, and it decreased in both study groups after the WRP. The L-MDA concentration decreased in both groups and was lower in the HG group after the WRP. There were no differences in the α- and γ-tocopherol concentrations and the α/γ-tocopherol ratio before or after the WRP between the NG and HG groups. Simultaneously, the concentration of α-tocopherol decreased in the NG group, while γ-tocopherol and the α/γ-tocopherol ratio decreased in the HG group after the WRP.

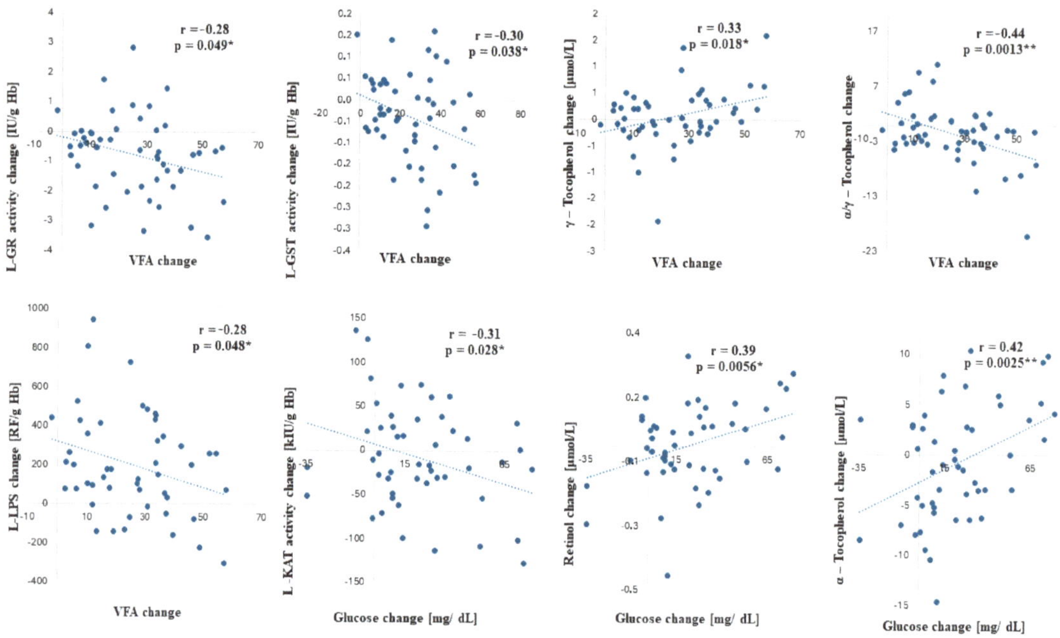

Figure 1. The correlations between selected parameters. Pearson's coefficient r expresses the strength of the correlation. The correlation is significant if $p < 0.05$ (*) and $p < 0.005$ (**). **Legend**: VFA—visceral fat area; L-GST—hemolysate glutathione transferase; L-GR—hemolysate glutathione reductase; L-KAT—hemolysate catalase; and L-LPS—hemolysate lipofuscin.

Table 4. Comparative analysis in normo- (NG) and hyperglycemia (HG) groups of people with obesity (PWO). Data are expressed as mean ± SD. For within-group (p (X_1 vs. X_2)) and between-group (#p) comparisons of the differences between measurements from before the diet program (**X_1**) and after the diet program (**X_2**), a *t*-test was used.

Parameter	X	Normoglycemia NG (N = 18)	Hyperglycemia HG (N = 35)	#p
Weight loss		12.4 ± 10.2	9.2 ± 5.8	#0.24
VFA loss		21.1 ± 18.4	21.2 ± 21.6	#0.83
TBW loss		2.69 ± 3.1	2.26 ± 3.91	#0.81
SMM loss		2.11 ± 2.67	2.07 ± 3.70	#0.95
L-SOD	1	179.56 ± 16.41	179.28 ± 15.06	#0.95
	2	184.56 ± 21.00	173.13 ± 15.75	**#0.033**
p (X_1 vs. X_2)		0.41	0.11	
L-GPX	1	57.55 ± 10.08	64.03 ± 10.37	**#0.038**
	2	48.58 ± 14.87	54.51 ± 15.93	#0.20
p (X_1 vs. X_2)		**0.0034**	**6.3*10^{-7}**	

Table 4. *Cont.*

Parameter	X	Normoglycemia NG (N = 18)	Hyperglycemia HG (N = 35)	#p
L-GST	1	0.23 ± 0.09	0.21 ± 0.11	#0.56
	2	0.33 ± 0.11	0.23 ± 0.13	**#0.013**
p (X_1 vs. X_2)		**0.008**	0.33	
L-KAT	1	496.0 ± 77.7	460.9 ± 72.4	#0.12
	2	496.4 ± 78.57	477.4 ± 64.9	#0.36
p (X_1 vs. X_2)		0.82	0.11	
L-GR	1	8.79 ± 2.70	8.23 ± 1.76	#0.38
	2	9.40 ± 1.93	9.00 ± 1.84	#0.48
p (X_1 vs. X_2)		**0.064**	**0.0007**	
L-LPS	1	1232.1 ± 239.6	1037.1 ± 207.6	**#0.004**
	2	1031.2 ± 187.0	857.0 ± 236.0	**#0.010**
p (X_1 vs. X_2)		**0.024**	**0.00012**	
L-MDA	1	0.41 ± 0.10	0.35 ± 0.05	#0.21
	2	0.35 ± 0.05	0.32 ± 0.05	**#0.031**
p (X_1 vs. X_2)		**0.045**	**0.0004**	
Retinol	1	0.58 ± 0.16	0.64 ± 0.16	#0.20
	2	0.58 ± 0.17	0.63 ± 0.15	#0.34
p (X_1 vs. X_2)		0.84	0.59	
α-tocopherol	1	25.72 ± 7.93	26.51 ± 7.58	#0.74
	2	28.18 ± 10.09	26.27 ± 7.92	#0.45
p (X_1 vs. X_2)		**0.038**	0.81	
γ-tocopherol	1	1.53 ± 0.55	1.65 ± 0.62	#0.49
	2	1.74 ± 0.67	1.45 ± 0.51	#0.14
p (X_1 vs. X_2)		0.18	**0.015**	
α/γ-tocopherol ratio	1	18.26 ± 6.80	17.08 ± 5.02	#0.54
	2	18.48 ± 11.20	19.04 ± 5.25	#0.81
p (X_1 vs. X_2)		0.90	**0.008**	

Legend: NG—normoglycemia group; HG—hyperglycemia group; X_1—before WRP; X_2—after WRP; weight [kg]; VFA—visceral fat area [cm^2]; TBW—total body water [L]; SMM—skeletal muscle mass [kg]; L-SOD—hemolysate superoxide dismutase [NU/mgHb]; L-GPX—hemolysate glutathione peroxidase [IU/gHb]; L-GST—hemolysate glutathione transferase [IU/gHb]; L-GR—hemolysate glutathione reductase [IU/gHb]; L-KAT—hemolysate catalase [kIU/gHb]; L-LPS—hemolysate lipofuscin [RF/gHb]; L-MDA—hemolysate malondialdehyde [μmol/gHb]; retinol [μmol/L]; α-tocopherol [μmol/L]; and γ- tocopherol [μmol/L].

4. Discussion

Obesity, often accompanied by metabolic disorders such as insulin resistance and hyperglycemia, significantly impacts cellular functions across the body, including RBCs. Some changes mimic the alterations induced by aging, but others have the opposite effect, possibly counteracting the adaptation of the organism to senescence [33]. Obesity induces a state of chronic oxidative stress, affecting erythrocytes' functionality by altering glucose metabolism pathways and increasing the production of ROS. However, existing human and animal studies show controversial results regarding obesity and antioxidant enzymes. Some studies have shown that antioxidant enzymes increase in obesity, but there are studies that found no significant difference in antioxidant enzyme concentrations in obese individuals [34]. The present study involved only people with obesity (PWO). We examined their responses after weight reduction from a few perspectives. We explored erythrocyte responses to oxidative stress in obese subjects undergoing weight loss, highlighting the differences between those with a normoglycemic state (NG) and those with a hyperglycemic state (HG) before the WRP. This study also describes the effect on RBC redox state in relation to different types of tissue loss during a weight-reduction program (WRP).

The interactions between glucose metabolism, oxidative state, and the integrity of erythrocyte cell membranes are highly complex. The most important processes involving RBC (Figure 2a) and their associations to redox state (Figure 2b) are shown in Figure 2.

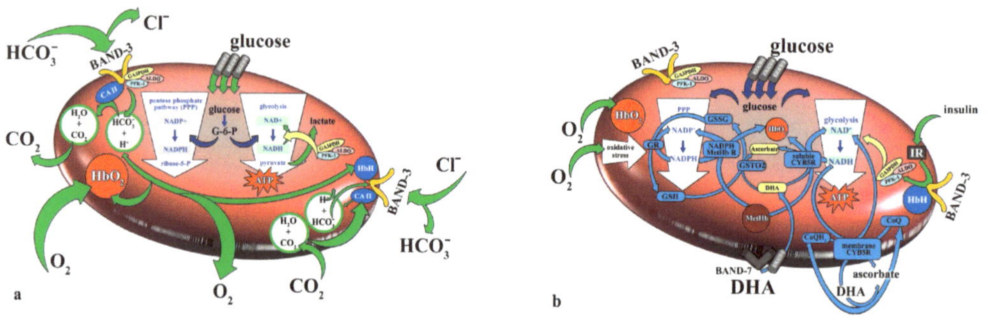

Figure 2. (**a**,**b**). Mutual interactions between glucose metabolism, oxidative state, and cell membrane in mature RBCs. RBCs are important glucose-consuming cells, and there are few metabolic pathways involved in glucose consumption. Glycolysis is a producer of NADH, lactate for other cells, and is the only source of ATP. The pentose phosphate pathway (PPP) is a source of NADPH, vital for the regeneration of glutathione. Under physiological conditions, about 7% of glucose is used in the PPP. Increased glucose influx into RBCs can be a reason for PPP increase. Band-3, the anion exchanger of HCO_3^-/Cl^-, binds carbonic anhydrase II (CA II) and deoxygenated hemoglobin HbH. After hemoglobin oxygenation (HbO_2), glycolytic enzymes are bound to band-3 instead of HbH. Besides glucose-6-phosphate dehydrogenase activity, this is the second regulatory mechanism which helps to shift glucose-6-phosphate to glycolysis (in the case of HbH prevalence) or PPP (in the case of HbO_2 prevalence). Insulin-independent glucose transporter 1 (GLUT1) interacts with various membrane proteins, such as band-7, dematin, adducin, and band-3. GLUT1 is also able to transport dehydroascorbic acid (DHA). Band-7's interaction with GLUT1 causes a change in the substrate preference of the transported molecule from glucose to DHA. Ascorbate's involvement in antioxidant activity involves cooperation with both PPP and glycolysis. In hyperglycemic individuals, there can be a reduced level of ascorbate in RBCs, resulting from the competitive effect of DHA and glucose or due to the changed expression of GLUT1. The actin cytoskeleton complexes around GLUT1 may be chemically modified in diabetic individuals' RBCs. The dynamics of the cytoskeleton/actin can regulate glucose uptake through GLUT1. The human RBC membrane also contains the insulin receptor (IR). Insulin activates PFK (6-phosphofructo-1-kinase) via phosphorylation, causing it to detach from band-3 and leading to the redistribution of the enzyme within RBCs, thus supporting glycolysis. The enzymes glutathione reductase (GR), glutathione-S-transferase (GSTO2), NADPH methemoglobine reductase (NADPH MetHb R), and cytochrome B5 reductase (CYB5R) are closely dependent on metabolic pathways [35–49].

The glucose uptake rate, enzyme activity, and production and utilization of intermediate metabolites and ATP in the erythrocytes of obese patients with hyperglycemia can be altered [46,50]. Insulin resistance, a common feature of obesity, influences RBC function by modulating GLUT1 expression and glucose uptake. On one hand, increased glucose metabolism in RBCs helps to consume excess glucose from the blood and thus reduces the formation of glycated end-products (AGEs) in the serum. On the other hand, it intensifies glycation processes in RBCs, leading to oxidative stress and causing lipid and protein oxidation, resulting in their denaturation and changes in the structure and function of RBCs. These changes lead to a progressive decline in RBC deformability, associated with increased fragility and shortened lifespan [48,51,52]. Insulin-mediated glucose metabolism in erythrocytes highlights the complex interaction between insulin sensitivity and RBC health.

RBC defensive mechanisms are crucial in managing oxidative stress in erythrocytes and may have significant implications for the treatment and management of diabetes, where controlling hyperglycemia and protecting against lipid peroxidation can significantly impact the reduction in complications [53,54].

In response to lipid peroxidation, RBCs utilize several antioxidant enzymes that help protect the cells from damage. Glutathione peroxidase (GPX) is a key enzyme that reduces lipid peroxides to alcohols, using reduced glutathione (GSH) as an electron donor. The byproduct of this reaction is oxidized glutathione (GSSG), which is then reduced back to GSH by glutathione reductase (GR) in the presence of NADPH, ensuring the continuation of the antioxidant cycle. GST also plays a critical role in RBC protection. The peroxidase activity of GST, crucial for breaking down lipid peroxides, not only relies on selenium but also requires the presence of GSH. GSH's coupling reaction with electrophilic products of lipid peroxidation leads to the neutralization of harmful aldehydes and other peroxidation products. Thus, GST helps reduce toxicity and improve the solubility of these products in water, facilitating their elimination from the cell [46,50]. In the present study, regardless of the type of tissue loss, WRP led to a significant reduction in L-GPX and a significant increase in L-GR, without significant changes in L-SOD or L-CAT, although more drastic weight loss was accompanied with a decrease in L-SOD and an increase in L-CAT.

Weight loss is usually not only accompanied by losses of fat but also reductions in other aspects, including body water and muscle mass. However, losing lean body mass has a number of negative health consequences, which is why a lot of research is focused on strategies to maintain muscle mass during weight loss, including the use of a wide range of nutritional supplements [55]. In our previous publications, we showed that obese patients with persistent hyperglycemia are resistant to fat tissue reductions even after an extensive WRP based on balanced nutrition and exercise [56]. Moreover, we showed different lipid metabolism responses depending on the extent of muscle loss in PWO. Greater muscle loss was associated with a decrease in triacylglycerols, reflecting the loss of fat stores, but smaller muscle losses were linked to an increase in beneficial HDL-cholesterol. In addition, different patterns of changes in superoxide dismutase (SOD) isoenzyme activity were observed. Patients with lower mitochondrial MnSOD activity in serum at the outset were more likely to lose more non-fat tissue, and mitochondrial MnSOD activity in serum significantly increased after glycemia improvement. On the contrary, lower cytosolic CuZnSOD activity in serum at the outset was found to be related to smaller losses of non-fat tissues and significantly increased after the WRP in normoglycemic patients, but this did not occur in the hyperglycemia patients [16]. Isoenzyme SOD activity responses may also vary in different skeletal muscle fiber types upon stimuli, as has been described, for example, in aging [57]. Circulating cytosolic SOD, derived from both hemolysis and peripheral tissues as a result of cellular secretion [58], is the major SOD isoenzyme in RBCs, and in this study, it did not experience any changes in general during the WRP. However, patients losing more muscle finished with significantly lower L-SOD values compared to patients with less muscle loss after the WRP, highlighting the importance of preventing muscle loss due to the impact on RBCs' antioxidant defense.

The mechanisms underlying muscle mass reduction during weight loss are not fully understood. Evidence indicates that a short-term (up to 21 days) calorie restriction (about 35% energy deficit/day) can decrease the postprandial rate of muscle protein synthesis and reduce the basal rate of muscle protein synthesis [59]. Contrastingly, in another study, it was indicated that a prolonged moderate calorie restriction and 5–10% weight loss increased the rate of muscle protein synthesis [60]. Overall, muscle mass reduction during a prolonged moderate calorie restriction is believed to be mediated by increased muscle proteolysis rather than suppressed muscle protein synthesis, and it is viewed as an undesirable effect. It appears that increased physical activity during weight loss may protect against this unfavorable process [61]. The results of the current study indicate that other factors that affect the type of tissue being reduced during weight loss should also be taken into account. One of these factors could be hyperglycemia state.

Also, glycation was involved in the change in L-SOD during the WRP. In this study, before the WRP, there was no significant difference between SOD levels in the NG and HG patients, which is in accordance with Turpin et al. (2020) [62], who showed no differences in SOD levels in glycated RBCs compared to non-glycated RBCs. However, after the WRP, in NG PWO, there was a slight increase in L-SOD activity (non-significant), and in HG PWO, there was a slight decrease in L-SOD activity (non-significant), finally leading to significantly lower levels in the HG group compared to the NG group, thus pointing to the importance of glycemia in obesity as an important factor negatively affecting WRPs. Turpin et al. (2020) [62] also showed no difference in glycated RBCs compared to non-glycated RBCs in terms of catalase levels. Similarly, in the present study, no difference was detected in normo- vs. hyperglycemic PWO before the WRP. Regarding other enzymes, Turpin et al. also measured peroxidase activity, which was decreased in glycated RBCs, differing from what we found in our study. However, these other authors measured total peroxidase activity and focused on the relation to diabetes regardless of obesity.

Following the WRP, the RBCs of the NG individuals showed improved antioxidant defense, indicated by a decrease in the activity of glutathione peroxidase (L-GPX) and increased activities of glutathione reductase (L-GR) and, notably, glutathione S-transferase (L-GST). This was also observed in those who lost more than 15% of their VFA. This contrasts with the results for the HG individuals, for which only the L-GPx and L-GR activities were changed after the WRP. Interestingly, the correlation analysis showed that glucose correction is related to a reduction in lipid-soluble antioxidant vitamins and an increase in the activity of L-KAT, which, before the diet program, was lower in the HG group (though not significantly so) compared to the NG group. L-KAT and L-GPX act on the same substrate, and according to Gaetani et al. [63,64], both are equally active in the detoxification of hydrogen peroxide, but L-GPX was, contrary to catalase, significantly higher in terms of activity in the HG group before the WRP compared to the NG group. This inverse relation between L-KAT and L-GPX in the HG group is probably related to the action of NADPH, which is essential for glutathione regeneration but also directly binds L-KAT and protects the molecule against oxidative damage and denaturation [64,65]. The higher the consumption of NADPH for glutathione reduction, necessary for L-GPX activity, the less NADPH available to bind and protect L-KAT. Another explanation is that L-GPX shows a higher affinity for hydrogen peroxide than L-KAT and therefore plays a more important role in most physiological situations when the amount of hydrogen peroxide produced is not too high. Insufficient L-KAT activity is thus compensated for by an increase in L-GPX activity, and conversely, reduced peroxidase activity is compensated for by an increase in L-KAT activity [65]. Ramezanipour et al. (2014) [66] studied the effect of weight reduction on antioxidant enzymes in a group of 30 obese women, specifically focusing on values from the beginning of the study and after a 3-month intervention. In terms of their results, the glutathione reductase and catalase activities showed a significant increase after weight reduction, but no significant changes were seen in the superoxide dismutase and glutathione peroxidase activities, as partially corroborated by our study.

Moreover, generally, glutathione-related enzyme activities showed different patterns between groups. In the NG group, L-GPX significantly decreased, and L-GR and L-GST significantly increased, contrary to the HG group, in which there was no change in L-GST activity. Additionally, correlation analysis showed that the changes in L-GR and L-GST were not related to glucose improvement, but their increased activities in RBCs are associated with VFA reduction. The increase in L-GST activity observed in the NG obese individuals (Figure 3) indicates the potential for better antioxidant protection for the normoglycemic subjects after weight reduction and highlights the different metabolic adaptations of RBCs to weight loss in obese patients with normo- and hyperglycemia. Our results also indirectly point to an improvement in NADPH availability after glycemia correction.

Regarding vitamins, the most significant changes were present in patients who lost more weight, more VFA, and more TBW. We also found different vitamin E levels in normo- vs. hyperglycemia patients. Both vitamins are proven to have significant effects on RBCs. Vitamin A is involved in strengthening RBC membrane integrity, but with higher doses, it may exert an undesirable hemolytic effect [67], while vitamin E cooperates with vitamin C in RBC protection [68]. Both vitamin A and E exert antioxidant properties, but while vitamin E is a direct antioxidant preventing lipid peroxidation, vitamin A acts indirectly via the regulation of the transcription of a number of genes involved in mediating the body's canonical antioxidant responses [68]. Although our study was without significant differences in retinol levels, retinol levels decrease with a decrease in glucose, thus implying an inter-relationship. This correlation between retinol and glucose levels was also described by Gök et al. [69] in their work studying antioxidant enzymes and vitamins in women with uterine myoma and endometrial cancer. Different levels of both vitamins A and E, relative to the control group, were also described by Darenskaya et al. [70], who reported lower values of retinol and α-tocopherol in those with diabetic nephropathy (both stage 1 and 2). They focused on men with T1DM, and both α-tocopherol and retinol show high sensitivity (ROC and AUC) in the earlier stage of diabetic nephropathy but not in the later stage.

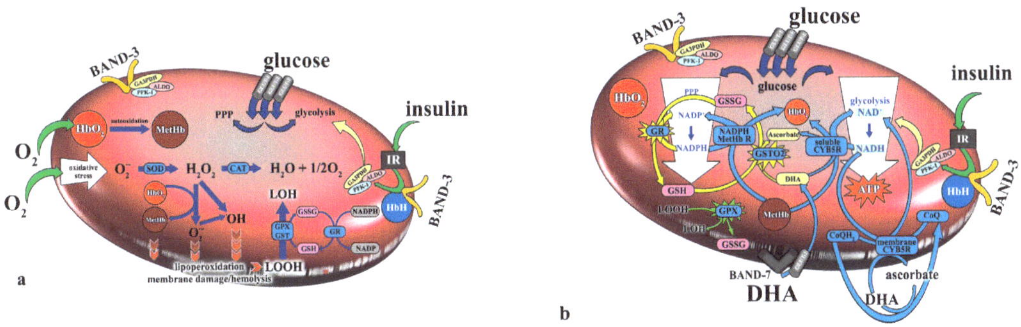

Figure 3. (**a**,**b**). The formation of ROS in RBCs and their processing and scavenging by enzymes. The most important enzymes are superoxide dismutase (SOD), catalase (CAT), glutathione peroxidase (GPX), and glutathione transferase (GST). Glutathione reductase (GR) is vital for glutathione's recovery from the oxidized form (GSSG) to the reduced form (GSH). Glutathione-dependent enzymes are involved in protecting against lipid peroxidation via reducing lipoperoxide (LOOH) to alcohol (LOH). Additionally, GR is dependent on the availability of NADPH from the PPP (pentose phosphate pathway). Weight reduction itself results in a decrease in GPX activity and an increase in GR activity regardless of the type of tissue loss or glycemia level. A different pattern was found for GST, which increased only in the normoglycemia patients and did not experience change in the hyperglycemia patients. Water-soluble ascorbate seems to be a factor involved in GST changes in normo- and hyperglycemia patients, and this could also be connected to the type of tissue loss, as GST only increased in patients who lost less than 5% of their total body and skeletal muscle mass.

The results regarding L-LPS are also interesting. LPS in RBCs was significantly lower in the HG group both before the WRP and after the WRP. However, a decrease in L-LPS does not correlate with a decrease in glucose, but there is a relation between VFA and L-LPS, which is inverse and shows that the higher the decrease in VFA, the lower the decrease in LPS in RBCs. Similar results were also found in our compartment loss analysis, as L-LPS decreased the most in those who lost less weight, less VFA, more TBW, and more muscle. In RBCs, much of the lipid oxidation is catalyzed by HbO_2 and is inhibited when hemoglobin is present as MetHb [71]. Also, hypercapnia exerts a protective effect against LPS formation [72]. Thus, in the HG group, the lower content of oxygenated hemoglobin was probably the cause of the low LPS levels.

5. Conclusions

This comprehensive analysis underscores the complex interactions between glucose metabolism, oxidative state, and erythrocyte membrane integrity, which are critical for understanding metabolic processes in diabetes and their management. This study highlights the significant metabolic adaptability of erythrocytes in response to systemic changes induced by obesity and hyperglycemia, suggesting avenues for future research to explore potential therapeutic targets in order to improve metabolic health in obese individuals.

6. Study Limitation

Despite the admission of 300 eligible patients to the program, compliance was met only by 53, which were included in the study. This was due to the following reasons: (a) the majority of the originally preselected patients appeared to be taking medications affecting glucose and lipid metabolism; (b) we experienced patient dropouts, as some patients who started the program stopped attending follow-up visits before the program was terminated; and (c) despite declaring they would do so, some patients did not follow the prescribed diet and/or physical activity regimen.

Author Contributions: Conceptualization, B.S. and J.Z.-F.; methodology, B.S. and A.G.-G.; formal analysis, A.B.; investigation, B.S., P.Z.-F., and M.D.; resources, J.Z.-F.; data curation, A.B.; writing—original draft preparation, B.S., A.B., and B.Č.; writing—review and editing, A.B., B.Č., and J.Z.-F.; funding acquisition, J.Z.-F. All authors have read and agreed to the published version of the manuscript.

Funding: This research was funded by the Medical University of Silesia in Katowice (PCN-1-018/K/1/O).

Institutional Review Board Statement: This study was conducted in accordance with the Declaration of Helsinki and approved by the Medical University of Silesia Ethics Committee in Katowice, Poland, No. KNW/0022/KB1/19/I/16.

Informed Consent Statement: Written informed consent to participate in the study and publish the research results was obtained from all subjects. To avoid bias, for biochemical tests and statistical analysis, the samples were anonymized and numbered.

Data Availability Statement: The database of aggregated statistics ready for analysis is stored in a secure, confidential, and password-protected repository on the server of the Medical University of Silesia. The data were anonymized. Completely non-identifiable records may be made available to interested persons/organizations upon request to jzalejskafiolka@sum.edu.pl.

Acknowledgments: The authors would like to acknowledge the support they received from Michal Toborek (Department of Biochemistry and Molecular Biology, University of Miami) and the Metropolis GZM Science and Education Support Fund under the program "Metropolitan Fund for Science Support" for 2021 (RW/13/2021). Special thanks to Irena Gryc and the laboratory staff from the clinic in Miasteczko Śląskie for allowing visiting patients and for their help in blood collection.

Conflicts of Interest: The authors declare no conflicts of interest.

References

1. Available online: https://www.who.int/news-room/fact-sheets/detail/obesity-and-overweight (accessed on 2 February 2024).
2. Available online: https://ec.europa.eu/eurostat/statistics-explained/index.php?title=Overweight_and_obesity_-_BMI_statistics (accessed on 2 February 2024).
3. Blüher, M. Obesity: Global epidemiology and pathogenesis. *Nat. Rev. Endocrinol.* **2019**, *15*, 288–298. [CrossRef] [PubMed]
4. De Lorenzo, A.; Soldati, L.; Sarlo, F.; Calvani, M.; Di Lorenzo, N.; Di Renzo, L. New obesity classification criteria as a tool for bariatric surgery indication. *World J. Gastroenterol.* **2016**, *22*, 681–703. [CrossRef]
5. Páscoa, R.; Teixeira, A.; Henriques, T.S.; Monteiro, H.; Monteiro, R.; Martins, C. Characterization of an obese population: A retrospective longitudinal study from real-world data in northern Portugal. *BMC Prim. Care* **2024**, *24*, 99. [CrossRef] [PubMed]
6. Manna, P.; Jain, S.K. Obesity, Oxidative Stress, Adipose Tissue Dysfunction, and the Associated Health Risks: Causes and Therapeutic Strategies. *Metab. Syndr. Relat. Disord.* **2015**, *13*, 423–444. [CrossRef] [PubMed]
7. Pilch, W.; Wyrostek, J.; Piotrowska, A.; Czerwińska-Ledwig, O.; Zuziak, R.; Sadowska-Krępa, E.; Maciejczyk, M.; Żychowska, M. Blood pro-oxidant/antioxidant balance in young men with class II obesity after 20 sessions of whole body cryostimulation: A preliminary study. *Redox Rep.* **2021**, *26*, 10–17. [CrossRef] [PubMed]
8. de Oliveira, S.; Saldanha, C. An overview about erythrocyte membrane. *Clin. Hemorheol. Microcirc.* **2010**, *44*, 63–74. [CrossRef]
9. Kostara, C.E.; Tsiafoulis, C.G.; Bairaktari, E.T.; Tsimihodimos, V. Altered RBC membrane lipidome: A possible etiopathogenic link for the microvascular impairment in Type 2 diabetes. *J. Diabetes Complicat.* **2021**, *35*, 107998. [CrossRef]
10. Bryk, A.H.; Wiśniewski, J.R. Quantitative Analysis of Human Red Blood Cell Proteome. *J. Proteome Res.* **2017**, *16*, 2752–2761. [CrossRef]
11. Barshtein, G. Biochemical and Biophysical Properties of Red Blood Cells in Disease. *Biomolecules* **2022**, *12*, 923. [CrossRef]
12. Kuhn, V.; Diederich, L.; Keller, T.C.S., 4th; Kramer, C.M.; Lückstädt, W.; Panknin, C.; Suvorava, T.; Isakson, B.E.; Kelm, M.; Cortese-Krott, M.M. Red Blood Cell Function and Dysfunction: Redox Regulation, Nitric Oxide Metabolism, Anemia. *Antioxid. Redox Signal.* **2017**, *26*, 718–742. [CrossRef]

13. Maurya, P.K.; Kumar, P.; Chandra, P. Biomarkers of oxidative stress in erythrocytes as a function of human age. *World J. Methodol.* **2015**, *5*, 216–222. [CrossRef] [PubMed]
14. Möller, M.N.; Orrico, F.; Villar, S.F.; López, A.C.; Silva, N.; Donzé, M.; Thomson, L.; Denicola, A. Oxidants and Antioxidants in the Redox Biochemistry of Human Red Blood Cells. *ACS Omega* **2022**, *8*, 147–168. [CrossRef] [PubMed]
15. Marseglia, L.; Manti, S.; D'Angelo, G.; Nicotera, A.; Parisi, E.; Di Rosa, G.; Gitto, E.; Arrigo, T. Oxidative stress in obesity: A critical component in human diseases. *Int. J. Mol. Sci.* **2014**, *16*, 378–400. [CrossRef]
16. Szlachta, B.; Birková, A.; Wielkoszyński, T.; Gospodarczyk, A.; Hubková, B.; Dydoń, M.; Zalejska-Fiolka, J. Serum Oxidative Status in People with Obesity: Relation to Tissue Losses, Glucose Levels, and Weight Reduction. *Antioxidants* **2023**, *12*, 1923. [CrossRef]
17. Ryan, D.H.; Yockey, S.R. Weight Loss and Improvement in Comorbidity: Differences at 5%, 10%, 15%, and Over. *Curr. Obes. Rep.* **2017**, *6*, 187–194. [CrossRef]
18. Nowak, W.S.; Skotnicki, A.B. Podstawy diagnostyki hematologicznej. In *Diagnostyka Laboratoryjna z Elementami Biochemii Klinicznej*; Dembińska-Kieć, A., Naskalski, J.W., Solnica, B., Eds.; Edra Urban & Partner: Wrocław, Poland, 2017; Volume IV, pp. 275–276.
19. Oyanagui, Y. Reevaluation of assay methods and establishment of kit for superoxide dismutase activity. *Anal. Biochem.* **1984**, *142*, 290–296. [CrossRef] [PubMed]
20. Johansson, L.H.; Borg, L.A. A spectrophotometric method for determination of catalase activity in small tissue samples. *Anal. Biochem.* **1988**, *174*, 331–336. [CrossRef] [PubMed]
21. Paglia, D.E.; Valentine, W.N. Studies on the quantitative and qualitative characterization of erythrocyte hemolysate glutathione peroxidase. *J. Lab. Clin. Med.* **1967**, *70*, 158–169. [PubMed]
22. Richterich, R. *Chemia Kliniczna*; PZWL: Warszawa, Poland, 1971.
23. Habig, W.H.; Jakoby, W.B. Assays for differentiation of glutathione S-transferases. *Methods Enzymol.* **1981**, *77*, 398–405.
24. Jain, S.K. In vivo externalization of phosphatidylserine and phosphatidylethanolamine in the membrane bilayer and hypercoagulability by the lipid peroxidation of erythrocyte hemoerythrocytes in rats. *J. Clin. Investig.* **1985**, *76*, 281–286. [CrossRef]
25. Ohkawa, H.; Ohishi, N.; Yagi, K. Assay for lipid peroxides in animal tissues by thiobarbituric acid reaction. *Anal. Biochem.* **1979**, *95*, 351–358. [CrossRef] [PubMed]
26. Shearer, M.J. Vitamins. In *HPLC of Small Molecules*; Lim, C.K., Ed.; IRL Press: Oxford, UK, 1986.
27. Chung, Y.J.; Kim, E.Y. Usefulness of bioelectrical impedance analysis and ECW ratio as a guidance for fluid management in critically ill patients after operation. *Sci. Rep.* **2021**, *11*, 12168. [CrossRef] [PubMed]
28. *ISO 9001:2015*. Available online: https://www.iso.org/obp/ui/#iso:std:iso:9001:ed-5:v1:en (accessed on 4 August 2024).
29. *ISO 13485:2016*. Available online: https://www.iso.org/obp/ui/en/#iso:std:iso:13485:ed-3:v1:en (accessed on 4 August 2024).
30. *EN60601-1*. Available online: https://www.bsigroup.com/en-GB/capabilities/medical-devices/en-60601-medical-electrical-equipment-and-systems/ (accessed on 4 August 2024).
31. *EN60601-1-2*. Available online: https://www.eurofins.com/media/12153172/en60601-1-2-implications-of-4th-edition-iwp-d01.pdf (accessed on 4 August 2024).
32. Directive 93/42/EEC. Available online: https://eur-lex.europa.eu/legal-content/EN/TXT/PDF/?uri=CELEX:31993L0042 (accessed on 4 August 2024).
33. Domingo-Ortí, I.; Lamas-Domingo, R.; Ciudin, A.; Hernández, C.; Herance, J.R.; Palomino-Schätzlein, M.; Pineda-Lucena, A. Metabolic footprint of aging and obesity in red blood cells. *Aging* **2021**, *13*, 4850–4880. [CrossRef] [PubMed]
34. Bali, M.; Thomas, S.R. Activation vers l'avant de la glycolyse de l'erythrocyte humain: Une etude par modelisation mathematique. *Comptes Rendus l'Acad. Sci.* **2001**, *324*, 185–199.
35. Grundy, J.C. Aspects of blood cell biochemistry: Erythrocytes, platelets, and stem cells. In *Functional Metabolism: Regulation and Adaptation*; Storey, K.B., Ed.; John Wiley: Hoboken, NJ, USA, 2004; pp. 505–528.
36. Delgado, T.C.; Castro, M.M.; Geraldes, C.F.; Jones, J.G. Quantitation of erythrocyte pentose pathway flux with [2–13C] glucose and 1H NMR analysis of the lactate methyl signal. *Magn. Reson. Med.* **2004**, *51*, 1283–1286. [CrossRef] [PubMed]
37. Bravi, M.C.; Armiento, A.; Laurenti, O.; Cassone-Faldetta, M.; De Luca, O.; Moretti, A.; De Mattia, G. Insulin decreases intracellular oxidative stress in patients with type 2 diabetes mellitus. *Metabolism* **2006**, *55*, 691–695. [CrossRef] [PubMed]
38. Puckeridge, M.; Chapman, B.E.; Conigrave, A.D.; Grieve, S.M.; Figtree, G.A.; Kuchel, P.W. Stoichiometric relationship between Na(+) ions transported and glucose consumed in human erythrocytes: Bayesian analysis of (23)Na and (13)C NMR time course data. *Biophys. J.* **2013**, *104*, 1676–1684. [CrossRef] [PubMed]
39. Rungaldier, S.; Oberwagner, W.; Salzer, U.; Csaszar, E.; Prohaska, R. Stomatin interacts with GLUT1/SLC2A1, band 3/SLC4A1, and aquaporin-1 in human erythrocyte membrane domains. *Biochim. Biophys. Acta* **2013**, *1828*, 956–966. [CrossRef] [PubMed]
40. Sage, J.M.; Carruthers, A. Human erythrocytes transport dehydroascorbic acid and sugars using the same transporter complex. *Am. J. Phys. Cell Physiol.* **2014**, *306*, 910–917. [CrossRef]
41. Holman, G.D. Chemical biology probes of mammalian GLUT structure and function. *Biochem. J.* **2018**, *475*, 3511–3534. [CrossRef]
42. Khan, A.A.; Hanada, T.; Mohseni, M.; Jeong, J.J.; Zeng, L.; Gaetani, M.; Li, D.; Reed, B.C.; Speicher, D.W.; Chishti, A.H. Demain and adducin provide a novel link between the spectrin cytoskeleton and human erythrocyte membrane by directly interacting with glucose transporter-1. *J. Biol. Chem.* **2008**, *283*, 14600–14609. [CrossRef] [PubMed]
43. Rumsey, S.C.; Kwon, O.; Xu, G.W.; Burant, C.F.; Simpson, I.; Levine, M. Glucose transporter isoforms GLUT1 and GLUT3 transport dehydroascorbic acid. *J. Biol. Chem.* **1997**, *272*, 18982–18989. [CrossRef] [PubMed]

44. Vera, J.C.; Rivas, C.I.; Fischbarg, J.; Golde, D.W. Mammalian facilitative hexose transporters mediate the transport of dehydroascorbic acid. *Nature* **1993**, *364*, 79–82. [CrossRef] [PubMed]
45. Lapatsina, L.; Brand, J.; Poole, K.; Daumke, O.; Lewin, G.R. Stomatin-domain proteins. *Eur. J. Cell Biol.* **2012**, *91*, 240–245. [CrossRef] [PubMed]
46. Zancan, P.; Sola-Penna, M. Calcium influx: A possible role for insulin modulation of intracellular distribution and activity of 6-phosphofructo-1-kinase in human erythrocytes. *Mol. Genet. Metab.* **2005**, *86*, 392–400. [CrossRef] [PubMed]
47. Bakhtiari, N.; Hosseinkhani, S.; Larijani, B.; Mohajeri-Tehrani, M.R.; Fallah, A. Red blood cell ATP/ADP & nitric oxide: The best vasodilators in diabetic patients. *J. Diabetes Metab. Disord.* **2012**, *11*, 9. [PubMed]
48. Tu, H.; Li, H.; Wang, Y.; Niyyati, M.; Leshin, J.; Levine, M. Low red blood cell vitamin C concentrations induce red blood cell fragility: A link to diabetes via glucose, glucose transporters, and Dehydroascorbic acid. *EBioMedicine* **2015**, *2*, 1735–1750. [CrossRef]
49. Simic, T.; Savic-Radojevic, A.; Pljesa-Ercegovac, M.; Matic, M.; Sasic, T.; Dragicevic, D.; Mimic-Oka, J. The role of glutathione S-transferases in urinary tract tumors. *J. Med. Bioch.* **2008**, *27*, 360–366. [CrossRef]
50. Porter-Turner, M.M.; Skidmore, J.C.; Khokher, M.A.; Singh, B.M.; Rea, C.A. Relationship between erythrocyte GLUT1 function and membrane glycation in type 2 diabetes. *Br. J. Biomed. Sci.* **2016**, *68*, 203–207. [CrossRef]
51. Zhang, H.; Shen, Z.; Hogan, B.; Barakat, A.I.; Misbah, C. ATP release by red blood cells under flow: Model and simulations. *Biophys. J.* **2018**, *115*, 2218–2229. [CrossRef]
52. Symeonidis, A.; Athanassiou, G.; Psiroyannis, A.; Kyriazopoulou, V.; Kapatais-Zoumbos, K.; Missirlis, Y.; Zoumbos, N. Impairment of erythrocyte viscoelasticity is correlated with levels of glycosylated haemoglobin in diabetic patients. *Clin. Lab. Haematol.* **2001**, *23*, 103–109. [CrossRef] [PubMed]
53. Benedik, P.S.; Hamlin, S.K. The physiologic role of erythrocytes in oxygen delivery and implications for blood storage. *Crit. Care Nurs. Clin. N. Am.* **2014**, *26*, 325–335. [CrossRef] [PubMed]
54. Clemens, M.R.; Waller, H.D. Lipid peroxidation in erythrocytes. *Chem. Phys. Lipids* **1987**, *45*, 251–268. [CrossRef] [PubMed]
55. Willoughby, D.; Hewlings, S.; Kalman, D. Body Composition Changes in Weight Loss: Strategies and Supplementation for Maintaining Lean Body Mass, a Brief Review. *Nutrients* **2018**, *10*, 1876. [CrossRef]
56. Zalejska-Fiolka, J.; Birková, A.; Hubková, B.; Wielkoszyński, T.; Čižmárová, B.; Szlachta, B.; Fiolka, R.; Błaszczyk, U.; Wylęgała, A.; Kasperczyk, S.; et al. Successful correction of hyperglycemia is critical for weight loss and a decrease in cardiovascular risk in obese patients. *J. Nutr. Biochem.* **2022**, *106*, 109021. [CrossRef] [PubMed]
57. Hollander, J.; Bejma, J.; Ookawara, T.; Ohno, H.; Ji, L.L. Superoxide dismutase gene expression in skeletal muscle: Fiber-specific effect of age. *Mech. Ageing Dev.* **2000**, *116*, 33–45. [CrossRef] [PubMed]
58. Mondola, P.; Bifulco, M.; Serù, R.; Annella, T.; Ciriolo, M.R.; Santillo, M. Presence of CuZn superoxide dismutase in human serum lipoproteins. *FEBS Lett.* **2000**, *467*, 57–60. [CrossRef] [PubMed]
59. Pasiakos, S.M.; Cao, J.J.; Margolis, L.M.; Sauter, E.R.; Whigham, L.D.; McClung, J.P.; Rood, J.C.; Carbone, J.W.; Combs, C.F., Jr.; Young, A.J. Effects of high-protein diets on fat-free mass and muscle protein synthesis following weight loss: A randomized controlled trial. *FASEB J.* **2013**, *27*, 3837–3847. [CrossRef] [PubMed]
60. Villareal, D.T.; Smith, G.; Shah, K.; Mittendorfer, B. Effect of Weight Loss on the Rate of Muscle Protein Synthesis During Fasted and Fed Conditions in Obese Older Adults. *Obesity* **2012**, *20*, 1780–1786. [CrossRef]
61. Thomas, D.T.; Erdman, K.A.; Burke, L.M. Position of the Academy of Nutrition and Dietetics, Dietitians of Canada, and the American College of Sports Medicine: Nutrition and Athletic Performance. *J. Acad. Nutr. Diet.* **2016**, *116*, 501–528. [CrossRef]
62. Turpin, C.; Catan, A.; Guerin-Dubourg, A.; Debussche, X.; Bravo, S.B.; Álvarez, E.; Van Den Elsen, J.; Meilhac, O.; Rondeau, P.; Bourdon, E. Enhanced oxidative stress and damage in glycated erythrocytes. *PLoS ONE* **2020**, *15*, e0235335. [CrossRef] [PubMed]
63. Gaetani, G.F.; Galiano, S.; Canepa, L.; Ferraris, A.M.; Kirkman, H.N. Catalase and glutathione peroxidase are equally active in detoxification of hydrogen peroxide in human erythrocytes. *Blood* **1989**, *73*, 334–339. [CrossRef]
64. Gaetani, G.F.; Kirkman, H.N.; Mangerini, R.; Ferraris, A.M. Importance of catalase in the disposal of hydrogen peroxide within human erythrocytes. *Blood* **1994**, *84*, 325–330. [CrossRef] [PubMed]
65. Cattani, L.; Ferri, A. The function of NADPH bound to Catalase. *Boll. Soc. Ital. Biol. Sper.* **1994**, *70*, 75–82. [PubMed]
66. Ramezanipour, M.; Jalali, M.; Sadrzade-Yeganeh, H.; Keshavarz, S.A.; Eshraghian, M.R.; Bagheri, M.; Emami, S.S. Effect of weight reduction on antioxidant enzymes and their association with dietary intake of vitamins A, C and E. *Bras. Endocrinol.* **2014**, *58*, 744–749. [CrossRef] [PubMed]
67. Glauert, A.M.; Daniel, M.R.; Lucy, J.A.; Dingle, J.T. Studies on the mode of action of excess of vitamin A. VII. Changes in the fine structure of erythrocytes during haemolysis by vitamin A. *J. Cell Biol.* **1963**, *17*, 111–121. [CrossRef]
68. Pallavi, M.; Rajashekaraiah, V. Synergistic activity of vitamin-C and vitamin-E to ameliorate the efficacy of stored erythrocytes. *Transfus. Clin. Biol.* **2023**, *30*, 87–95. [CrossRef] [PubMed]
69. Gök, F.; Ekin, S.; Karaman, E.; Erten, R.; Yıldız, D.; Bakır, A. Total Sialic Acid, Antioxidant Enzyme Activities, Trace Elements, and Vitamin Status before and after Surgery in Women with Uterine Myoma and Endometrial Cancer. *Reprod. Sci.* **2023**, *30*, 2743–2757. [CrossRef] [PubMed]

70. Darenskaya, M.; Chugunova, E.; Kolesnikov, S.; Semenova, N.; Michalevich, I.; Nikitina, O.; Lesnaya, A.; Kolesnikova, L. Receiver Operator Characteristic (ROC) Analysis of Lipids, Proteins, DNA Oxidative Damage, and Antioxidant Defense in Plasma and Erythrocytes of Young Reproductive-Age Men with Early Stages of Type 1 Diabetes Mellitus (T1DM) Nephropathy in the Irkutsk Region, Russia. *Metabolites* **2022**, *12*, 1282. [CrossRef]
71. Orrico, F.; Laurance, S.; Lopez, A.C.; Lefevre, S.D.; Thomson, L.; Möller, M.N.; Ostuni, M.A. Oxidative Stress in Healthy and Pathological Red Blood Cells. *Biomolecules* **2023**, *13*, 1262. [CrossRef]
72. Skoumalová, A.; Herget, J.; Wilhelm, J. Hypercapnia protects erythrocytes against free radical damage induced by hypoxia in exposed rats. *Cell Biochem. Funct.* **2008**, *26*, 801–807. [CrossRef] [PubMed]

antioxidants

MDPI

Article

Stratification of $\beta^S\beta^+$ Compound Heterozygotes Based on L-Glutamine Administration and RDW: Focusing on Disease Severity

Aimilia Giannaki [1,†], Hara T. Georgatzakou [1,†], Sotirios P. Fortis [1], Alkmini T. Anastasiadi [1,2], Efthimia G. Pavlou [1,3], Efrosyni G. Nomikou [3], Maria P. Drandaki [4], Angeliki Kotsiafti [4], Aikaterini Xydaki [4], Christina Fountzoula [5], Effie G. Papageorgiou [1], Vassilis L. Tzounakas [2] and Anastasios G. Kriebardis [1,*]

[1] Laboratory of Reliability and Quality Control in Laboratory Hematology (HemQcR), Department of Biomedical Sciences, School of Health & Caring Sciences, University of West Attica (UniWA), 12243 Egaleo, Greece; agiannaki@uniwa.gr (A.G.); cgeorgatz@uniwa.gr (H.T.G.); sfortis@uniwa.gr (S.P.F.); aanastasiadi@uniwa.gr (A.T.A.); epavlou@uniwa.gr (E.G.P.); efipapag@uniwa.gr (E.G.P.)

[2] Department of Biochemistry, School of Medicine, University of Patras, 26504 Patras, Greece; vtzounakas@upatras.gr

[3] Blood Bank and Hemophilia Unit, Hippokration Hospital, 11527 Athens, Greece; aima@hippocratio.gr

[4] Thalassemia and Sickle Cell Unit, Expertise Center of Hemoglobinopathies and Their Complications, Hippokration General Hospital, 11527 Athens, Greece; thalassemia@chaniahospital.gr (M.P.D.); ankotsiafti@uth.gr (A.K.); xydaki@hippocratio.gr (A.X.)

[5] Laboratory of Chemistry, Biochemistry and Cosmetic Science (ChemBiochemCosm), Department of Biomedical Sciences, School of Health & Caring Sciences, University of West Attica (UniWA), 12243 Egaleo, Greece; chfountz@uniwa.gr

* Correspondence: akrieb@uniwa.gr; Tel.: +30-210-538-5813

† These authors contributed equally to this work.

Citation: Giannaki, A.; Georgatzakou, H.T.; Fortis, S.P.; Anastasiadi, A.T.; Pavlou, E.G.; Nomikou, E.G.; Drandaki, M.P.; Kotsiafti, A.; Xydaki, A.; Fountzoula, C.; et al. Stratification of $\beta^S\beta^+$ Compound Heterozygotes Based on L-Glutamine Administration and RDW: Focusing on Disease Severity. *Antioxidants* **2023**, *12*, 1982. https://doi.org/10.3390/antiox12111982

Academic Editors: Reto Asmis and Miriam M. Cortese-Krott

Received: 6 October 2023
Revised: 30 October 2023
Accepted: 7 November 2023
Published: 8 November 2023

Abstract: Sickle cell disease (SCD) is heterogeneous in terms of manifestation severity, even more so when in compound heterozygosity with beta-thalassemia. The aim of the present study was to stratify $\beta^S\beta^+$ patient blood samples in a severity-dependent manner. Blood from thirty-two patients with HbS/β-thalassemia compound heterozygosity was examined for several parameters (e.g., hemostasis, inflammation, redox equilibrium) against healthy controls. Additionally, SCD patients were a posteriori (a) categorized based on the L-glutamine dose and (b) clustered into high-/low-RDW subgroups. The patient cohort was characterized by anemia, inflammation, and elevated coagulation. Higher-dose administration of L-glutamine was associated with decreased markers of inflammation and oxidation (e.g., intracellular reactive oxygen species) and an altered coagulation profile. The higher-RDW group was characterized by increased hemolysis, elevated markers of inflammation and stress erythropoiesis, and oxidative phenomena (e.g., membrane-bound hemoglobin). Moreover, the levels of hemostasis parameters (e.g., D-Dimers) were greater compared to the lower-RDW subgroup. The administration of higher doses of L-glutamine along with hydroxyurea seems to attenuate several features in SCD patients, probably by enhancing antioxidant power. Moreover, anisocytosis may alter erythrocytes' coagulation processes and hemolytic propensity. This results in the disruption of the redox and pro-/anti-inflammatory equilibria, creating a positive feedback loop by inducing stress erythropoiesis and, thus, the occurrence of a mixed erythrocyte population.

Keywords: sickle cell disease; L-glutamine; RDW; coagulation; inflammation; oxidative stress

1. Introduction

Sickle cell disease (SCD) is one of the most common and severe monogenic disorders worldwide and is estimated to affect ~300,000 infants every year. SCD, which Pauling characterized as a "molecular disease" in 1949, occurs due to a point mutation in the β-globin gene that leads to the production of hemoglobin S (HbS). Due to its altered biophysical properties, the latter can polymerize under hypoxic conditions, and the fibers formed change red blood cell (RBC) deformability features [1]. This is followed by a cascade of events, including vessel occlusion and hemolysis, which are responsible for the symptomatology of the disease, such as pain crises, and even organ damage [2]. Despite its monogenic basis, some clinical phenotypes of SCD present extreme variability, compelling scientists to propose that it concerns a monogenic disease with a polygenic phenotype [3]. The most frequent and most studied form of SCD is the homozygous state for the β^s mutation, while other forms include compound heterozygosity for HbS and thalassemias. Nevertheless, even within the distinct genotypes, there is a wide range of manifestation severity.

The variable symptom severity has been linked to several distinct blood parameters. RBCs with low deformability present increased adhesive properties, with both features being indisputable contributors to the disease sequelae [4–6]. The levels of microRNAs have been also linked to patient symptomatology [7], while free heme has been similarly discussed as a potential biomarker [8]. Of note, a cluster of seventeen circulating molecules, including hematological parameters (e.g., monocytes, MCV) and serum markers (e.g., lactate dehydrogenase, bilirubin), has been found to correlate with morbidity and mortality in SCD [9]. More recently, the presence of mitochondria-retaining RBCs and reticulocytes, as well as free mitochondrial DNA, has been associated with well-known contributors to the disease manifestations, including hemolysis and inflammatory reactions [10–12].

As stated above, the β^S mutation can be combined with other mutations in compound heterozygosity. While HbS/β^0 thalassemia clinically resembles the homozygous sickle cell disease state, HbS/β^+ thalassemia's manifestations vary depending on the specific allele [13], and, although it is considered a mild disease, this is not always the case [14]. The present study aimed to find a stratifying parameter to evaluate the variability in disease-manifestation-related parameters in the blood of HbS-β^+ thalassemia compound heterozygotes, which compose a less studied SCD subgroup.

2. Materials and Methods

2.1. Subjects

Thirty-two patients with HbS/β-thalassemia compound heterozygosity (HbS-β^+) and twenty age- and sex-matched healthy controls were included in the study. All HbS-β^+ patients were receiving folate and most were under treatment with hydroxyurea with/without L-glutamine, while some of them were transfusion-dependent. L-glutamine was consumed by the patients as a food supplement (dose: 16 ± 6 g/day; Glutamine DB EXTRA supplement). The exclusion criteria were as follows: vaso-occlusive crisis 1 month prior to onset of L-glutamine, malignancy, known peptic ulcer, and intolerance to any of the ingredients. Whole blood samples from both patients and controls were collected in ethylenediaminetetraacetic acid (EDTA) and 3.2% sodium citrate blood collection tubes (BD Vacutainer Blood Collection Tubes, BD Biosciences, San Jose, CA, USA).

2.2. Material Supplies

All materials and common chemicals were obtained from Sigma-Aldrich (Munich, Germany) unless otherwise stated. Antibodies used for flow cytometry experiments were obtained from BD Biosciences (San Jose, CA, USA). The Zymuphen™ EV activity kit (Hyphen Biomed, Neuville-sur-Oise, France) was used for the measurement of extracellular vesicles' procoagulant activity. The ECL Western blot detection kit was from GE Healthcare. Antibodies against Hb (CR8000GAP) were from Europa Bioproducts, (Ipswich, UK) while the antibody against 4.1 R was kindly provided by Prof. J. Delaunay (Laboratoire d' Hématologie, d'Immunologie et de Cytogénétique, Hopital de Bicetre, Le Kremlin-Bicetre, France). HRP-conjugated antibodies to rabbit IgGs were from GE Healthcare (Chicago, IL, USA) and those to goat IgGs were from Sigma-Aldrich (St. Louis, MO, USA).

2.3. Hematological and Biochemical Analysis

Classical hematological analysis was performed using a Siemens Advia 2120i Hematology Analyzer. Biochemical analysis of serum components (urea, creatinine, uric acid, glucose, cholesterol, triglycerides, calcium, phosphorus, potassium, sodium, chlorine, magnesium, iron, ferritin, B12, folate, proteins, albumin, serum glutamyl oxalate transaminase, serum glutamyl pyruvate transaminase, gamma-glutamyl transferase, alkaline phosphatase, total, direct and indirect bilirubin, total creatine phosphokinase, amylase, lactate dehydrogenase, and vitamin D) was conducted using the automatic Clinical Chemistry Analyzer ARCHITECT C16000 (Abbott, Chicago, IL, USA). C-reactive protein levels were determined in the Architect C8000 analyzer by a commercial kit (Abbott Laboratories (Hellas), Athens, Greece). HbS and HbF levels were evaluated by the fully automated VARIANT II Hemoglobin Testing System (BioRad, Hercules, CA, USA).

2.4. Hemolysis and Redox Parameters

Spontaneous hemolysis (levels of plasma Hb) was calculated by spectrophotometry using Harboe's method, followed by Allen's correction and normalization to hematocrit and intracellular Hb levels. The propensity of erythrocytes for osmotic lysis was evaluated upon exposure to ascending NaCl concentrations. The antioxidant capacity of plasma (total, TAC; uric acid-dependent, UAdAC; and uric acid-independent, UAiAC) was measured via the method of Benzie and Strain [15]. Briefly, plasma with and without uricase treatment, mixed with the ferric reducing/antioxidative power (FRAP) solution, was incubated at 37 °C for 4 min and absorbance was measured at 593 nm. Intracellular mean fluorescence and the percentage of reactive oxygen species (ROS)-positive RBCs were detected in a FACSCanto II Cytometer (BD Biosciences, San Jose, CA, USA) by using the fluorescent probe CM-H_2DCFDA (Invitrogen, Molecular Probes, Eugene, OR, USA) [16].

2.5. Hemostasis Parameters

A secondary hemostasis screening test, including the measurement of prothrombin time (PT-INR), activated partial thromboplastin clotting time (APTT), and fibrinogen and D-dimer values, as well as the levels of von Willebrand and factor VIII, was performed as previously described [16]. Quantitative determination of the thrombin/antithrombin III complex (TAT) in human plasma was assessed by enzyme immunoassays (Siemens Healthineers, Erlangen, Germany). Thrombin time (TT) was measured by STA Thrombin reagent (Diagnostica Stago, Parsippany, NJ, USA).

2.6. Membrane Isolation and Immunoblotting

RBC membranes were isolated by hypotonic (5 mmol/L sodium phosphate buffer) lysis of RBCs. Membrane aliquots were immune-probed for integral and membrane-bound proteins by using horseradish-peroxidase-conjugated secondary antibodies and enhanced chemiluminescence development. Semi-quantification of protein bands was performed by scanning densitometry (Gel Analyzer v.1.0, Biosure, Athens, Greece).

2.7. Statistical Analysis

For statistical analysis, the Statistical Package for Social Sciences (IBM SPSS Software; version 26.0 for Windows IBM Corp., Armonk, NY, USA; administrated by UniWA and the University of Patras) was used. Patients were a posteriori categorized using the information regarding L-glutamine consumption or a two-step cluster analysis (in the case of RDW). The differences between groups were evaluated by parametric and non-parametric tests according to the distribution profile of each parameter. Significance was accepted at $p < 0.05$.

3. Results

3.1. Variation from Controls

The first step of the study was to ensure that our patient cohort presented the anticipated divergence from the healthy controls. As expected, SCD patients were characterized as anemic due to decreased RBC counts, intracellular Hb, and hematocrit, but an increased red cell distribution width (RDW) (Table 1). Besides the higher levels of HbF and HbS, more reticulocytes were found in the SCD samples. While the complete number of white blood cells did not differ from that in the controls, the proportion of monocytes and basophils was found increased. The biochemical analysis showed that markers of hemolysis, like bilirubin and lactate dehydrogenase, along with markers of inflammation (ferritin, C-reactive protein) and liver function (e.g., alkaline phosphatase and gamma-glutamyl transferase), also presented higher values in the serum of SCD patients compared to controls. Several hemostasis and coagulation parameters, including the levels of D-dimers and von Willebrand factor, and the procoagulant activity of EVs, were elevated in the disease cohort. Lastly, the free Hb and plasma antioxidant capacity were higher in SCD patients, along with the ROS cargo of their RBCs, while osmotic hemolysis was lower (Table 1).

Table 1. Variations in SCD patients from healthy controls.

	Patients	Controls	Normal Range
Age (years)	49.8 ± 11.9	44.7 ± 7.6	
	General Blood Test		
White blood cells (×10^3/μL)	7.6 ± 1.7	6.4 ± 1.3	5.2–12.4
Neutrophils (%)	57.4 ± 9.0	56.8 ± 7.1	40.0–74.0
Lymphocytes (%)	29.3 ± 7.3	31.3 ± 6.5	19.0–48.0
Monocytes (%)	**7.5 ± 3.1** *	5.8 ± 1.3	3.4–9.0
Eosinophils (%)	**2.5 ± 1.3** *	3.6 ± 1.8	0.0–7.0
Basophils (%)	**1.0 ± 0.5** *	0.7 ± 0.3	0.0–1.5
Neutrophil/lymphocyte ratio	2.2 ± 1.0	1.9 ± 0.5	-
Red blood cells (×10^6/μL)	**3.8 ± 0.9** *	5.0 ± 0.4	4.2–6.1
Hemoglobin (g/dL)	**10.0 ± 1.5** *	14.2 ± 1.1	12.0–18.0
Hematocrit (%)	**31.8 ± 4.4** *	44.0 ± 3.6	37.0–52.0
MCV(fL)	86.4 ± 12.0	88.9 ± 4.0	80.0–99.0
MCH (pg)	27.1 ± 4.2	28.6 ± 1.3	27.0–31.0
MCHC (gr/dL)	**31.3 ± 1.3** *	32.2 ± 0.7	33.0–37.0
RDW (%)	**19.3 ± 2.3** *	13.1 ± 0.9	11.5–14.5
Platelets (×10^3/μL)	310.7 ± 189.2	273.4 ± 40.0	130.0–400.0
Mean platelet volume (MPV; fL)	10.0 ± 1.2	9.5 ± 1.4	7.2–11.1
Reticulocyte count (%)	**8.0 ± 4.8** *	1.5 ± 0.3	0.5–2.0
HbS (%)	**65.9 ± 14.3** *	0.0 ± 0.0	0
HbF (%)	**14.3 ± 9.6** *	0.3 ± 0.1	0.8–2
	Serum Biochemical Analysis		
Glucose (mg/dL)	**89.3 ± 10.1** *	81.5 ± 10.8	70–105
Urea (mg/dL)	28.8 ± 13.7	27.2 ± 5.1	18.0–55.0
Creatinine (mg/dL)	0.78 ± 0.18	0.83 ± 0.09	0.72–1.25
Uric acid (mg/dL)	5.1 ± 1.0	4.7 ± 0.8	3.5–7.2
Cholesterol (mg/dL)	**149.4 ± 25.4** *	183.1 ± 22.6	0.0–200.0
Triglycerides (mg/dL)	124.7 ± 53.1	99.7 ± 37.0	0.0–150.0
Calcium (mg/dL)	9.2 ± 0.6	9.3 ± 0.4	8.4–10.2
Phosphorus (mg/dL)	3.4 ± 0.5	3.4 ± 0.5	2.4–4.7
Potassium (mmol/L)	4.3 ± 0.3	4.2 ± 0.2	3.5–5.1
Sodium (mmol/L)	138.6 ± 1.8	139.4 ± 1.8	136.0–145.0
Chlorine(mmol/L)	105.7 ± 1.9	105.8 ± 1.8	98.0–107.0
Magnesium (mg/dL)	2.0 ± 0.3	2.0 ± 0.1	1.60–2.60
Iron (mg/dL)	130.0 ± 96.1	109.4 ± 51.0	50–150 (F); 60–160 (M)
Ferritin (ng/mL)	**359.9 ± 215.5** *	61.0 ± 42.3	14.0–233.0 (F); 16.4–293.3 (M)

Table 1. *Cont.*

	Patients	Controls	Normal Range
B12 (pg/mL)	336.3 ± 146.2	371.4 ± 185.0	179.0–1162.0
Folate (ng/mL)	**24.1 ± 15.0 ***	6.4 ± 2.5	2.5–17.0
Proteins (mg/dL)	7.4 ± 0.6	7.3 ± 0.4	6.40–8.30
Albumin (gr/dL)	4.4 ± 0.3	4.4 ± 0.3	3.50–5.00
SGOT (U/L)	**32.4 ± 11.5 ***	19.0 ± 6.3	5.0–34.0
SGPT (U/L)	28.8 ± 25.0	22.3 ± 12.6	0.0–55.0
Gamma-glutamyl transferase (U/L)	**34.4 ± 24.1 ***	19.2 ± 9.6	12.0–64.0
Alkaline phosphatase (ALP; U/L)	**78.5 ± 23.1 ***	63.1 ± 11.9	40.0–150.0
HDL (mg/dL)	**39.9 ± 8.7 ***	55.4 ± 14.1	>50
LDL (mg/dL)	**84.7 ± 20.9 ***	107.8 ± 18.9	<110
Immunoglobulins (g/dL)	3.0 ± 0.7	2.9 ± 0.7	
Total bilirubin (mg/dL)	**2.2 ± 1.1 ***	0.6 ± 0.2	0.2–1.2
Indirect bilirubin (mg/dL)	**1.4 ± 0.9 ***	0.3 ± 0.1	0.01–0.9
Direct bilirubin (mg/dL)	**0.7 ± 0.3 ***	0.2 ± 0.1	0.00–0.30
Lactate dehydrogenase (IU/L)	**337.9 ± 102.1 ***	184.5 ± 31.7	125.0–220.0
Creatine phosphokinase total (IU/L)	**41.8 ± 29.7 ***	107.7 ± 80.9	30.0–200.0
Vitamin D (ng/mL)	24.3 ± 11.1	21.4 ± 8.2	30.0–100.0
C-reactive protein (mg/L)	**6.0 ± 5.9 ***	1.7 ± 1.6	0.0–5.0
	Hemostasis–Coagulation Parameters		
Prothrombin time INR	**1.1 ± 0.4 ***	1.0 ± 0.1	0.8–1.1
APTT (s)	29.7 ± 5.0	29.2 ± 2.9	<36
Fibrinogen (mg/dL)	306.2 ± 115.5	327.9 ± 74.1	180–350
D-Dimer (μg/L)	**2552.6 ± 2186.0 ***	261.9 ± 98.4	<500
Factor VIII (%)	**90.6 ± 31.0 ***	123.3 ±17.6	60–140
von Willebrand factor (%)	**166.3 ± 82.4 ***	110 ± 20.8	60–140
TAT complex (μg/L)	**7.0 ± 3.7 ***	3.2 ± 0.6	2.0–4.2
EV procoagulant activity (nM PS)	**28.0 ± 12.2 ***	20.3 ± 8.5	
	Hemolysis and Redox Status		
Hemolysis (%)	**0.20 ± 0.09 ***	0.09 ± 0.08	
Osmotic fragility (% [NaCl])	**0.33 ± 0.04 ***	0.46 ± 0.02	
Intracellular ROS (MFI)	**744.1 ± 258.5 ***	480.6 ± 211.2	
Plasma TAC (μM Fe^{2+})	**782.6 ± 184.9 ***	506.3 ± 109.4	
Plasma UAdAC (μM Fe^{2+})	**438.2 ± 106.9 ***	306.2 ± 125.1	
Plasma UAiAC (μM Fe^{2+})	**344.4 ± 166.2 ***	200.0 ± 59.8	

Values are presented as mean ± SD. Bold, (*): $p < 0.05$ patients vs. controls. APTT: activated partial thromboplastin time; A.U.: arbitrary units; EV: extracellular vesicles; HDL: high-density lipoproteins; LDL: low-density lipoproteins; MCH: mean corpuscular hemoglobin; MCHC: mean corpuscular hemoglobin concentration; MCV: mean corpuscular volume; MFI: mean fluorescence intensity; PS: phosphatidylserine; RDW: red cell distribution width; ROS: reactive oxygen species; SGOT: serum glutamic-oxaloacetic transaminase; SGPT: serum glutamate-pyruvate transaminase; TAC: total antioxidant capacity; TAT complex: thrombin–antithrombin complex; UAdAC: uric-acid-dependent antioxidant capacity; UAiAC: uric-acid-independent antioxidant capacity.

3.2. Glutamine-Based Categorization

The patient cohort varied in terms of therapy; therefore, the next step of the present study was to evaluate the effect of the dose of orally consumed L-glutamine to hydroxyurea-treated SCD patient samples. Transfusion-dependent patients were excluded from this categorization so as to consider only the effect of the dose of L-glutamine supplementation. The subgroup that consumed 15 g or more of glutamine per day presented elevated neutrophils and monocytes but decreased lymphocytes (Table 2). Regarding RBC values, mean corpuscular hemoglobin (MCH) and mean corpuscular volume (MCV) were reduced in the high-intake group, along with the presence of nucleated erythrocytes. The levels of inflammation-related molecules, like iron, ferritin, and immunoglobulins, were diminished in the same group, while markers of hemolysis, including lactate dehydrogenase and hemolysis percentage, presented a non-statistically significant trend for lower levels ($0.05 < p < 0.10$). It should be noted that the levels of HbF were decreased after supplementation with >15 g glutamine. Interestingly, the thrombin–antithrombin complex, fibrinogen, and D-dimers were decreased in the high-intake cohort, a result that also arose in oxidative stress parameters, like ROS accumulation and membrane-bound hemoglobin dimers (Table 2).

Table 2. Variations in SCD hydroxyurea-administered patients upon additional supplementation with glutamine of less or more than 15 g/day.

	<15 g/day (*n* = 10)	**≥15 g/day (*n* = 9)**
Age (years)	50.5 ± 14.7	40.3 ± 14.7
General Blood Test		
White blood cells (×10^3/μL)	5.7 ± 1.7	6.1 ± 2.0
Neutrophils (%)	**55.4 ± 7.0** *	65.8 ± 7.3
Lymphocytes (%)	**33.9 ± 6.3** *	22.8 ± 5.8 (33%)
Monocytes (%)	**5.2 ± 1.1** *	6.7 ± 1.1
Eosinophils (%)	2.5 ± 1.0	2.4 ± 1.4
Basophils (%)	0.7 ± 0.2	0.8 ± 0.3
Neutrophil/lymphocyte ratio	**1.7 ± 0.7** *	3.1 ± 1.2
Red blood cells (×10^6/μL)	3.4 ± 0.3 (100%)	4.0 ± 1.2 (55%)
Hemoglobin (g/dL)	10.2 ± 1.1 (90%)	10.2 ± 2.2 (66%)
Hematocrit (%)	32.5 ± 2.4 (100%)	32.4 ± 6.6 (66%)
Mean corpuscular volume (MCV; fL)	**97.0 ± 11.3** * (30%)	82.8 ± 9.8 (66%)
Mean corpuscular hemoglobin (MCH; pg)	**30.5 ± 4.5** * (40%)	26.0 ± 2.8 (66%)
MCH concentration (MCHC; gr/dL)	31.4 ± 1.2 (80%)	31.5 ± 1.6 (77%)
Red cell distribution width (RDW; %)	19.4 ± 2.4 (100%)	18.6 ± 2.7 (100%)
Platelets (×10^3/μL)	351.7 ± 263.3 (20%)	306.9 ± 201.9 (77%)
Mean platelet volume (MPV; fL)	10.0 ± 1.2 (10%)	9.6 ± 1.2
Reticulocyte count (%)	7.1 ± 1.6 (100%)	8.4 ± 6.7 (100%)
Nucleated red blood cells (%)	**19.8 ± 9.8** *	9.8 ± 7.3
HbS (%)	68.9 ± 6.4 (100%)	73.5 ± 9.3 (100%)
HbF (%)	**23.2 ± 7.9** * (100%)	10.3 ± 9.8 (100%)
Serum Biochemical Analysis		
Glucose (mg/dL)	88.7 ± 9.6	84.4 ± 6.8
Urea (mg/dL)	27.5 ± 17.2 (20%)	21.6 ± 7.7 (33%)
Creatinine (mg/dL)	0.78 ± 0.12 (20%)	0.73 ± 0.13 (22%)
Uric acid (mg/dL)	5.2 ± 0.7	5.3 ± 1.2 (11%)
Cholesterol (mg/dL)	156.9 ± 24.4	153.3 ± 27.9
Triglycerides (mg/dL)	**136.0 ± 30.2** * (20%)	91.2 ± 47.1 (11%)
Calcium (mg/dL)	**9.6 ± 0.8** * (20%)	8.9 ± 0.5 (11%)
Phosphorus (mg/dL)	3.5 ± 0.3	3.4 ± 0.4
Potassium (mmol/L)	4.4 ± 0.3	4.2 ± 0.3
Sodium (mmol/L)	139.1 ± 1.4	140.0 ± 1.2
Chlorine (mmol/L)	105.8 ± 1.5	106.3 ± 1.4
Magnesium (mg/dL)	2.0 ± 0.6 (10%)	2.0 ± 0.1
Iron (mg/dL)	**106.5 ± 29.1** *	69.4 ± 26.5 (22%)
Ferritin (ng/mL)	**338.8 ± 152.5** * (70%)	56.5 ± 36.0 (11%)
B12 (pg/mL)	269.1 ± 85.0 (20%)	311.2 ± 98.7
Folate (ng/mL)	20.5 ± 16.6 (50%)	25.1 ± 14.5 (55%)
Proteins (mg/dL)	**7.8 ± 0.3** *	7.0 ± 0.4
Albumin (g/dL)	4.7 ± 0.2	4.5 ± 0.4
Serum glutamic-oxaloacetic transaminase (SGOT; U/L)	27.8 ± 6.8 (10%)	26.4 ± 14.2 (11%)
Serum glutamate-pyruvate transaminase (SGPT; U/L)	19.8 ± 7.3	22.4 ± 15.2
Gamma-glutamyl transferase (U/L)	20.1 ± 14.1	34.5 ± 15.6
Alkaline phosphatase (ALP; U/L)	76.1 ± 14.6	86.7 ± 24.8
High-density lipoproteins (HDL; mg/dL)	41.5 ± 7.8 (80%)	39.7 ± 9.0 (66%)
Low-density lipoproteins (LDL; mg/dL)	88.5 ± 20.1 (10%)	95.6 ± 22.3 (33%)
Immunoglobulins (g/dL)	**3.2 ± 0.4** *	2.5 ± 0.6
Total bilirubin (mg/dL)	1.9 ± 0.9 (80%)	1.7 ± 0.8 (66%)

Table 2. *Cont.*

	<15 g/day (n = 10)	≥15 g/day (n = 9)
Indirect bilirubin (mg/dL)	1.3 ± 0.8 (50%)	1.1 ± 0.7 (44%)
Direct bilirubin (mg/dL)	0.65 ± 0.13 (100%)	0.58 ± 0.16 (100%)
Lactate dehydrogenase (IU/L)	284.5 ± 57.9 (80%)	251.8 ± 27.8 (77%)
Creatine phosphokinase total (IU/L)	33.4 ± 19.8 (50%)	47.0 ± 35.6 (11%)
Vitamin D (ng/mL)	21.7 ± 6.8 (70%)	22.9 ± 7.8 (66%)
C-reactive protein (mg/L)	4.3 ± 2.4 (10%)	6.7 ± 6.4 (33%)
Hemostasis–Coagulation Parameters		
Prothrombin time INR	1.02 ± 0.06 (10%)	1.08 ± 0.12 (22%)
Activated partial thromboplastin time (APTT; s)	28.2 ± 2.0	29.1 ± 2.9
Fibrinogen (mg/dL)	**364.0 ± 79.5** * (40%)	257.1 ± 104.6 (55%)
D-dimer (μg/L)	**1187.6 ± 360.9** * (100%)	693.1 ± 307.1 (55%)
Factor VIII (%)	78.2 ± 23.4 (10%)	75.3 ± 33.4 (33%)
von Willebrand factor (%)	126.8 ± 21.2 (20%)	141.1 ± 84.2 (33%)
Thrombin–antithrombin complex (μg/L)	**5.8 ± 1.4** * (50%)	4.3 ± 1.3 (22%)
EV procoagulant activity (nM PS)	22.5 ± 7.6	27.2 ± 6.3
Hemolysis and Redox Status		
Hemolysis (%)	0.21 ± 0.08	0.15 ± 0.05
Osmotic fragility (% [NaCl])	0.33 ± 0.06	0.34 ± 0.02
Intracellular reactive oxygen species (MFI)	**880.5 ± 194.9** *	585.4 ± 231.4
Plasma TAC (μM Fe^{2+})	778.7 ± 177.5	745.2 ± 211.2
Plasma UadAC (μM Fe^{2+})	417.1 ± 99.7	435.0 ± 95.0
Plasma UaiAC (μM Fe^{2+})	361.5 ± 158.2	310.2 ± 225.9
Membrane-bound hemoglobin dimers (A.U.)	**1.51 ± 0.89** *	0.64 ± 0.46

Values are presented as mean ± SD. Bold, (*): $p < 0.05$ glutamine administration <15 g/day vs. glutamine administration ≥15 g/day. The percentages in parentheses indicate the proportion of patient samples with values outside the normal range. A.U.: arbitrary units; EV: extracellular vesicles; MFI: mean fluorescence intensity; PS: phosphatidylserine; TAC: total antioxidant capacity; UAdAC: uric-acid-dependent antioxidant capacity; UAiAC: uric-acid-independent antioxidant capacity.

3.3. RDW-Based Categorization

Through clustering analyses, we sought to categorize all SCD patients, regardless of the choice of therapy, based on one of the parameters measured. The analysis outcome revealed that RDW had good potential in dichotomizing our heterogeneous patient group (Figure 1). A cut-off value of RDW = 19% split our cohort into two distinct subgroups with minimal overlap (Figure 1) and significant differences. To initially examine the effect of L-glutamine consumption upon RDW stratification, we checked the presence of high- and low-dose glutamine recipients in the new cohorts. There was representation of both L-glutamine groups in both RDW categories (six low- and four high-dose in the high-RDW group vs. four low- and five high-dose in the low-RDW group, dose of glutamine: 13.5 ± 6.3 vs. 16.7 ± 7.1 g, respectively, p = 0.315).

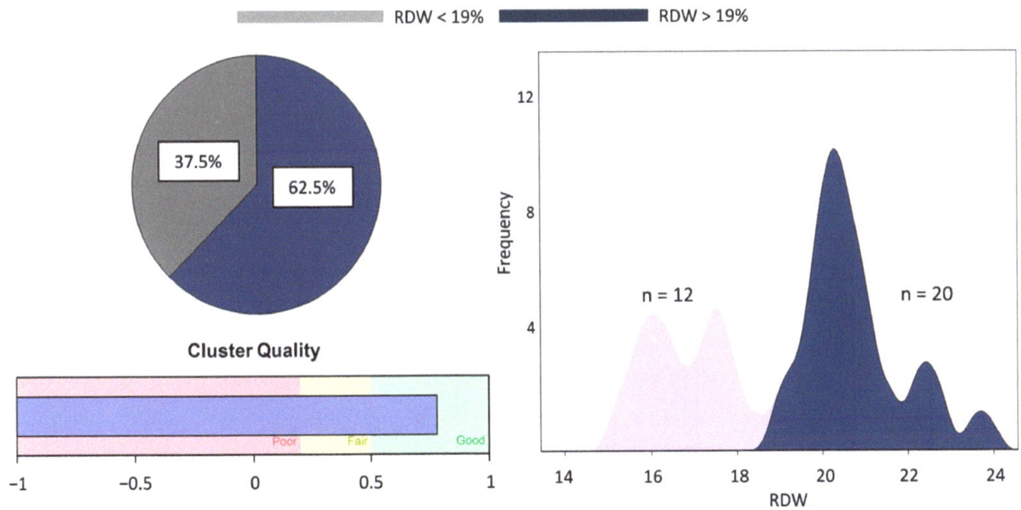

Figure 1. Cluster analysis revealed two subgroups stratified by RDW values.

Regarding RBC parameters, the levels of intracellular Hb were lower in the group of RDW > 19% (10.8 ± 1.8 vs. 9.5 ± 1.1 g/dL, lower RDW vs. higher RDW, $p = 0.017$), in contrast to the percentages of reticulocytes and nucleated RBCs, which exhibited the opposite result (reticulocytes: 4.3 ± 1.1 vs. 9.4 ± 4.2%; nucleated RBCs: 5.4 ± 5.1 vs. 24.7 ± 15.2%, lower RDW vs. higher RDW, $p = 0.01$ and 0.005, respectively). While bilirubin did not differ between the two groups, the levels of hemolysis and serum lactate dehydrogenase were higher in the patients with increased RDW (Figure 2A). Osmotic hemolysis presented lower values in the same group (Figure 2A), while extracellular K^+ was elevated (4.1 ± 0.2 vs. 4.4 ± 0.3 mmol/L, lower RDW vs. higher RDW, $p = 0.006$). Oxidative stress markers, like intracellular ROS accumulation and the binding of Hb to the membrane and the cytoskeleton, were also elevated in patients of higher RDW compared to their counterparts (Figure 2B). Nonetheless, plasma antioxidant capacity was similar between the two groups (Figure 2B).

Hemostasis and coagulation parameters were notably dissimilar between the two groups, with most of them, including platelets, von Willebrand factor, fibrinogen, and D-dimers, exhibiting increased values in the higher-RDW group (Figure 3). Of note, neither the prothrombin time nor the activated partial thromboplastin time were altered (Figure 3).

The white blood cell count was higher in the group with increased RDW, with monocytes following the same pattern but the neutrophil to lymphocyte ratio being inferior compared to the lower-RDW group (Figure 4). In the same context, markers of inflammation, like C-reactive protein and globulins, were elevated in the group of RDW > 19%, even though interleukin-6 was unaffected (Figure 4). Finally, the levels of serum albumin were lower in the same group (4.7 ± 0.3 vs. 4.3 ± 0.3 g/dL, lower RDW vs. higher RDW, $p = 0.003$).

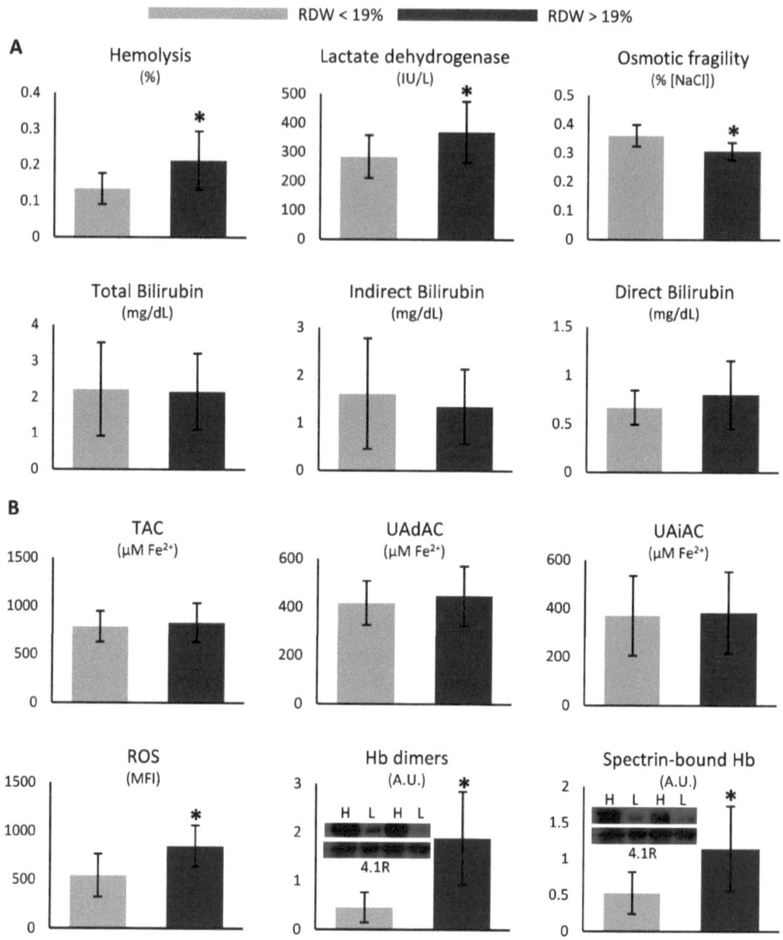

Figure 2. Differences in (**A**) hemolysis markers and (**B**) redox parameters between SCD samples stratified by their RDW values. Inserts: immunoblots of selected samples for high (H) and low (L) RDW subgroups. The 4.1R protein was used as an internal loading control. TAC: total antioxidant capacity; UAdAC: uric-acid-dependent antioxidant capacity; UAiAC: uric-acid-independent antioxidant capacity; ROS: reactive oxygen species; MFI: mean fluorescence intensity; Hb: hemoglobin. (*) $p < 0.05$.

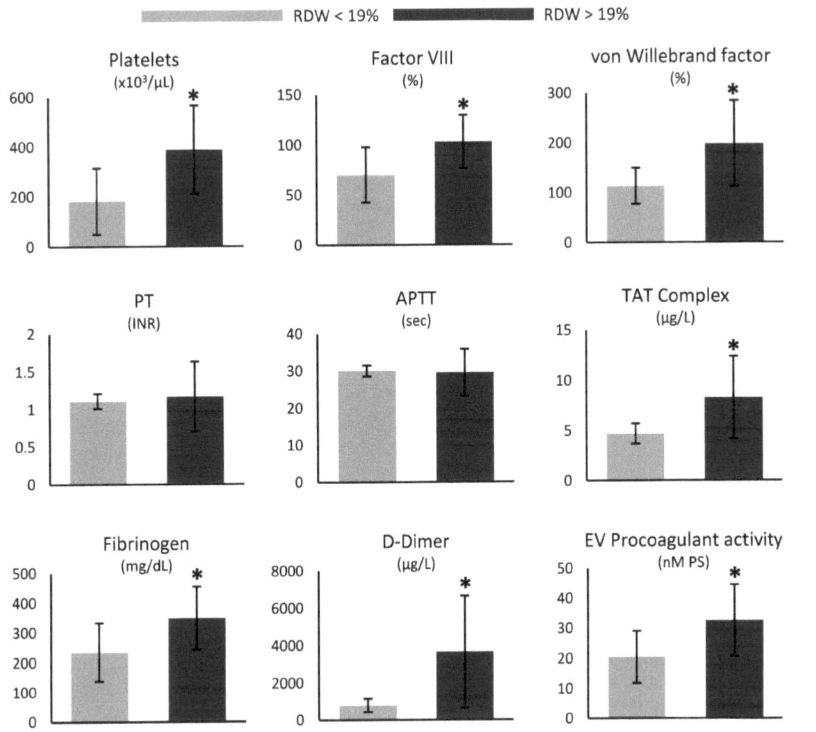

Figure 3. Differences in hemostasis and coagulation parameters between SCD samples stratified by their RDW values. PT: prothrombin time; INR: international normalized ratio; APTT: activated partial thromboplastin time; TAT: thrombin–antithrombin; EV: extracellular vesicles; PS: phosphatidylserine. (*) $p < 0.05$.

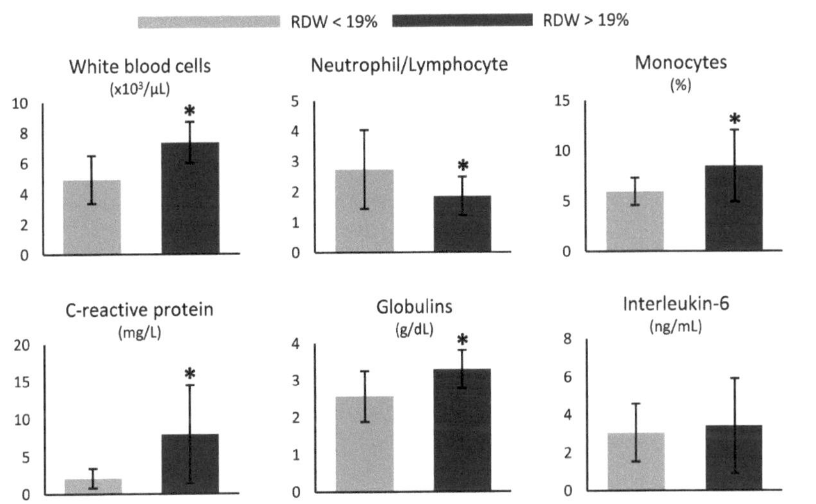

Figure 4. Differences in indicators of inflammation between SCD samples stratified by their RDW values. (*) $p < 0.05$.

4. Discussion

Compound heterozygotes are a highly heterogeneous group of SCD patients, mainly owing to the nature of the second mutation in β-genes that accompanies β^S. The combination of the alleles is expected to influence both hallmarks of the disease's manifestations, as well as the overall clinical picture. In this context, we hereby present (a) data regarding the impact of the dose of glutamine consumption upon markers of oxidation, inflammation, and coagulation and (b) the potential of RDW to categorize heterogeneous patient cohorts with regard to familiar features of SCD-related phenomena, such as free hemoglobin, oxidative stress, hemostasis, and pro-inflammatory molecules. Both analyses of the research were carried out in a compact patient cohort that was distinguished from healthy controls as expected.

4.1. Higher Dose of L-Glutamine Is Protective in Terms of Oxidation, Coagulation, and Inflammation

The first drug to be approved for SCD was hydroxyurea, which leads to the elevation of HbF levels, diluting, in this way, HbS and decreasing the probability of its polymerization [17]. In 2017, L-glutamine was also approved as an SCD medication, and it attenuates the disease's symptoms when administered either alone or in combination with hydroxyurea [18]. Glutamine can strengthen the erythrocytic antioxidant power by contributing to glutathione synthesis, as well as the synthesis of nicotinamide adenine dinucleotide (NAD^+) and its reduced form, NADH [19]. In this context, the currently presented low oxidative markers in the group that was characterized by higher doses of L-glutamine were rather anticipated.

While the introduction of L-glutamine as an SCD therapeutic agent was based on its potential antioxidant effect, it should be noted that L-glutamine is a quite versatile amino acid. The orally administered L-glutamine supports L-arginine production, the bioavailability of which is reduced in SCD patients but is vital for the formation of nitric oxide (NO) [19]. Among its other roles, NO inhibits hemostatic activation and adhesion molecules [20]. Interestingly, in our study, some parameters of coagulation and fibrinolysis, like the thrombin–antithrombin complex and D-dimers, were reduced after higher L-glutamine supplementation. Accordingly, SCD patients treated with oral L-glutamine present reduced adhesion of sickle erythrocytes to endothelial cells [21]. In a different disease setting, namely hypertension, there are drugs that lead to a D-dimer decrease and, at the same time, NO elevation [22]. Whether the production of NO after glutamine administration is responsible for the current observation is not yet clear.

It should not be omitted that glutamine is an important amino acid for the metabolism of immune cells and a significant modulator of leukocyte function. Among white blood cells, neutrophils –increased in the current study– consume glutamine at the highest rates [23], to support their energy and redox metabolism. Nonetheless, glutamine is also known for its anti-inflammatory roles, since it suppresses the NFκB and STAT pro-inflammatory pathways [24]; therefore, in a disease characterized by a steady inflammatory state, like SCD, its administration could be beneficial in terms of inflammatory phenotypes. Indeed, our results support a slightly differentiated inflammation profile when glutamine is present in higher doses, as evidenced by the lower levels of lymphocytes, and immunoglobulins, which might both be increased in the specific disease setting [25,26], as well as the decreased levels of inflammation markers like ferritin.

While hemolysis parameters did not differ significantly between the two subgroups, there was a trend for improved lysis profiles after high-dose glutamine consumption, a finding consistent with a recent study that examined the effect of prolonged glutamine administration on the clinical, hematological, and biochemical features of SCD patients [27]. Another indicator of disease severity that was significantly reduced was the presence of nucleated RBCs, which reflects increased erythropoietic drive or bone marrow necrosis [28]. On the other hand, the observed lower levels of HbF in the higher-dose group are a finding that needs further investigation, since, to our knowledge, there is not a reported relation between glutamine administration and HbF production. Of note, the secondary outcomes of a recent clinical trial [18] regarding glutamine supplementation/HbF production are still anticipated.

4.2. Increased RDW Is Linked to Markers of Disease Severity

RDW is a hematological parameter that seems to characterize systemic abnormalities, since it has been suggested as a potential biomarker in several pathological states, including cardiovascular diseases [29], autoimmunity [30], and cancer [31]. Sickle cell disease is not absent from this list, since, in $\beta^S\beta^S$ patients, low RDW has been found to be associated with milder clinical manifestations and improved hemoglobin parameters [32]. It was recently suggested that oxidative stress affects RDW [33]; therefore, redox disequilibrium could lie behind the constant emergence of the biomarker potential of this parameter.

In fact, in the current study, oxidative stress, as evidenced by ROS accumulation and Hb binding to the membrane, was increased in the higher-RDW group. Intracellular ROS have a significant impact on the physiology of RBCs. The membrane and the cytoskeleton are the two main sites of oxidative stress manifestation, through the carbonylation of proteins, peroxidation of lipids, and attachment of hemichromes [34]. All these make RBCs more prone to lysis, in consistency with our current hemolysis findings. Notably, osmotically induced hemolysis was inversely associated with the levels of RDW. It is well known that in hemoglobinopathies, such as thalassemic syndromes and SCD, the affected cells, mostly in the disease state but also in heterozygous conditions, are less fragile following osmotic stress. Moreover, at least in the beta-thalassemia trait, the higher the severity of the mutation, the more osmotically resistant RBCs are [35], while the irreversibly sickled cells have been linked to lower osmotic fragility in SCD due to their excessive damage [36].

The RDW index was recently suggested to be associated with the interactions of RBCs with vessels [37]. The size of erythrocytes, as indicated by RDW, may contribute to the platelet margination effect by altering the dynamic behavior in the bulk flow [37,38]. Platelets can then interact with adhesive proteins, including von Willebrand factor, aggregate, and become activated. The initiation of the coagulation cascade, supported by the elevated levels of factor VIII, thrombin, and fibrinogen, can then lead to the formation of thrombus, which can be then lysed via fibrinolysis, as indicated by the increased levels of D-dimers. In addition, the higher hemolysis in those with increased anisocytosis can lead to the generation of free heme [39], which can act in a prothrombotic way, as recently presented in detail [40]. Lastly, the increased levels of extracellular vesicles that expose phosphatidylserine on their exterior can also activate the coagulation cascade due to the negatively charged surface that enhances the activation of prothrombin to thrombin [41].

In previous studies, RDW has been closely associated with inflammation markers, like interleukins and C-reactive protein [42]. Accordingly, in the current study, the high-RDW group was associated with a slight increase in some inflammation markers, like leukocyte counts, C-reactive protein, and immunoglobulins. Monocytes, which have been reported to be elevated in SCD and were also increased in the currently presented high-RDW subgroup, are often modulated into inflammatory monocytes [43], which intensifies the production of pro-inflammatory molecules in SCD. Inflammation and the altered macrophages can alter both erythropoiesis and erythrocyte survival, which can lead to a mixed RBC population in circulation, mirrored by RDW [44]. Consequently, the modified inflammatory profile could be attributed to the increased extracellular hemoglobin (which acts as a damage-associated molecular pattern [45]), the high oxidative stress [46], or the activation of hemostasis [47] in the high-anisocytosis group, and its occurrence might positively feedback the observed size heterogeneity in the circulating RBCs [48]. To support this, the increased presence of nucleated RBCs and reticulocytes in the high-RDW group is suggestive of stress erythropoiesis [44].

Our study has several limitations. First of all, the a posteriori stratification used both for L-glutamine and RDW renders our findings essentially descriptive. In fact, and since the presented study was not designed as a clinical trial, the examined groups were not constructed beforehand but the patients were only studied for a period during which some of them were already under oral L-glutamine consumption, hydroxyurea treatment, or transfusion therapy. In this context, the fact that there are no data regarding the period before the start of L-glutamine consumption limits the potential to better (a) appreciate the effect of glutamine administration and (b) evaluate dose-dependent outcomes. Such information would be of great value for future studies, and especially randomized clinical trials, since it would support the evaluation of the patients' cohorts both before and after supplement administration, instead of reporting a snapshot. However, we strongly believe that, in light of the few studies that focus on L-glutamine in the context of SCD, and especially the less studied group of $\beta^S\beta^+$ compound heterozygotes, the reported data remain valuable and challenging and can give insight into the biochemical and physiological profiles of these patients in relation to food supplement consumption.

5. Conclusions

The present study suggests that glutamine in combination with hydroxyurea has a beneficial effect on $\beta^S\beta^+$ blood parameters, probably due to its implication in enhancing the antioxidant defenses. Moreover, despite the high variability in several parameters, RDW seems to adequately subcategorize patients, with its levels being related to well-known contributors to the disease, such as hemolysis, oxidative stress, inflammation, and coagulation. While it is true that "RDW cannot be regarded as the 'panacea' of the new century", as perfectly stated by Lippi and Plebani [49], it seems to be an informative parameter in heterogeneous conditions such as sickle cell disease.

Author Contributions: Conceptualization, A.G.K.; methodology, H.T.G., S.P.F. and V.L.T.; validation, S.P.F., A.T.A., V.L.T. and A.G.K.; formal analysis, A.G., H.T.G., A.T.A., E.G.P. (Effie G. Papageorgiou) and V.L.T.; investigation, A.G., H.T.G., S.P.F. and E.G.P. (Efthimia G. Pavlou); resources, A.G., E.G.P. (Efthimia G. Pavlou), E.G.N., M.P.D., A.K., A.X. and A.G.K.; writing—original draft preparation, A.G., H.T.G., S.P.F., A.T.A., V.L.T. and A.G.K.; writing—review and editing, E.G.P. (Efthimia G. Pavlou), E.G.N., M.P.D., A.K., A.X., C.F. and E.G.P. (Effie G. Papageorgiou); visualization, A.T.A. and V.L.T.; supervision, C.F., E.G.P. (Effie G. Papageorgiou) and A.G.K.; project administration, A.G.K. All authors have read and agreed to the published version of the manuscript.

Funding: The "Biomedical Methods and Technology in Diagnosis" graduate program funded part of the consumables used in this research (3/02-12-2022).

Institutional Review Board Statement: This study was approved by the Research Ethics Committee of the University of West Attica (UniWA) (approval number: 51108/01-07-2021). The research was conducted ethically, with all study procedures being performed in accordance with the requirements of the World Medical Association's Declaration of Helsinki.

Informed Consent Statement: Informed consent was obtained from all subjects involved in the study.

Data Availability Statement: All data presented in this study are available upon request.

Acknowledgments: The authors would like to thank all patients that voluntarily participated in this study; Marinos Akridas, Sante Plus, Kallithéa, for the kind offer of the glutamine supplement; Maria Sioumala, Stavroula Daskalaki, Areti Giannioti and Maria Liosi for their contribution to blood sampling; and Marianna Antonelou for the help that she offered regarding the immunoblotting experiments. The authors would also like to thank the postgraduate program "Pedagogy through innovative Technologies and Biomedical approaches" for supporting this research. Finally, all authors would like to thank Sophia Delicou, Clinical Hematologist, Thalassemia and Sickle Cell Unit, Expertise Center of Hemoglobinopathies and Their Complications, Hippokrateio General Hospital, for her help and critical comments on this publication.

Conflicts of Interest: The authors declare no conflict of interest.

References

1. Li, X.; Dao, M.; Lykotrafitis, G.; Karniadakis, G.E. Biomechanics and biorheology of red blood cells in sickle cell anemia. *J. Biomech.* **2017**, *50*, 34–41. [CrossRef] [PubMed]
2. Rees, D.C.; Williams, T.N.; Gladwin, M.T. Sickle-cell disease. *Lancet* **2010**, *376*, 2018–2031. [CrossRef] [PubMed]
3. Driss, A.; Asare, K.O.; Hibbert, J.M.; Gee, B.E.; Adamkiewicz, T.V.; Stiles, J.K. Sickle Cell Disease in the Post Genomic Era: A Monogenic Disease with a Polygenic Phenotype. *Genom. Insights* **2009**, *2009*, 23–48. [CrossRef]
4. Alapan, Y.; Matsuyama, Y.; Little, J.A.; Gurkan, U.A. Dynamic deformability of sickle red blood cells in microphysiological flow. *Technology* **2016**, *4*, 71–79. [CrossRef]
5. Knisely, M.R.; Tanabe, P.J.; Walker, J.K.L.; Yang, Q.; Shah, N.R. Severe Persistent Pain and Inflammatory Biomarkers in Sickle Cell Disease: An Exploratory Study. *Biol. Res. Nurs.* **2022**, *24*, 24–30. [CrossRef] [PubMed]
6. White, J.; Callaghan, M.U.; Gao, X.; Liu, K.; Zaidi, A.; Tarasev, M.; Hines, P.C. Longitudinal assessment of adhesion to vascular cell adhesion molecule-1 at steady state and during vaso-occlusive crises in sickle cell disease. *Br. J. Haematol.* **2022**, *196*, 1052–1058. [CrossRef]
7. Sangokoya, C.; Telen, M.J.; Chi, J.T. microRNA miR-144 modulates oxidative stress tolerance and associates with anemia severity in sickle cell disease. *Blood* **2010**, *116*, 4338–4348. [CrossRef]
8. Gbotosho, O.T.; Kapetanaki, M.G.; Kato, G.J. The Worst Things in Life are Free: The Role of Free Heme in Sickle Cell Disease. *Front. Immunol.* **2020**, *11*, 561917. [CrossRef]
9. Du, M.; Van Ness, S.; Gordeuk, V.; Nouraie, S.M.; Nekhai, S.; Gladwin, M.; Steinberg, M.H.; Sebastiani, P. Biomarker signatures of sickle cell disease severity. *Blood Cells Mol. Dis.* **2018**, *72*, 1–9. [CrossRef]
10. Tumburu, L.; Ghosh-Choudhary, S.; Seifuddin, F.T.; Barbu, E.A.; Yang, S.; Ahmad, M.M.; Wilkins, L.H.W.; Tunc, I.; Sivakumar, I.; Nichols, J.S.; et al. Circulating mitochondrial DNA is a proinflammatory DAMP in sickle cell disease. *Blood* **2021**, *137*, 3116–3126. [CrossRef]
11. Moriconi, C.; Dzieciatkowska, M.; Roy, M.; D'Alessandro, A.; Roingeard, P.; Lee, J.Y.; Gibb, D.R.; Tredicine, M.; McGill, M.A.; Qiu, A.; et al. Retention of functional mitochondria in mature red blood cells from patients with sickle cell disease. *Br. J. Haematol.* **2022**, *198*, 574–586. [CrossRef] [PubMed]
12. Esperti, S.; Nader, E.; Stier, A.; Boisson, C.; Carin, R.; Marano, M.; Robert, M.; Martin, M.; Horand, F.; Cibiel, A.; et al. Increased retention of functional mitochondria in mature sickle red blood cells is associated with increased sickling tendency, hemolysis and oxidative stress. *Haematologica* **2023**, *108*, 11. [CrossRef] [PubMed]
13. Serjeant, G.R.; Serjeant, B.E.; Fraser, R.A.; Hambleton, I.R.; Higgs, D.R.; Kulozik, A.E.; Donaldson, A. Hb S-beta-thalassemia: Molecular, hematological and clinical comparisons. *Hemoglobin* **2011**, *35*, 1–12. [CrossRef] [PubMed]
14. Notarangelo, L.D.; Agostini, A.; Casale, M.; Samperi, P.; Arcioni, F.; Gorello, P.; Perrotta, S.; Masera, N.; Barone, A.; Bertoni, E.; et al. HbS/beta+ thalassemia: Really a mild disease? A National survey from the AIEOP Sickle Cell Disease Study Group with genotype-phenotype correlation. *Eur. J. Haematol.* **2020**, *104*, 214–222. [CrossRef]
15. Benzie, I.F.; Strain, J.J. The ferric reducing ability of plasma (FRAP) as a measure of "antioxidant power": The FRAP assay. *Anal. Biochem.* **1996**, *239*, 70–76. [CrossRef]
16. Pavlou, E.G.; Georgatzakou, H.T.; Fortis, S.P.; Tsante, K.A.; Tsantes, A.G.; Nomikou, E.G.; Kapota, A.I.; Petras, D.I.; Venetikou, M.S.; Papageorgiou, E.G.; et al. Coagulation Abnormalities in Renal Pathology of Chronic Kidney Disease: The Interplay between Blood Cells and Soluble Factors. *Biomolecules* **2021**, *11*, 1309. [CrossRef]
17. McGann, P.T.; Ware, R.E. Hydroxyurea therapy for sickle cell anemia. *Expert Opin. Drug Saf.* **2015**, *14*, 1749–1758. [CrossRef]
18. Niihara, Y.; Miller, S.T.; Kanter, J.; Lanzkron, S.; Smith, W.R.; Hsu, L.L.; Gordeuk, V.R.; Viswanathan, K.; Sarnaik, S.; Osunkwo, I.; et al. A Phase 3 Trial of l-Glutamine in Sickle Cell Disease. *N. Engl. J. Med.* **2018**, *379*, 226–235. [CrossRef]
19. Sadaf, A.; Quinn, C.T. L-glutamine for sickle cell disease: Knight or pawn? *Exp. Biol. Med.* **2020**, *245*, 146–154. [CrossRef]
20. Goncalves, R.P. Nitric oxide status in sickle cell anemia. *Rev. Bras. Hematol. Hemoter.* **2012**, *34*, 255. [CrossRef]
21. Niihara, Y.; Matsui, N.M.; Shen, Y.M.; Akiyama, D.A.; Johnson, C.S.; Sunga, M.A.; Magpayo, J.; Embury, S.H.; Kalra, V.K.; Cho, S.H.; et al. L-glutamine therapy reduces endothelial adhesion of sickle red blood cells to human umbilical vein endothelial cells. *BMC Blood Disord.* **2005**, *5*, 4. [CrossRef] [PubMed]
22. Zhang, W.R.; Sun, M.; Luo, J.K. Serum nitric oxide and D-dimer before and after administering antihypertensive drugs in essential hypertension. *Hunan Yi Ke Da Xue Xue Bao Hunan Yike Daxue Xuebao Bull. Hunan Med. Univ.* **2003**, *28*, 382–384.
23. Pithon-Curi, T.C.; De Melo, M.P.; Curi, R. Glucose and glutamine utilization by rat lymphocytes, monocytes and neutrophils in culture: A comparative study. *Cell Biochem. Funct.* **2004**, *22*, 321–326. [CrossRef] [PubMed]
24. Liboni, K.C.; Li, N.; Scumpia, P.O.; Neu, J. Glutamine modulates LPS-induced IL-8 production through IkappaB/NF-kappaB in human fetal and adult intestinal epithelium. *J. Nutr.* **2005**, *135*, 245–251. [CrossRef]
25. Balandya, E.; Reynolds, T.; Obaro, S.; Makani, J. Alteration of lymphocyte phenotype and function in sickle cell anemia: Implications for vaccine responses. *Am. J. Hematol.* **2016**, *91*, 938–946. [CrossRef]
26. Cherif-Alami, S.; Hau, I.; Arnaud, C.; Kamdem, A.; Coulon, B.; Idoux, E.; Bechet, S.; Creidy, R.; Bernaudin, F.; Epaud, R.; et al. Serum Immunoglobulin Levels in Children with Sickle Cell Disease: A Large Prospective Study. *J. Clin. Med.* **2019**, *8*, 1688. [CrossRef]
27. Elenga, N.; Loko, G.; Etienne-Julan, M.; Al-Okka, R.; Adel, A.M.; Yassin, M.A. Real-World data on efficacy of L-glutamine in preventing sickle cell disease-related complications in pediatric and adult patients. *Front. Med.* **2022**, *9*, 931925. [CrossRef]

28. Ballantine, J.D.; Kwon, S.; Liem, R.I. Nucleated Red Blood Cells in Children With Sickle Cell Disease Hospitalized for Pain. *J. Pediatr. Hematol. Oncol.* **2019**, *41*, e487–e492. [CrossRef] [PubMed]
29. Arkew, M.; Gemechu, K.; Haile, K.; Asmerom, H. Red Blood Cell Distribution Width as Novel Biomarker in Cardiovascular Diseases: A Literature Review. *J. Blood Med.* **2022**, *13*, 413–424. [CrossRef]
30. Vaya, A.; Alis, R.; Hernandez, J.L.; Calvo, J.; Mico, L.; Romagnoli, M.; Ricarte, J.M. RDW in patients with systemic lupus erythematosus. Influence of anaemia and inflammatory markers. *Clin. Hemorheol. Microcirc.* **2013**, *54*, 333–339. [CrossRef]
31. Huang, D.P.; Ma, R.M.; Xiang, Y.Q. Utility of Red Cell Distribution Width as a Prognostic Factor in Young Breast Cancer Patients. *Medicine* **2016**, *95*, e3430. [CrossRef] [PubMed]
32. Thame, M.; Grandison, Y.; Mason, K.; Thompson, M.; Higgs, D.; Morris, J.; Serjeant, B.; Serjeant, G. The red cell distribution width in sickle cell disease--is it of clinical value? *Clin. Lab. Haematol.* **1991**, *13*, 229–237. [CrossRef]
33. Joosse, H.J.; van Oirschot, B.A.; Kooijmans, S.A.A.; Hoefer, I.E.; van Wijk, R.A.H.; Huisman, A.; van Solinge, W.W.; Haitjema, S. In-vitro and in-silico evidence for oxidative stress as drivers for RDW. *Sci. Rep.* **2023**, *13*, 9223. [CrossRef] [PubMed]
34. Wang, Q.; Zennadi, R. The Role of RBC Oxidative Stress in Sickle Cell Disease: From the Molecular Basis to Pathologic Implications. *Antioxidants* **2021**, *10*, 1608. [CrossRef] [PubMed]
35. Anastasiadi, A.T.; Tzounakas, V.L.; Dzieciatkowska, M.; Arvaniti, V.Z.; Papageorgiou, E.G.; Papassideri, I.S.; Stamoulis, K.; D'Alessandro, A.; Kriebardis, A.G.; Antonelou, M.H. Innate Variability in Physiological and Omics Aspects of the Beta Thalassemia Trait-Specific Donor Variation Effects. *Front. Physiol.* **2022**, *13*, 907444. [CrossRef]
36. Figueiredo, M.S.; Zago, M.A. The role of irreversibly sickled cells in reducing the osmotic fragility of red cells in sickle cell anemia. *Acta Physiol. Pharmacol. Latinoam. Organo La Asoc. Latinoam. Cienc. Fisiol. Y La Asoc. Latinoam. Farmacol.* **1985**, *35*, 49–56.
37. Ananthaseshan, S.; Bojakowski, K.; Sacharczuk, M.; Poznanski, P.; Skiba, D.S.; Prahl Wittberg, L.; McKenzie, J.; Szkulmowska, A.; Berg, N.; Andziak, P.; et al. Red blood cell distribution width is associated with increased interactions of blood cells with vascular wall. *Sci. Rep.* **2022**, *12*, 13676. [CrossRef]
38. Weisel, J.W.; Litvinov, R.I. Red blood cells: The forgotten player in hemostasis and thrombosis. *J. Thromb. Haemost. JTH* **2019**, *17*, 271–282. [CrossRef]
39. Vallelian, F.; Buehler, P.W.; Schaer, D.J. Hemolysis, free hemoglobin toxicity, and scavenger protein therapeutics. *Blood* **2022**, *140*, 1837–1844. [CrossRef]
40. Mubeen, S.; Domingo-Fernandez, D.; Diaz Del Ser, S.; Solanki, D.M.; Kodamullil, A.T.; Hofmann-Apitius, M.; Hopp, M.T.; Imhof, D. Exploring the Complex Network of Heme-Triggered Effects on the Blood Coagulation System. *J. Clin. Med.* **2022**, *11*, 5975. [CrossRef]
41. Nader, E.; Garnier, Y.; Connes, P.; Romana, M. Extracellular Vesicles in Sickle Cell Disease: Plasma Concentration, Blood Cell Types Origin Distribution and Biological Properties. *Front. Med.* **2021**, *8*, 728693. [CrossRef]
42. Lippi, G.; Targher, G.; Montagnana, M.; Salvagno, G.L.; Zoppini, G.; Guidi, G.C. Relation between red blood cell distribution width and inflammatory biomarkers in a large cohort of unselected outpatients. *Arch. Pathol. Lab. Med.* **2009**, *133*, 628–632. [CrossRef]
43. Liu, Y.; Zhong, H.; Vinchi, F.; Mendelson, A.; Yazdanbakhsh, K. Patrolling monocytes in sickle cell hemolytic conditions. *Transfus. Clin. Biol. J. De La Soc. Fr. De Transfus. Sang.* **2019**, *26*, 128–129. [CrossRef] [PubMed]
44. Sesti-Costa, R.; Costa, F.F.; Conran, N. Role of Macrophages in Sickle Cell Disease Erythrophagocytosis and Erythropoiesis. *Int. J. Mol. Sci.* **2023**, *24*, 6333. [CrossRef]
45. Bozza, M.T.; Jeney, V. Pro-inflammatory Actions of Heme and Other Hemoglobin-Derived DAMPs. *Front. Immunol.* **2020**, *11*, 1323. [CrossRef]
46. Lugrin, J.; Rosenblatt-Velin, N.; Parapanov, R.; Liaudet, L. The role of oxidative stress during inflammatory processes. *Biol. Chem.* **2014**, *395*, 203–230. [CrossRef]
47. Foley, J.H.; Conway, E.M. Cross Talk Pathways Between Coagulation and Inflammation. *Circ. Res.* **2016**, *118*, 1392–1408. [CrossRef]
48. Paulson, R.F.; Ruan, B.; Hao, S.; Chen, Y. Stress Erythropoiesis is a Key Inflammatory Response. *Cells* **2020**, *9*, 634. [CrossRef]
49. Lippi, G.; Plebani, M. Red blood cell distribution width (RDW) and human pathology. One size fits all. *Clin. Chem. Lab. Med.* **2014**, *52*, 1247–1249. [CrossRef]

antioxidants

Review

The Role of Ergothioneine in Red Blood Cell Biology: A Review and Perspective

Tiffany A. Thomas [1], Richard O. Francis [1], James C. Zimring [2], Joseph P. Kao [3], Travis Nemkov [4] and Steven L. Spitalnik [1,*]

1 Laboratory of Transfusion Biology, Department of Pathology and Cell Biology, Columbia University Irving Medical Center, New York, NY 10032, USA; tt2254@cumc.columbia.edu (T.A.T.)
2 Department of Pathology, University of Virginia School of Medicine, Charlottesville, VA 22903, USA
3 Center for Biomedical Engineering, Department of Physiology, University of Maryland School of Medicine, Baltimore, MD 21201, USA
4 Department of Biochemistry and Molecular Genetics, University of Colorado Anschutz Medical Campus, Denver, CO 80203, USA
* Correspondence: ss2479@cumc.columbia.edu

Abstract: Oxidative stress can damage tissues and cells, and their resilience or susceptibility depends on the robustness of their antioxidant mechanisms. The latter include small molecules, proteins, and enzymes, which are linked together in metabolic pathways. Red blood cells are particularly susceptible to oxidative stress due to their large number of hemoglobin molecules, which can undergo auto-oxidation. This yields reactive oxygen species that participate in Fenton chemistry, ultimately damaging their membranes and cytosolic constituents. Fortunately, red blood cells contain robust antioxidant systems to enable them to circulate and perform their physiological functions, particularly delivering oxygen and removing carbon dioxide. Nonetheless, if red blood cells have insufficient antioxidant reserves (e.g., due to genetics, diet, disease, or toxin exposure), this can induce hemolysis in vivo or enhance susceptibility to a "storage lesion" in vitro, when blood donations are refrigerator-stored for transfusion purposes. Ergothioneine, a small molecule not synthesized by mammals, is obtained only through the diet. It is absorbed from the gut and enters cells using a highly specific transporter (i.e., SLC22A4). Certain cells and tissues, particularly red blood cells, contain high ergothioneine levels. Although no deficiency-related disease has been identified, evidence suggests ergothioneine may be a beneficial "nutraceutical." Given the requirements of red blood cells to resist oxidative stress and their high ergothioneine content, this review discusses ergothioneine's potential importance in protecting these cells and identifies knowledge gaps regarding its relevance in enhancing red blood cell circulatory, storage, and transfusion quality.

Keywords: ergothioneine; antioxidant; RBC; erythrocyte; ROS; nutraceutical; supplement

Citation: Thomas, T.A.; Francis, R.O.; Zimring, J.C.; Kao, J.P.; Nemkov, T.; Spitalnik, S.L. The Role of Ergothioneine in Red Blood Cell Biology: A Review and Perspective. *Antioxidants* **2024**, *13*, 717. https://doi.org/10.3390/antiox13060717

Academic Editor: Brandon Reeder

Received: 16 May 2024
Revised: 3 June 2024
Accepted: 10 June 2024
Published: 13 June 2024

1. Introduction

Red blood cells (RBCs) are critically important for organismal biology by, for example, delivering oxygen to multiple tissues and removing carbon dioxide [1]. To transport oxygen, every human RBC contains ~250 million hemoglobin molecules [2], each of which contains four heme groups, with each of those containing one atom of the ferrous form of iron caged within a porphyrin ring. Because oxygen binds directly to ferrous iron in hemoglobin, oxygen-rich hemoglobin is at risk of undergoing auto-oxidation, whereby oxygen extracts an extra electron, thus converting ferrous iron (i.e., Fe^{+2}) to its ferric (i.e., Fe^{+3}) state, which yields met-hemoglobin, while simultaneously producing superoxide. The latter sparks a series of reactions, exacerbated by the abundance of iron inside RBCs, to produce multiple reactive oxygen species (ROS) and other toxic metabolites, including the hydroxyl radical and hydrogen peroxide. Thus, during their circulatory lifespan (e.g., ~120 days in humans), RBCs are continuously exposed to significant oxidative stresses as a result of their normal biological function. Therefore, to protect their cytosolic components and plasma membrane integrity, and to reduce non-oxygen-transporting met-hemoglobin back to its functional form, evolution provided RBCs with multiple mechanisms to protect themselves from oxidative stress. These include small molecules (e.g., vitamin C, vitamin E, reduced glutathione (i.e., GSH)) and proteins (e.g., superoxide dismutase, glutathione peroxidase, catalase, peroxiredoxin 2), which can eliminate ROS by participating in specific metabolic pathways. In addition, multiple redundant pathways repair RBC lipids and proteins damaged by oxidant stress (for reviews, see [3–5]), and enzymes that maintain antioxidant capacity continually replenish these pathways (e.g., glutathione reductase). To this end, the relevant small molecules, proteins, enzymes, reactions, and metabolic pathways have been studied intensively for decades in an effort, not only to understand how RBCs experience and respond to oxidant stress in health and disease, but also to develop approaches to enable RBCs to avoid and resist oxidant stress and enhance their abilities to ameliorate it [6].

Among the myriad independent, overlapping, and redundant pathways that handle oxidant stress in RBCs, an interesting and unusual molecule was identified more than a century ago: ergothioneine (i.e., ESH; see Figure 1) [7]. Its primary function appears to be as an antioxidant, and it is found in multiple organisms, tissues, and cells. Yet, despite many studies of its structure, location, and function, multiple mysteries still remain. Although it is the subject of excellent general reviews [8–10], this current contribution focuses on its potential relevance and importance for mammalian, primarily human, RBC biology, for which, to the best of our knowledge, no specific and comprehensive review has yet been published.

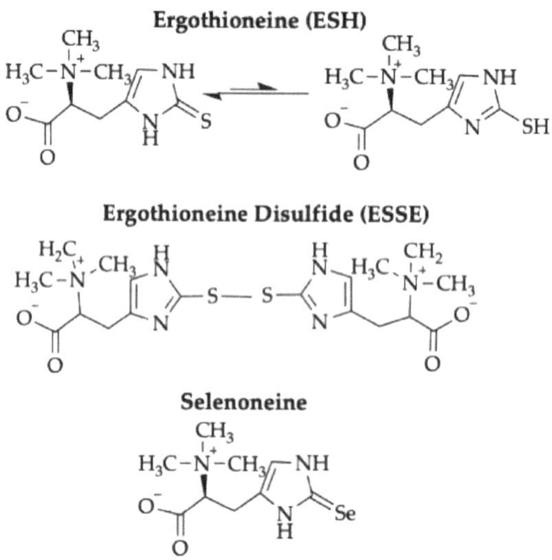

Figure 1. Structure of ergothioneine and related compounds. ESH has two tautomeric forms—thione (on the left) and thiol (on the right). Under physiological conditions, ESH primarily exists as its thione tautomer. The disulfide form, ESSE, is produced by the oxidation of ESH. Selenoneine, a selenium-containing analogue of ESH, is abundant in certain marine animals, is found in human RBCs, and may function as an antioxidant.

ESH is not synthesized by humans or other mammals, but rather, is incorporated from multiple dietary sources, including mushrooms, which have particularly high amounts (for a general review, see [11]). Nonetheless, high levels of ESH are found in multiple human tissues, including mature RBCs; this finding, along with the existence of a transporter, SLC22A4, which transports ESH with high specificity (see below), suggests that ESH has significant biological importance. However, no known disease-inducing deficiency states have yet been identified in humans or in animal models, although, even when ESH levels are "undetectable" in various animal models (see below), it is not clear that the compound is completely absent from the relevant tissues. Nonetheless, because no such deficiency disease has yet been identified, ESH is not classified as a "vitamin", although it may be characterized as a "nutraceutical" (see below). Indeed, given its apparent safety as a dietary supplement for humans [12,13], it has been proposed to have clinical utility in various settings [10], including aging [14,15], cognitive dysfunction [16], and infection [17], thereby suggesting its potential relevance as a nutraceutical. Its efficacy in such settings has been demonstrated in animal models, as expanded on below.

2. ESH in the Diet and Dietary Supplementation

ESH is not synthesized de novo by animals [18], but rather by fungi [19], principally certain species of mushrooms (e.g., king bolete and oyster mushrooms), although it can also be found in lower concentrations in multiple other foods including beans (e.g., black turtle beans, red kidney beans), grains, eggs, and meat [10,20]. In addition to ESH being obtained by dietary means, purified preparations of ESH are available that are suitable for human use [21].

Dietary supplementation with ESH has been studied in various animal models. For example, in mice, it increases levels in "whole blood" and other tissues [22]. In rats supplemented with tritium-labeled ESH, this compound becomes rapidly detectable in plasma and then undetectable; however, its levels are stable in "corpuscles", presumably RBCs (see below) [23]. In addition, little, or no, tritiated ESH disappeared from the blood in vivo after 1 week of fasting [23]. Moreover, dietary ESH in rats protected their kidneys and liver against Fenton reaction-derived oxidative damage, particularly against lipid peroxidation [24].

The initial studies exploring ESH dietary supplementation in humans, particularly in the context of a nutraceutical, examined ESH amounts in various foods [11,25] and focused on healthy human volunteers. For example, one human clinical trial demonstrated the bioavailability of ESH from eating mushrooms [26]. Another showed the "slow" accumulation of pure ESH into "whole blood", in contrast to plasma, consistent with the possible incorporation of ESH into RBCs during erythropoiesis [27]. To begin evaluating the potential therapeutic benefits of pure dietary ESH, another human clinical trial found that plasma levels increased, and then decreased rapidly, whereas RBC levels began increasing ~7 days after initiating ESH administration, again consistent with its incorporation during erythropoiesis [27]. This study also found decreasing trends in oxidative stress and inflammation, suggestive of a therapeutic effect [27]. More recently, a description of the design of the "first" randomized human clinical trial was published, which aims to study the therapeutic effects of ESH supplementation in patients with the metabolic syndrome [28]. The goal of this "pilot" trial is to understand the magnitude of any potential effects, thereby allowing for power calculations to be determined for a subsequent definitive trial [28]; however, this trial has not yet enrolled any patients [29].

Even though no therapeutic benefits have been proven to date in humans, dietary supplementation with ESH seems safe for human use. For example, regulatory agencies in Europe [12] and the United States [13] have deemed it safe as a dietary supplement or food additive, at least up to 30 mg/day for adults and 20 mg/day for children. Indeed, this even seems to hold for fetuses and breast-feeding neonates [30].

Nonetheless, some caution is required when developing inclusion and exclusion criteria during the design of clinical trials, because there is, at least, the suggestion that ESH may exacerbate cancer and certain infections (e.g., tuberculosis) by preventing the oxidative stress that may be protective in these pathogenic processes [8]. In addition, ESH supplementation may affect immune function via the potentiation of toll-like receptor (TLR) signaling or through other mechanisms [20,31]. Thus, potential adverse effects should be considered when designing clinical trials of dietary supplementation with ESH in specific clinical settings.

Finally, several recent studies examined selenoneine, a selenium-containing molecule that is structurally similar to ESH (Figure 1) and may have similar biological effects. Selenoneine is present in high amounts in the skin of beluga whales and in RBCs among the Inuit who consume this food [32]. By analogy with ESH, selenoneine may function as an antioxidant. Although it appears to be transported into human cells by SLC22A4, the ESH transporter (see below) [33], and may be important in protection against methylmercury toxicity, its biology is still not well understood.

3. ESH Mechanism(s) of Action: What Does It Do and How Does It Work?

3.1. *ESH Detoxifies ROS and Binds Metal Cations to Protect Cells from Damage*

At a molecular level, a major mechanism by which ESH exerts its antioxidant function(s) involves the direct inactivation of various ROS and free radicals [10], including the hydroxyl radical [34] and singlet oxygen [35], the latter of which can be produced by neutrophils during inflammatory processes [36]. In contrast, ESH does not seem to react directly with hydrogen peroxide or superoxide anion [37,38]. Moreover, although it may not directly detoxify hydrogen peroxide [37,38], ESH can protect cells from hydrogen peroxide-mediated damage [39]. In addition, ESH, which can alternate between its tautomeric forms (i.e., thiol or thione (Figure 1)), may primarily exist as a thione inside cells; as such, it is protected from auto-oxidation, making it more stable than GSH, which, as a thiol, is readily oxidized and, thereby, inactivated [39].

Interestingly, ESH can also directly interact with various metal cations, potentially including iron [34]. Thus, in a physiological way, ESH may enhance ferrochelatase-mediated iron incorporation into protoporphyrin IX to produce heme during normal hemoglobin synthesis, perhaps by helping maintain iron in its reduced, ferrous (Fe^{+2}) state [40]. In addition, ESH may neutralize highly reactive ferryl (Fe^{+4}) hemoglobin [38], thereby preventing met-hemoglobin production and free radical-catalyzed lipid peroxidation [41]. Given the close homology between myoglobin in muscle cells and hemoglobin in RBCs, and their corresponding roles in oxygen delivery, it is perhaps not surprising that ESH can also reduce the ferryl (Fe^{+4}) form of myoglobin [37]. As one potential clinical application of ESH's interactions with metal cations, it ameliorated iron-induced liver toxicity in rats, potentially due to its ability to scavenge ROS and potentially bind iron [42].

3.2. *How Is ESH Regenerated after Its Oxidation to the ESSE Disulfide?*

Some uncertainty remains regarding the exact metabolic pathways and reaction mechanisms by which ESH exerts its antioxidant effects and is then regenerated and/or catabolized [22,43]. For example, Servillo et al. state that "...in the presence of oxidants, ESH can form the disulfide according to the usual pattern, but this disulfide (ESSE) can behave very different from alkylthiol disulfides, in that, being unstable at physiological pH, it undergoes a progressive decomposition by disproportion. Our data show that, from 2 mol of ESSE, 3 mol of ESH and 1 mol of EH are formed. *Notably, the partial ESH regeneration from ESSE does not require reducing substances, the process being a disproportion*" (emphasis added; Figure 2) [44]. Therefore, these results suggest that providing reducing equivalents, such as NADPH derived from the pentose phosphate pathway, may not be required to maintain reasonable amounts of the reduced form of ESH in cells. Nonetheless, the slow kinetics of this disproportionation reaction raises questions about its biological relevance. In addition, these authors suggest that this process could occur inside cells, not just in vitro, as when exposing endothelial cells to oxidative stress caused by high glucose levels, paraquat, superoxide, or hydrogen peroxide [45], although the reaction kinetics may not favor this process in vivo. Nonetheless, there is not universal agreement regarding this proposed mechanism [46]. For example, an alternate proposal suggests that when ESH is oxidized by, for example, the hydroxyl radical, it can be "repaired" (i.e., regenerated) by interacting with vitamin C, which reduces oxidized ESH and simultaneously generates the ascorbyl radical [47]; of course, in cells, the ascorbyl radical would then need to be reduced [48]. In addition, thioredoxin reductase can directly reduce oxidized ESH by itself, without requiring GSH [49]. Finally, when human RBCs were subjected to oxidative stress using arsenicals, Reglinski et al. postulated a novel and unusual mechanism in which ESH may exert its protective effect by undergoing an "environmental", but not a chemical, change, suggesting that it is not "consumed" when playing this role [50].

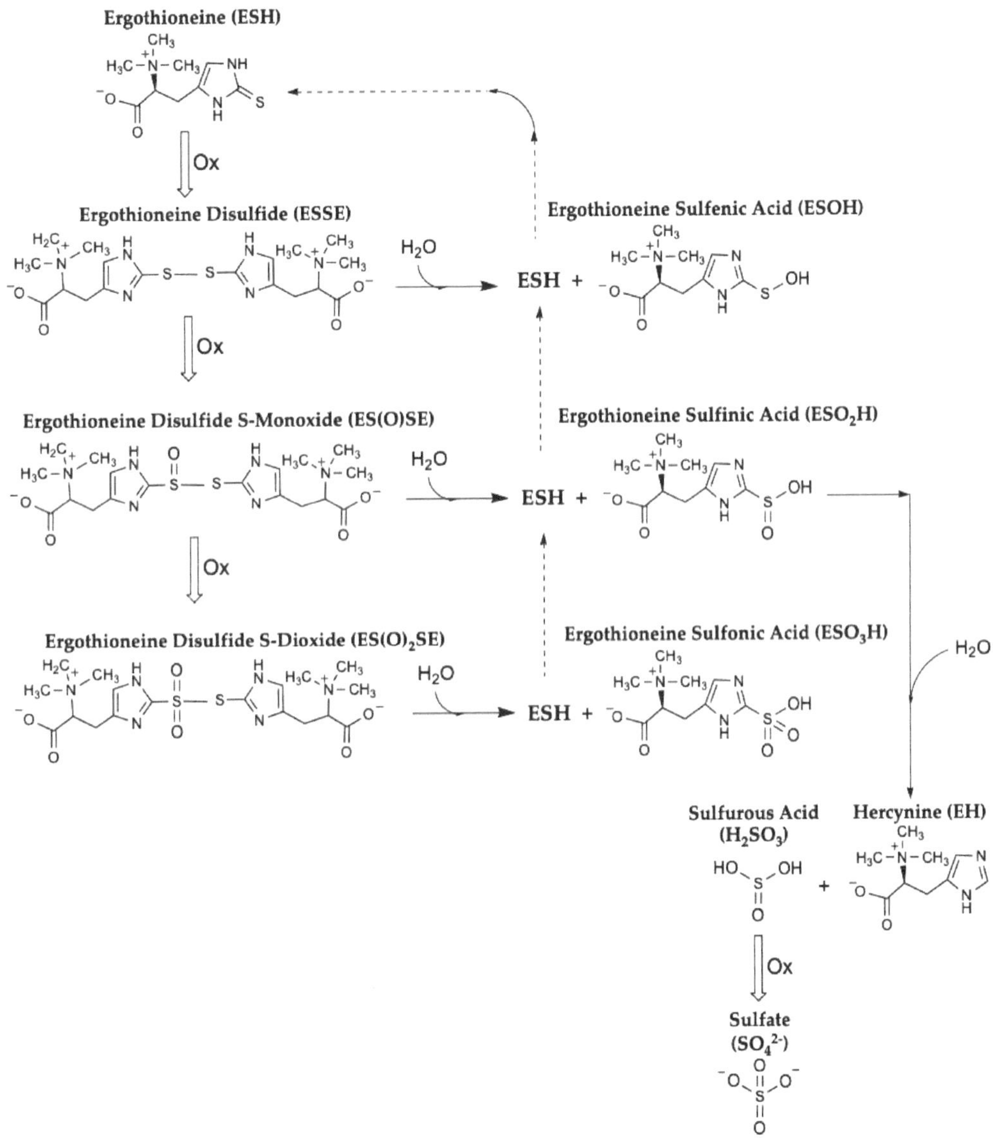

Figure 2. One proposed pathway of ESH oxidation with subsequent regeneration of reduced ESH. This pathway does not require reducing equivalents provided by NADPH (figure inspired by Ref. [44]). Nonetheless, other pathways have been proposed, including ones involving GSH to regenerate reduced ESH [46,47,49,51].

In contrast, Oumari et al. described an elegant mechanism, which does involve GSH, whereby ESH is regenerated after detoxifying singlet oxygen [35]. In addition, when ferryl (Fe^{+4}) hemoglobin is produced by nitrite, it can be reduced by ESH, the thione of which is then "consumed" by forming the ESSE disulfide; ESH can then be regenerated (i.e., reduced) by interacting with thiol-containing cysteine and, perhaps, GSH [38]. Similarly, GSH may be involved in reducing oxidized ESH (i.e., the ESSE disulfide) after ESH participates in reducing ferryl (Fe^{+4}) myoglobin [37].

Thus, there remains some debate whether, inside RBCs and other cells, oxidized ESH can be recycled without requiring reducing equivalents derived from, for example, the pentose phosphate pathway (i.e., NADPH). Therefore, it would be interesting to test this concept using RBCs from individuals with enzymatically diminished forms of glucose-6-phosphate dehydrogenase (i.e., G6PD), the rate-limiting step in the pentose phosphate pathway. Because G6PD-deficient RBCs cannot generate large amounts of NADPH, decreasing their ability to respond to oxidative stress because of defective regeneration of GSH, it would be interesting to determine whether or not oxidized ESH (i.e., ESSE disulfide) is efficiently regenerated in these RBCs.

3.3. *How Does ESH Work in Cells and How Do Cells Modulate/Amplify Its Function(s)?*

Using rats as a model organism, dietary supplementation with ESH protected kidney and liver tissues, and the polyunsaturated fatty acids (PUFAs) in their cellular membranes, from Fenton reaction-derived damage and lipid peroxidation [24]. In addition, ESH ameliorated iron-induced liver toxicity, perhaps by both scavenging ROS and binding iron [42]. Finally, in the setting of experimental diabetes, ESH protected rat endothelial cells against oxidative damage, thereby maintaining their normal function [52].

In the central nervous system, ESH improved antioxidant status (including the GSH/GSSG ratio) in the brains of mice [53]. In addition, ESH protected a human neuronal hybridoma cell line against hydrogen peroxide- and peroxynitrite-mediated toxicity, including preventing DNA damage [54]. Similarly, ESH inhibited hydrogen peroxide-mediated DNA damage and cell death in the PC12 rat neural cell line [55].

Interestingly, several studies suggest that one of the mechanisms, at least, by which ESH protects cells from oxidative damage is by upregulating signaling through the Nrf2 antioxidant pathway [56]. These studies examined oxidative damage induced by ultraviolet light in keratinocyte cell lines in vitro [57,58] and in a rat model of diabetes in vivo [59]. In addition, the latter provided evidence that ESH could also function by inhibiting inflammation mediated by the NFκB signaling pathway [59].

Given that a major effector of inflammation is the production of oxidant stress, it is perhaps not surprising that, in an effort to maintain homeostasis, inflammation could induce increases in intracellular ESH levels by enhancing its transport by SLC22A4, the highly specific ESH transporter (see just below). Indeed, *SLC22A4* mRNA levels in CD14+ cells were upregulated by proinflammatory cytokines, such as TNFα [60]. In addition, multiple pro-inflammatory signaling molecules, including, IL1β, TNFα, and NFκB, were each involved in upregulating SLC22A4 expression in human synoviocytes, which has potential relevance for the pathophysiology of rheumatoid arthritis [61].

4. How Does ESH Enter Organisms and Cells?

4.1. *Identifying ESH Transporters and Their Cell/Tissue Locations*

A major advance in the field of ESH biology was the discovery that the OCTN1 molecule ("Organic Cation Transporter-1"), which had previously been characterized as an "orphan" transporter, putatively for small organic cations and zwitterions, was actually the highly specific cell membrane transporter of ESH [41]. Although this cell surface molecule had been given several names (e.g., OCTN1, ETT ("ergothioneine transporter")), it is now generally agreed that its official name is SLC22A4 (i.e., "solute carrier family 22 member 4"); this is the descriptor we will use herein.

SLC22A4 protein and mRNA are highly expressed in various tissues, including the intestine [62], bone marrow [63], and fetal liver [63]. Nonetheless, there is some disagreement regarding whether or not it is highly expressed in adult liver [63,64]. Importantly, for the purposes of this review, in addition to its bone marrow expression, SLC22A4 is highly expressed by K562 cells [63], a human leukemic cell line that is often used to study the basic mechanisms involved in erythropoiesis [65].

More recent studies, which require further elaboration, suggest that a separate and distinct ESH transporter, SLC22A15, is required for ESH to enter the brain [66], even though relatively little ESH is found in mouse brain and SLC22A15 is less effective than SLC22A4 in transporting ESH [67]. In addition, the SLC22A15 transporter function is a bit more promiscuous, in that it can effectively transport carnitine and carnosine [66]. Although the tissue distribution of SLC22A15 overlaps with that of SLC22A4, it is much more prominent in the brain [66]. Nonetheless, studies suggest that no ESH is found in the brains of *SLC22A4* knockout mice (see below) [68,69]. Thus, more work remains regarding the biology of SLC22A15, and studies with single- and double-knockout (i.e., both *SLC22A4* and *SLC22A15*) mice could be very revealing.

4.2. Structure–Function Information about ESH Transporters

SLC22A4 is highly specific for transporting ESH and requires sodium ion to exert its effects [70]. Interestingly, analogous to ESH, carnitine, also a small zwitterion, is transported into cells by OCTN2, a highly homologous transporter, now officially designated as SLC22A5. Although SLC22A4 and SLC22A5 are highly homologous, they exhibit exquisite selectivity for ESH and carnitine, respectively, and recent, elegant molecular modeling studies helped explain these specificities [71]. In addition, site-directed mutagenesis studies of SLC22A4 and SLC22A5 were performed in an effort to help elucidate these specificities; however, none of the engineered SLC22A4 single point mutations allowed for carnitine transport; instead each maintained, or improved, ESH transport [72]. It is hoped that a complete elucidation of these specificities will become available in the future, which should also explain the apparent ability of SLC22A4 to transport metformin, among other small molecules [73]. Finally, at this time, less is known about the molecular mechanisms involved in ESH transport in the brain by SLC22A15 [66].

4.3. Elucidation of ESH Function Using Genetically Modified Organisms

The availability of *SLC22A4* knockouts in various model organisms has shed important light on ESH function and its biological importance. The first such model described *SLC22A4* knockout mice, in which ESH was not detected in any tissue examined [68]. In addition, although dietary ESH supplementation of these mice led to slight, transient increases in circulating ESH levels, no tissue incorporation was apparent. Given that few, if any, abnormalities were definitively detected in these mice at baseline, including no detectable hematological abnormalities, this suggests that an "ESH deficiency state" is not analogous to a vitamin deficiency. This concept is supported by an older study of rats fed an ESH-deficient diet, in that no abnormalities were seen in these rats or in their offspring, which were also maintained on an ESH-deficient diet [74]. Nonetheless, these knockout mice did exhibit greater intestinal susceptibility to oxidative stress. Indeed, a subsequent study of streptozotocin-induced diabetes in these mice identified increased oxidative stress and kidney fibrosis [75]. Thus, for the particular focus of this review, even though no hematological abnormalities were found in these mice at baseline, it is possible that they may have ineffective responses to disease-related stress. For example, crossing these mice with G6PD-deficient mice [76] might produce offspring exhibiting exacerbated oxidant-induced hemolysis (see below). Finally, although a SLC22A15 knockout mouse (i.e., Slc22a15$^{em1(IMPC)J}$) is listed in the JAX catalog [77], to our knowledge, no papers have yet been published describing its biology. Nonetheless, studies with this mouse, along with those of a *SLC22A4/SLC22A15* double knockout, should be of interest.

Analogous ESH transporter constructs were also made in "simple" organisms, such as zebrafish [78] and roundworms [79]. Although these approaches will be useful for obtaining additional structure–function information and for elucidating the role(s) of ESH in the biology of organismal oxidative stress, there is currently no information regarding the effects of this "absolute" ESH deficiency on RBC biology in zebrafish, and *C. elegans* is not useful for this purpose.

4.4. Human ESH Transporter Variants

Multiple SLC22A4 polymorphisms were identified and characterized in human populations. Studying these naturally occurring variants is useful for elucidating transporter activity and substrate specificity [80,81], although their effects, if any, on RBC structure and function are not yet understood. In addition, there may be specific disease-relevant variants, which can influence intracellular ESH concentrations, such as in the context of Crohn's disease [60]. Finally, although perhaps not relevant for RBC biology, SLC22A15 variants were also identified [66].

5. ESH and RBC Biology

5.1. Introduction

As described above, RBCs in vivo are continuously exposed to oxidant stress and, as such, use multiple mechanisms to resist these attacks and protect and/or repair oxidant damage to their cytoplasmic contents and cellular membrane [3,6]. From the perspective of transfusion medicine, refrigerated storage is necessary to maintain an appropriate, readily available inventory of donor RBC units so that they can be transfused into patients expeditiously, when required. During storage, these RBCs are exposed to additional oxidative stress, which, one can argue, is the "prime mover" in causing RBC "storage lesion" [82–84]. The latter encompasses a plethora of physiological and morphological effects [85,86], which ultimately lead to decreasing RBC storage quality and transfusion quality as the refrigerated storage interval increases; this results in individual patients receiving highly variable doses of therapeutically effective RBCs from individual donor units [87]. Interestingly, recent studies suggest that the "exposome", as experienced by individual donors due to variations in diet, habits, pharmaceutical exposures, and the microbiome, may be quite relevant for the resistance or susceptibility of their donated units to the storage lesion [88]. Given that dietary supplementation with antioxidants can modulate the effects of the RBC storage lesion [84], and given that ESH is enriched in certain diets and was identified in the RBC exposome [88], it is not unreasonable to determine whether specific supplementation with ESH can affect RBC properties when circulating in vivo and during refrigerated storage in vitro, especially given ESH's ability to interact directly with ROS and its possible interactions with free iron [9].

5.2. RBC ESH Levels

ESH, first identified more than a century ago, was also detected in the blood of humans and other animals not long thereafter [7]. Interestingly, the vast majority of "blood" ESH is found in RBCs (1285 $\pm$ 1363 ng/mL) [89], with only low levels in plasma (107.4 $\pm$ 20.5 ng/mL) [89], white blood cells, and platelets [7,68,90–93]. Indeed, *SLC22A4* knockout mice did not have any detectable ESH in their RBCs or plasma [68].

Multiple studies measured RBC ESH levels in various human populations and ethnicities, and in various disease settings [94,95]. Although RBC ESH levels vary significantly between individuals, there is little daily variation in serial measures in particular individuals [89,96]; nonetheless, it is possible to determine a reference range for humans [97]. Although RBC antioxidant potential appears to decrease with organismal aging [98], there is a lack of certainty whether this is also true for RBC ESH levels [10,96]; nonetheless, in rats, no change in RBC ESH levels was observed with aging [99]. Interestingly, ESH levels in RBCs decrease throughout their circulatory lifespan in vivo [100], perhaps contributing to their senescent phenotype by making them more sensitive to oxidative damage.

In an interesting study, when mice were given phenylhydrazine, a severe oxidant stressor that produces massive hemolysis, the RBCs remaining in the circulation had significantly higher ESH levels [76], perhaps suggesting that RBCs with low ESH levels were more susceptible to oxidant damage and were rapidly cleared. In addition, RBC ESH levels varied significantly when measured in RBCs obtained from genetically inbred strains of mice [101], suggesting that these levels may be genetically determined. Indeed, ESH levels were higher in RBCs from C57Bl/6 mice, as compared to FVB mice. This is particularly interesting given that, in the context of RBC storage and transfusion, C57Bl/6 mice are "good storers" and FVB mice are "poor storers" [101]; therefore, this difference in RBC ESH content may be physiologically relevant.

Multiple studies have evaluated the effects of dietary supplementation on ESH levels in "blood" in general, and in RBCs in particular. Given that the vast majority of ESH in the circulation is in RBCs, we assume that when "blood" levels were measured [22], these actually reflected ESH amounts in RBCs. For example, in a rat model, ESH accumulated slowly in circulating RBCs following dietary supplementation [102]. In addition, using radiolabeled ESH, the compound rapidly appeared in plasma and then disappeared rapidly, but then was stably present in circulating "corpuscles" with little or no decrease in cellular ESH levels after one week of fasting [23]. Analogous studies in rabbits showed that RBC levels decreased when they were fed an ESH-deficient diet, which then reversed with dietary supplementation [103]. Finally, in mice receiving one oral dose of radiolabeled ESH, RBC levels of this compound continued to increase slowly over the ensuing days [68]. Taken together, these results in animal models suggest that dietary ESH rapidly enters, and then rapidly disappears from, plasma; it also enters the RBC compartment slowly and then remains stable over time.

In humans, the failure to detect ESH in plasma from fasting subjects argues for its rapid clearance from this compartment, whereas the lack of an apparent effect of dietary ESH on steady-state RBC levels suggests that only persistent changes in dietary ESH uptake over long time periods would induce significant changes in RBC intracellular ESH concentrations [104]. Supporting this concept, a study of the effects of changes in altitude on human RBC metabolism demonstrated increasing 2,3-diphosphoglycerate levels with increasing altitude, presumably to enhance oxygen offloading during relative hypoxia; nonetheless, RBC ESH levels did not change, presumably resulting from a stable diet during the study period [105]. These data also suggest that RBC ESH was not "consumed" due to the altitude changes during this study.

Finally, compounds related to ESH, including trimethyl histidine (i.e., hercynine), methyl-ESH, and selenoneine are highly present in human RBCs [93]. Given the potential role of selenoneine as an antioxidant, and its abundance in certain foods (see above), it is surprising that mice fed a selenoneine-enriched diet for 32 days exhibited no detectable selenoneine in their RBCs [106]. Thus, more work on this interesting molecule is eagerly anticipated.

5.3. Does ESH Enter Mature RBCs or Only Erythropoietic Progenitors?

Some of the studies described above document that dietary ESH rapidly appears in plasma and is then rapidly cleared, over minutes to hours; it then appears slowly in circulating RBCs, over days to weeks, and remains relatively stable in this compartment. These results are consistent with the concept that ESH enters the RBC compartment during erythropoiesis, but does not enter into, or efflux from, mature circulating RBCs. These results would also be consistent with the absence of functional SLC22A4 transporters, or the ability of these molecules to function effectively, on the surface of mature RBCs. They also suggest that there is little or no efflux out of mature RBCs, and, therefore, any changes in RBC levels would be due to intracellular "consumption" in response to oxidative stress. Data directly supporting, or refuting, this hypothesis are not entirely consistent and are described just below.

Although several studies support the concept that ESH can be transported into mature RBCs [92,107], other, possibly more convincing, studies demonstrate that ESH does not directly enter into, or efflux out of, mature RBCs [41,67,91]. Indeed, the functional expression of SLC22A4 was not detected on mature RBCs and reticulocytes, suggesting that ESH found in RBCs derives from its transport into erythropoietic progenitors, remaining there throughout their differentiation into mature RBCs [104]. To this end, erythropoietic progenitors express high levels of *SLC22A4* mRNA [41,107]. In addition, *SLC22A4* mRNA is detectable in human hematopoietic stem cells [41] and may be relevant in the context of polycythemia vera [108]. In addition, given that different types of reticulocytes are identifiable, and that they may vary in their antioxidant "machinery" [5], it would be interesting to evaluate *SLC22A4* mRNA and ESH levels in these cell types, particularly in disease settings that induce stress erythropoiesis. Finally, given the increasing sensitivity in identifying additional components of the human and murine RBC proteome [109], it would be important to determine definitively, using highly purified samples of mature RBCs, lacking leukocytes and platelets, whether or not the SLC22A4 transporter protein is present and functional on these cells.

An alternative approach to address this issue could use tissue culture in vitro. For example, the immortalized K562 human cell line expresses high levels of *SLC22A4* mRNA [63] and can be induced to differentiate into RBC-like, hemoglobin-containing cells [65]. In addition, silencing *SLC22A4* decreases both transporter expression and ESH uptake [110]. Thus, it would be interesting to measure SLC22A4 protein levels and transporter function in terminally differentiated K562 RBC-like cells. Analogous approaches could also be employed using mouse or human CD34+ hematopoietic stem cells, derived from various sources, by culturing them in vitro and inducing them to differentiate into, at least, enucleated reticulocytes [111].

5.4. What Are the Functions of ESH in RBCs?

Given the high amounts of ESH in circulating mature RBCs, and the expression of its transporter in the bone marrow and during erythropoiesis [63], it is reasonable to suggest that the major function of ESH is to protect erythropoietic progenitors and mature RBCs from oxidative stress induced by hemoglobin auto-oxidation. Indeed, silencing *SLC22A4* mRNA expression in K562 cells induces decreased expression of the transporter protein, decreased cellular uptake of ESH, and decreased proliferation and differentiation of these cells, potentially due to increased apoptosis [110]; these results suggest a role in protecting erythropoiesis from damaging oxidative stress. In addition, in rabbits fed an ESH-deficient diet, RBC ESH levels decreased, and these RBCs were more sensitive to sodium nitrite-induced methemoglobin formation; importantly, this defect was reversed by dietary supplementation with ESH [103]. Similar results for met-hemoglobin production were seen in rats fed a methionine-deficient diet; in addition to observing lower ESH levels, ESH was found to be more important than reduced glutathione for providing this protection [112]. Although ESH was also reported to protect human RBCs from oxidative damage due to arsenicals [50], some questions were raised about these results [113].

If, indeed, RBC ESH levels decrease with human aging, this may enhance their "senescent phenotype", which could potentially be ameliorated by dietary ESH supplementation [10]. Nonetheless, despite *SLC22A4* knockout mice having no detectable ESH in their circulating RBCs, they were not anemic, nor did they display hematological abnormalities suggestive of dysfunctional hematopoiesis [68]. However, these results did not derive from intensive studies of murine erythropoiesis in vivo or in vitro, nor were these animals subjected to a disease model (e.g., sepsis) that would induce stress erythropoiesis. As such, it would be interesting to breed these mice with those exhibiting an RBC disease phenotype of increased oxidative stress (e.g., G6PD deficiency, sickle cell disease), to study the relevance, or lack thereof, of ESH function in these disease contexts.

5.5. Clinical Implications of ESH in RBC Biology

Due to its antioxidant properties and apparent safety as a dietary supplement, increasing numbers of publications have suggested the potential benefits of ESH supplementation in various non-hematological disease settings, and as an anti-aging nutraceutical (for a general review, see [8]). In addition, in the context of non-hematological disorders, circulating RBCs can function as sensors, measuring and integrating oxidative stress throughout a given organism [3]. As such, measuring RBC ESH levels could identify a potential organismal deficit (or abundance) of antioxidant potential. As one example, RBC ESH levels were significantly elevated in patients with rheumatoid arthritis [60]. Because pro-inflammatory cytokines and cytokine signaling are prominent in this disease (e.g., IL1β, TNFα, NFκB) [114], and because these increase SLC22A4 expression in synoviocytes [61], this may provide a compensatory mechanism to provide protection by increasing ESH levels in these cells. If these cytokines similarly affected bone marrow erythropoiesis by increasing SLC22A4 expression and function, then elevated circulating RBC ESH levels could provide a marker of disease activity. Similarly, in pre-eclampsia [115], the pathogenesis of which may also involve oxidative stress [116], this proposed mechanism could also explain the presence of elevated RBC ESH levels, perhaps as part of a compensatory process [117]. Finally, it is important to consider the possibility that elevated ESH levels in various tissues, including RBCs, may exacerbate cancer and certain infections by ameliorating the oxidative stress with which the host could target the pathogenic process [8]. Thus, it is important to consider this potential adverse effect if a therapeutic approach aims to use dietary supplementation to raise RBC ESH levels.

In contrast, in the context of hematological disorders, ESH levels may modify disease severity in settings where RBCs are particularly susceptible to oxidative damage, thereby leading to intravascular and/or extravascular hemolysis. For example, low RBC ESH levels may exacerbate hemolytic crises in patients with G6PD deficiency or shorten RBC circulatory lifespan in patients with sickle cell disease. Nonetheless, it is important to note that, following a hemolytic crisis in G6PD deficiency, the remaining RBCs are typically younger, with elevated reticulocyte counts; both of these cell types have higher than average G6PD enzymatic activity (in the context of the average activity of G6PD-deficient RBCs) because it typically decreases during circulatory aging in vivo [118]. Similarly, RBC ESH levels presumably decrease throughout their circulatory lifespan [100]. Interestingly, although ESH was not identified in RBCs in a metabolomics investigation of sickle cell disease [119], it was found to be significantly decreased in an analogous study [120]; the authors of the latter suggested that dietary ESH supplementation could be therapeutically beneficial for these patients [120].

Although not a hematological disorder per se, experimental hemorrhage in a rat model led to increased RBC ESH levels following recovery [99]. Similar results were seen in humans in the more controlled and limited context of blood donation (see below) [94]. A potential explanation for this finding could be that erythropoietic recovery yields increased numbers of circulating (stress) reticulocytes and younger RBCs, which can have higher ESH levels in their cytoplasm.

Finally, the RBCs of patients with hereditary overhydrated stomatocytosis, due to variants in Rh-associated glycoprotein, have low ESH levels [121]; similar results were found in patients with chronic granulocytic leukemia [122]. In addition, erythropoietic progenitors in patients with polycythemia had an increased expression of *SLC22A4* mRNA [108]. Nonetheless, the mechanistic explanations for these results, and their pathologic implications, are not yet known.

5.6. Clinical Applications of ESH for RBC Transfusion and the RBC Storage Lesion

From the perspective of volunteer blood donation, acute phlebotomy of ~10% of a donor's circulating blood volume subsequently induces enhanced erythropoiesis in an effort to restore RBC mass to baseline levels. Assuming that there is sufficient ESH in the diet, one might predict that RBC ESH levels, on average, might increase, as a result of this burst of newly formed RBCs. Indeed, in an early study of healthy human volunteers, RBC ESH levels did increase at two weeks post-donation [94].

Following donation, human RBC concentrates, which can be refrigerator-stored for up to 42 days before transfusion, experience the "storage lesion", which is primarily due to oxidative stress (see above). Interestingly, in an early study using a rabbit model, RBC ESH levels decreased during storage [103]. In contrast, in a mouse model, RBC levels remained stable, or even increased, during storage [101]. However, whether or not ESH levels change during the storage of human RBCs is not yet known. Nonetheless, given the potential role of ESH in ameliorating oxidative stress during storage, thereby decreasing the severity of the storage lesion, one might expect that RBC ESH levels, and/or *SLC22A4* variants, would correlate with RBC transfusion quality (e.g., post-transfusion hemoglobin increments); however, no evidence for this effect has been described to date [87,123]. Nonetheless, results from mouse studies (see above) demonstrate differences in RBC levels when comparing inbred strains; in particular, levels in C57Bl/6 RBCs were significantly higher than those in FVB RBCs [101]. Given that FVB and C57Bl/6 mice are "poor storers" and "good storers", respectively [101], and that their respective *SLC22A4* sequences differ somewhat, this is an intriguing hypothesis to pursue.

6. Conclusions

In summary, ESH, an unusual and interesting antioxidant, is obtained from the diet, through the mediation of a highly specific and evolutionarily conserved transporter, and is present in high amounts in circulating mature RBCs. Although there are no known deficiency states requiring its definition as a "vitamin", there is increasing evidence that ESH could be therapeutically beneficial as a "nutraceutical". Given that oxidant stress is important in the pathophysiology of multiple hematological disorders, and is a major determinant of RBC lifespan in vivo and in vitro, there not only remains much to be learned regarding ESH function during erythropoiesis and in RBC biology, but there also is the potential that it could enhance the function, health, and longevity of human RBCs (Table 1).

Table 1. Unanswered questions.

1. **At which developmental stages during erythropoiesis can ESH be imported into cells?**
Can use ESH supplementation at different developmental stages during differentiation of K562 cells or CD34+ cells in vitro.
2. **How do *SLC22A4* gene transcription, SLC22A4 protein expression, and SLC22A4 transport function change during erythropoiesis?**
Can use highly sensitive proteomics methods and differentiation of CD34+ cells in vitro, to determine whether the transporter, or its fragments, can be identified at different cell stages.
3. **How do RBCs "normally" accumulate ESH? Can ESH enter mature RBCs circulating in vivo and/or "in the bag" in vitro?**
Can incubate isotopically labeled ESH in vitro (using radioactive or stable isotopes) with highly purified populations of reticulocyte-poor, mature human and mouse RBCs. Can use ESH dietary manipulation of mice, along with RBC biotinylation in vivo, to label and then isolate RBC populations of defined circulatory age.
4. **Do *SLC22A4* sequence variants affect the "efficiency" of ESH uptake during erythropoiesis (and/or in mature RBCs), thereby affecting RBC ESH levels?**
Can study RBCs obtained from human volunteers with *SLC22A4* sequence variants, inbred mouse stains with known SLC22A4 sequence variants, or genetically modified mice constructed to express specific *SLC22A4* sequence variants of interest. Can study differentiation of CD34+ cells obtained from human volunteers with *SLC22A4* sequence variants.

Table 1. *Cont.*

5.	**Does "stress" and/or disease modulate RBC ESH levels?**
a.	Does organismal "stress" in vivo (e.g., sepsis) or cellular stress in vitro (e.g., endotoxin exposure) affect SLC22A4 transporter levels and/or function and RBC ESH levels?
b.	Are there differential effects on reticulocytes, stress reticulocytes, and/or mature RBCs in hematological disorders, such as sickle cell disease?
c.	Do responses to anemia per se (e.g., hemorrhage resuscitation, iron therapy to treat iron-deficiency anemia, repetitive phlebotomy, and blood donation in healthy volunteers) affect SLC22A4 expression and function during erythropoiesis and ESH levels in mature RBCs?
d.	Does inflammation in previously healthy individuals, or in patients with inflammatory disorders, affect SLC22A4 expression and function during erythropoiesis and ESH levels in mature RBCs?
6.	**Is decreased SLC22A4 function, resulting in decreased RBC ESH levels, relevant in hematological settings?**
a.	Do *SLC22A4* knockout mice display any hematological pathology and/or defective erythropoiesis when placed under stress (e.g., inflammation, sepsis, hemorrhage resuscitation, iron deficiency, or immune-mediated hemolysis)?
b.	Does a defective SLC22A4 function (e.g., by knocking out *SLC22A4* expression) exacerbate hematological disorders, such as G6PD deficiency and sickle cell disease?
c.	Does knocking out both ESH transporters (i.e., SLC22A4 and SLC22A15) produce any additive effects in the settings describe above?
7.	**Do increased RBC ESH levels provide enhanced protection against oxidative stress in settings of NADPH "restriction" (e.g., G6PD deficiency)?**
8.	**Are RBC ESH levels independently relevant for modulating RBC "circulatory quality" (e.g., RBC lifespan, deformability, endothelial adherence, perfusion, oxygenation, and erythrophagocytosis)?**
9.	**What roles, if any, do RBC ESH levels play in refrigerated storage biology?**
a.	Do RBC ESH levels modulate the RBC storage lesion, thereby affecting RBC "storage quality"?
b.	Do *SLC22A4* variants in RBC donors affect RBC ESH levels, RBC "storage quality", and post-transfusion hemoglobin increments?
c.	If healthy blood donors are supplemented with dietary ESH, do their RBCs exhibit enhanced "circulatory quality" post-transfusion?

Author Contributions: Conceptualization, S.L.S.; writing—original draft preparation, S.L.S.; writing—review and editing, T.A.T., R.O.F., J.C.Z., J.P.K., T.N. and S.L.S. All authors have read and agreed to the published version of the manuscript.

Funding: This work was supported, in part, by a grant from the National Institutes of Health (NIH R01 HL148151).

Conflicts of Interest: S.L.S. is a member of the Scientific Advisory Boards of Hemanext, Inc. and Alcor, Inc. J.C.Z. is a founder of, and J.P.K. is a consultant for, Svalinn Therapeutics. T.N. is a founder of Omix Technologies, Inc. and Altis Biosciences LLC.

References

1. Anastasiadi, A.T.; Arvaniti, V.Z.; Hudson, K.E.; Kriebardis, A.G.; Stathopoulos, C.; D'Alessandro, A.; Spitalnik, S.L.; Tzounakas, V.L. Exploring unconventional attributes of red blood cells and their potential applications in biomedicine. *Protein Cell* **2024**, *15*, 315–330. [CrossRef] [PubMed]
2. Pierige, F.; Serafini, S.; Rossi, L.; Magnani, M. Cell-based drug delivery. *Adv. Drug Deliv. Rev.* **2008**, *60*, 286–295. [CrossRef] [PubMed]
3. Moller, M.N.; Orrico, F.; Villar, S.F.; Lopez, A.C.; Silva, N.; Donze, M.; Thomson, L.; Denicola, A. Oxidants and Antioxidants in the Redox Biochemistry of Human Red Blood Cells. *ACS Omega* **2023**, *8*, 147–168. [CrossRef] [PubMed]
4. Orrico, F.; Laurance, S.; Lopez, A.C.; Lefevre, S.D.; Thomson, L.; Moller, M.N.; Ostuni, M.A. Oxidative Stress in Healthy and Pathological Red Blood Cells. *Biomolecules* **2023**, *13*, 1262. [CrossRef] [PubMed]

5. Melo, D.; Ferreira, F.; Teles, M.J.; Porto, G.; Coimbra, S.; Rocha, S.; Santos-Silva, A. Reticulocyte Antioxidant Enzymes mRNA Levels versus Reticulocyte Maturity Indices in Hereditary Spherocytosis, beta-Thalassemia and Sickle Cell Disease. *Int. J. Mol. Sci.* **2024**, *25*, 2159. [CrossRef] [PubMed]
6. D'Alessandro, A.; Anastasiadi, A.T.; Tzounakas, V.L.; Nemkov, T.; Reisz, J.A.; Kriebardis, A.G.; Zimring, J.C.; Spitalnik, S.L.; Busch, M.P. Red Blood Cell Metabolism In Vivo and In Vitro. *Metabolites* **2023**, *13*, 793. [CrossRef] [PubMed]
7. Melville, D.B. Ergothioneine. *Vitam. Horm.* **1959**, *17*, 155–204.
8. Halliwell, B.; Cheah, I.K.; Tang, R.M.Y. Ergothioneine—A diet-derived antioxidant with therapeutic potential. *FEBS Lett.* **2018**, *592*, 3357–3366. [CrossRef] [PubMed]
9. Cheah, I.K.; Halliwell, B. Ergothioneine; antioxidant potential, physiological function and role in disease. *Biochim. Biophys. Acta* **2012**, *1822*, 784–793. [CrossRef]
10. Fu, T.T.; Shen, L. Ergothioneine as a Natural Antioxidant Against Oxidative Stress-Related Diseases. *Front. Pharmacol.* **2022**, *13*, 850813. [CrossRef]
11. Beelman, R.B.; Phillips, A.T.; Richie, J.P., Jr.; Ba, D.M.; Duiker, S.W.; Kalaras, M.D. Health consequences of improving the content of ergothioneine in the food supply. *FEBS Lett.* **2022**, *596*, 1231–1240. [CrossRef] [PubMed]
12. Turck, D.; Bresson, J.L.; Burlingame, B.; Dean, T.; Fairweather-Tait, S.; Heinonen, M.; Hirsch-ernst, K.I.; Mangelsdorm, I.; McArdle, H.; Naska, A.; et al. Safety of synthetic L-ergothioneine (Ergoneine®) as a novel food pursuant to Regulation (EC) No 258/97. *EFSA J.* **2016**, *14*, e04629. [CrossRef]
13. Adams, M.K.D. *GRAS Notice No. GRN 000734*; U.S. Food and Drug Administration: Silver Spring, MD, USA, 2018.
14. Apparoo, Y.; Phan, C.W.; Kuppusamy, U.R.; Sabaratnam, V. Ergothioneine and its prospects as an anti-ageing compound. *Exp. Gerontol.* **2022**, *170*, 111982. [CrossRef] [PubMed]
15. Tian, X.; Thorne, J.L.; Moore, J.B. Ergothioneine: An underrecognised dietary micronutrient required for healthy ageing? *Br. J. Nutr.* **2023**, *129*, 104–114. [CrossRef] [PubMed]
16. Wu, L.Y.; Kan, C.N.; Cheah, I.K.; Chong, J.R.; Xu, X.; Vrooman, H.; Hilal, S.; Venketasubramanian, N.; Chen, C.P.; Halliwell, B.; et al. Low Plasma Ergothioneine Predicts Cognitive and Functional Decline in an Elderly Cohort Attending Memory Clinics. *Antioxidants* **2022**, *11*, 1717. [CrossRef] [PubMed]
17. Cheah, I.K.; Halliwell, B. Could Ergothioneine Aid in the Treatment of Coronavirus Patients? *Antioxidants* **2020**, *9*, 595. [CrossRef]
18. Melville, D.B.; Horner, W.H.; Otken, C.C.; Ludwig, M.L. Studies on the origin of ergothioneine in animals. *J. Biol. Chem.* **1955**, *213*, 61–68. [CrossRef] [PubMed]
19. Stampfli, A.R.; Blankenfeldt, W.; Seebeck, F.P. Structural basis of ergothioneine biosynthesis. *Curr. Opin. Struct. Biol.* **2020**, *65*, 1–8. [CrossRef]
20. Ey, J.; Schomig, E.; Taubert, D. Dietary sources and antioxidant effects of ergothioneine. *J. Agric. Food Chem.* **2007**, *55*, 6466–6474. [CrossRef]
21. Qiu, Y.; Chen, Z.; Su, E.; Wang, L.; Sun, L.; Lei, P.; Xu, H.; Li, S. Recent Strategies for the Biosynthesis of Ergothioneine. *J. Agric. Food Chem.* **2021**, *69*, 13682–13690. [CrossRef]
22. Tang, R.M.Y.; Cheah, I.K.; Yew, T.S.K.; Halliwell, B. Distribution and accumulation of dietary ergothioneine and its metabolites in mouse tissues. *Sci. Rep.* **2018**, *8*, 1601. [CrossRef] [PubMed]
23. Mayumi, T.; Kawano, H.; Sakamoto, Y.; Suehisa, E.; Kawai, Y.; Hama, T. Studies on ergothioneine. V. Determination by high performance liquid chromatography and application to metabolic research. *Chem. Pharm. Bull.* **1978**, *26*, 3772–3778. [CrossRef] [PubMed]
24. Deiana, M.; Rosa, A.; Casu, V.; Piga, R.; Assunta Dessi, M.; Aruoma, O.I. L-ergothioneine modulates oxidative damage in the kidney and liver of rats in vivo: Studies upon the profile of polyunsaturated fatty acids. *Clin. Nutr.* **2004**, *23*, 183–193. [CrossRef] [PubMed]
25. Borodina, I.; Kenny, L.C.; McCarthy, C.M.; Paramasivan, K.; Pretorius, E.; Roberts, T.J.; van der Hoek, S.A.; Kell, D.B. The biology of ergothioneine, an antioxidant nutraceutical. *Nutr. Res. Rev.* **2020**, *33*, 190–217. [CrossRef] [PubMed]
26. Weigand-Heller, A.J.; Kris-Etherton, P.M.; Beelman, R.B. The bioavailability of ergothioneine from mushrooms (Agaricus bisporus) and the acute effects on antioxidant capacity and biomarkers of inflammation. *Prev. Med.* **2012**, *54*, S75–S78. [CrossRef] [PubMed]
27. Cheah, I.K.; Tang, R.M.; Yew, T.S.; Lim, K.H.; Halliwell, B. Administration of Pure Ergothioneine to Healthy Human Subjects: Uptake, Metabolism, and Effects on Biomarkers of Oxidative Damage and Inflammation. *Antioxid. Redox Signal.* **2017**, *26*, 193–206. [CrossRef] [PubMed]
28. Tian, X.; Cioccoloni, G.; Sier, J.H.; Naseem, K.M.; Thorne, J.L.; Moore, J.B. Ergothioneine supplementation in people with metabolic syndrome (ErgMS): Protocol for a randomised, double-blind, placebo-controlled pilot study. *Pilot Feasibility Stud.* **2021**, *7*, 193. [CrossRef] [PubMed]
29. Tian, X. Investigating the potential health benefits of ergothioneine supplementation for people with metabolic syndrome-Study Registry. *Pilot Feasibility Stud.* **2021**, *7*, 193. [CrossRef]
30. Cheah, I.K.; Halliwell, B. Ergothioneine, recent developments. *Redox. Biol.* **2021**, *42*, 101868. [CrossRef]
31. Yoshida, S.; Shime, H.; Funami, K.; Takaki, H.; Matsumoto, M.; Kasahara, M.; Seya, T. The Anti-Oxidant Ergothioneine Augments the Immunomodulatory Function of TLR Agonists by Direct Action on Macrophages. *PLoS ONE* **2017**, *12*, e0169360. [CrossRef]
32. Achouba, A.; Dumas, P.; Ouellet, N.; Little, M.; Lemire, M.; Ayotte, P. Selenoneine is a major selenium species in beluga skin and red blood cells of Inuit from Nunavik. *Chemosphere* **2019**, *229*, 549–558. [CrossRef] [PubMed]

33. Yamashita, M.; Yamashita, Y.; Suzuki, T.; Kani, Y.; Mizusawa, N.; Imamura, S.; Takemoto, K.; Hara, T.; Hossain, M.A.; Yabu, T.; et al. Selenoneine, a novel selenium-containing compound, mediates detoxification mechanisms against methylmercury accumulation and toxicity in zebrafish embryo. *Mar. Biotechnol.* **2013**, *15*, 559–570. [CrossRef] [PubMed]
34. Akanmu, D.; Cecchini, R.; Aruoma, O.I.; Halliwell, B. The antioxidant action of ergothioneine. *Arch. Biochem. Biophys.* **1991**, *288*, 10–16. [CrossRef]
35. Oumari, M.; Goldfuss, B.; Stoffels, C.; Schmalz, H.G.; Grundemann, D. Regeneration of ergothioneine after reaction with singlet oxygen. *Free Radic. Biol. Med.* **2019**, *134*, 498–504. [CrossRef] [PubMed]
36. Kiryu, C.; Makiuchi, M.; Miyazaki, J.; Fujinaga, T.; Kakinuma, K. Physiological production of singlet molecular oxygen in the myeloperoxidase-H_2O_2-chloride system. *FEBS Lett.* **1999**, *443*, 154–158. [CrossRef] [PubMed]
37. Arduini, A.; Eddy, L.; Hochstein, P. The reduction of ferryl myoglobin by ergothioneine: A novel function for ergothioneine. *Arch. Biochem. Biophys.* **1990**, *281*, 41–43. [CrossRef]
38. Arduini, A.; Mancinelli, G.; Radatti, G.L.; Hochstein, P.; Cadenas, E. Possible mechanism of inhibition of nitrite-induced oxidation of oxyhemoglobin by ergothioneine and uric acid. *Arch. Biochem. Biophys.* **1992**, *294*, 398–402. [CrossRef] [PubMed]
39. Paul, B.D.; Snyder, S.H. The unusual amino acid L-ergothioneine is a physiologic cytoprotectant. *Cell Death Differ.* **2010**, *17*, 1134–1140. [CrossRef]
40. Goldberg, A. The enzymic formation of haem by the incorporation of iron into protoporphyrin; importance of ascorbic acid, ergothioneine and glutathione. *Br. J. Haematol.* **1959**, *5*, 150–157. [CrossRef]
41. Grundemann, D.; Harlfinger, S.; Golz, S.; Geerts, A.; Lazar, A.; Berkels, R.; Jung, N.; Rubbert, A.; Schomig, E. Discovery of the ergothioneine transporter. *Proc. Natl. Acad. Sci. USA* **2005**, *102*, 5256–5261. [CrossRef]
42. Salama, S.A.; Omar, H.A. Modulating NF-kappaB, MAPK, and PI3K/AKT signaling by ergothioneine attenuates iron overload-induced hepatocellular injury in rats. *J. Biochem. Mol. Toxicol.* **2021**, *35*, e22729. [CrossRef] [PubMed]
43. Yadan, J.C. Matching chemical properties to molecular biological activities opens a new perspective on l-ergothioneine. *FEBS Lett.* **2022**, *596*, 1299–1312. [CrossRef]
44. Servillo, L.; Castaldo, D.; Casale, R.; D'Onofrio, N.; Giovane, A.; Cautela, D.; Balestrieri, M.L. An uncommon redox behavior sheds light on the cellular antioxidant properties of ergothioneine. *Free Radic. Biol. Med.* **2015**, *79*, 228–236. [CrossRef] [PubMed]
45. Servillo, L.; D'Onofrio, N.; Casale, R.; Cautela, D.; Giovane, A.; Castaldo, D.; Balestrieri, M.L. Ergothioneine products derived by superoxide oxidation in endothelial cells exposed to high-glucose. *Free Radic. Biol. Med.* **2017**, *108*, 8–18. [CrossRef] [PubMed]
46. Hartmann, L.; Seebeck, F.P.; Schmalz, H.G.; Grundemann, D. Isotope-labeled ergothioneine clarifies the mechanism of reaction with singlet oxygen. *Free Radic. Biol. Med.* **2023**, *198*, 12–26. [CrossRef] [PubMed]
47. Asmus, K.D.; Bensasson, R.V.; Bernier, J.L.; Houssin, R.; Land, E.J. One-electron oxidation of ergothioneine and analogues investigated by pulse radiolysis: Redox reaction involving ergothioneine and vitamin C. *Biochem. J.* **1996**, *315 Pt 2*, 625–629. [CrossRef]
48. May, J.M.; Cobb, C.E.; Mendiratta, S.; Hill, K.E.; Burk, R.F. Reduction of the ascorbyl free radical to ascorbate by thioredoxin reductase. *J. Biol. Chem.* **1998**, *273*, 23039–23045. [CrossRef] [PubMed]
49. Jenny, K.A.; Mose, G.; Haupt, D.J.; Hondal, R.J. Oxidized Forms of Ergothioneine Are Substrates for Mammalian Thioredoxin Reductase. *Antioxidants* **2022**, *11*, 185. [CrossRef] [PubMed]
50. Reglinski, J.; Smith, W.E.; Sturrock, R.D. Spin-echo 1H NMR detected response of ergothioneine to oxidative stress in the intact human erythrocyte. *Magn. Reson. Med.* **1988**, *6*, 217–223. [CrossRef] [PubMed]
51. Reglinski, J.; Hoey, S.; Smith, W.E.; Sturrock, R.D. Cellular response to oxidative stress at sulfhydryl group receptor sites on the erythrocyte membrane. *J. Biol. Chem.* **1988**, *263*, 12360–12366. [CrossRef]
52. Li, R.W.; Yang, C.; Sit, A.S.; Kwan, Y.W.; Lee, S.M.; Hoi, M.P.; Chan, S.W.; Hausman, M.; Vanhoutte, P.M.; Leung, G.P. Uptake and protective effects of ergothioneine in human endothelial cells. *J. Pharmacol. Exp. Ther.* **2014**, *350*, 691–700. [CrossRef] [PubMed]
53. Song, T.Y.; Lin, H.C.; Chen, C.L.; Wu, J.H.; Liao, J.W.; Hu, M.L. Ergothioneine and melatonin attenuate oxidative stress and protect against learning and memory deficits in C57BL/6J mice treated with D-galactose. *Free Radic. Res.* **2014**, *48*, 1049–1060. [CrossRef] [PubMed]
54. Aruoma, O.I.; Spencer, J.P.; Mahmood, N. Protection against oxidative damage and cell death by the natural antioxidant ergothioneine. *Food Chem. Toxicol.* **1999**, *37*, 1043–1053. [CrossRef] [PubMed]
55. Colognato, R.; Laurenza, I.; Fontana, I.; Coppede, F.; Siciliano, G.; Coecke, S.; Aruoma, O.I.; Benzi, L.; Migliore, L. Modulation of hydrogen peroxide-induced DNA damage, MAPKs activation and cell death in PC12 by ergothioneine. *Clin. Nutr.* **2006**, *25*, 135–145. [CrossRef] [PubMed]
56. Ngo, V.; Duennwald, M.L. Nrf2 and Oxidative Stress: A General Overview of Mechanisms and Implications in Human Disease. *Antioxidants* **2022**, *11*, 2345. [CrossRef] [PubMed]
57. Hseu, Y.C.; Lo, H.W.; Korivi, M.; Tsai, Y.C.; Tang, M.J.; Yang, H.L. Dermato-protective properties of ergothioneine through induction of Nrf2/ARE-mediated antioxidant genes in UVA-irradiated Human keratinocytes. *Free Radic. Biol. Med.* **2015**, *86*, 102–117. [CrossRef] [PubMed]
58. Ko, H.J.; Kim, J.; Ahn, M.; Kim, J.H.; Lee, G.S.; Shin, T. Ergothioneine alleviates senescence of fibroblasts induced by UVB damage of keratinocytes via activation of the Nrf2/HO-1 pathway and HSP70 in keratinocytes. *Exp. Cell Res.* **2021**, *400*, 112516. [CrossRef]
59. Dare, A.; Channa, M.L.; Nadar, A. L-ergothioneine and its combination with metformin attenuates renal dysfunction in type-2 diabetic rat model by activating Nrf2 antioxidant pathway. *Biomed. Pharmacother.* **2021**, *141*, 111921. [CrossRef]

60. Taubert, D.; Grimberg, G.; Jung, N.; Rubbert, A.; Schomig, E. Functional role of the 503F variant of the organic cation transporter OCTN1 in Crohn's disease. *Gut* **2005**, *54*, 1505–1506. [CrossRef]
61. Maeda, T.; Hirayama, M.; Kobayashi, D.; Miyazawa, K.; Tamai, I. Mechanism of the regulation of organic cation/carnitine transporter 1 (SLC22A4) by rheumatoid arthritis-associated transcriptional factor RUNX1 and inflammatory cytokines. *Drug Metab. Dispos.* **2007**, *35*, 394–401. [CrossRef]
62. Sugiura, T.; Kato, S.; Shimizu, T.; Wakayama, T.; Nakamichi, N.; Kubo, Y.; Iwata, D.; Suzuki, K.; Soga, T.; Asano, M.; et al. Functional expression of carnitine/organic cation transporter OCTN1/SLC22A4 in mouse small intestine and liver. *Drug Metab. Dispos.* **2010**, *38*, 1665–1672. [CrossRef] [PubMed]
63. Tamai, I.; Yabuuchi, H.; Nezu, J.; Sai, Y.; Oku, A.; Shimane, M.; Tsuji, A. Cloning and characterization of a novel human pH-dependent organic cation transporter, OCTN1. *FEBS Lett.* **1997**, *419*, 107–111. [CrossRef] [PubMed]
64. Wu, X.; George, R.L.; Huang, W.; Wang, H.; Conway, S.J.; Leibach, F.H.; Ganapathy, V. Structural and functional characteristics and tissue distribution pattern of rat OCTN1, an organic cation transporter, cloned from placenta. *Biochim. Biophys. Acta* **2000**, *1466*, 315–327. [CrossRef] [PubMed]
65. Ilboudo, D.P.; D'Alessandro, S.; Parapini, S.; Calvo Alvarez, E.; Misiano, P.; Taramelli, D.; Basilico, N. A rapid spectrophotometric method to identify inhibitors of human erythropoiesis. *J. Pharmacol. Toxicol. Methods* **2022**, *113*, 107134. [CrossRef]
66. Yee, S.W.; Buitrago, D.; Stecula, A.; Ngo, H.X.; Chien, H.C.; Zou, L.; Koleske, M.L.; Giacomini, K.M. Deorphaning a solute carrier 22 family member, SLC22A15, through functional genomic studies. *FASEB J.* **2020**, *34*, 15734–15752. [CrossRef] [PubMed]
67. Grundemann, D.; Hartmann, L.; Flogel, S. The ergothioneine transporter (ETT): Substrates and locations, an inventory. *FEBS Lett.* **2022**, *596*, 1252–1269. [CrossRef] [PubMed]
68. Kato, Y.; Kubo, Y.; Iwata, D.; Kato, S.; Sudo, T.; Sugiura, T.; Kagaya, T.; Wakayama, T.; Hirayama, A.; Sugimoto, M.; et al. Gene knockout and metabolome analysis of carnitine/organic cation transporter OCTN1. *Pharm. Res.* **2010**, *27*, 832–840. [CrossRef] [PubMed]
69. Ishimoto, T.; Nakamichi, N.; Nishijima, H.; Masuo, Y.; Kato, Y. Carnitine/Organic Cation Transporter OCTN1 Negatively Regulates Activation in Murine Cultured Microglial Cells. *Neurochem. Res.* **2018**, *43*, 116–128. [CrossRef]
70. Pochini, L.; Galluccio, M.; Scalise, M.; Console, L.; Pappacoda, G.; Indiveri, C. OCTN1: A Widely Studied but Still Enigmatic Organic Cation Transporter Linked to Human Pathology and Drug Interactions. *Int. J. Mol. Sci.* **2022**, *23*, 914. [CrossRef]
71. Pochini, L.; Barone, F.; Console, L.; Brunocilla, C.; Galluccio, M.; Scalise, M.; Indiveri, C. OCTN1 (SLC22A4) displays two different transport pathways for organic cations or zwitterions. *Biochim. Biophys. Acta Biomembr.* **2024**, *1866*, 184263. [CrossRef]
72. Bacher, P.; Giersiefer, S.; Bach, M.; Fork, C.; Schomig, E.; Grundemann, D. Substrate discrimination by ergothioneine transporter SLC22A4 and carnitine transporter SLC22A5: Gain-of-function by interchange of selected amino acids. *Biochim. Biophys. Acta* **2009**, *1788*, 2594–2602. [CrossRef] [PubMed]
73. Nakamichi, N.; Shima, H.; Asano, S.; Ishimoto, T.; Sugiura, T.; Matsubara, K.; Kusuhara, H.; Sugiyama, Y.; Sai, Y.; Miyamoto, K.; et al. Involvement of carnitine/organic cation transporter OCTN1/SLC22A4 in gastrointestinal absorption of metformin. *J. Pharm. Sci.* **2013**, *102*, 3407–3417. [CrossRef] [PubMed]
74. Kawano, H.; Higuchi, F.; Mayumi, T.; Hama, T. Studies on ergothioneine. VII. Some effects on ergothioneine on glycolytic metabolism in red blood cells from rats. *Chem. Pharm. Bull.* **1982**, *30*, 2611–2613. [CrossRef] [PubMed]
75. Makiishi, S.; Furuichi, K.; Yamamura, Y.; Sako, K.; Shinozaki, Y.; Toyama, T.; Kitajima, S.; Iwata, Y.; Sakai, N.; Shimizu, M.; et al. Carnitine/organic cation transporter 1 precipitates the progression of interstitial fibrosis through oxidative stress in diabetic nephropathy in mice. *Sci. Rep.* **2021**, *11*, 9093. [CrossRef] [PubMed]
76. D'Alessandro, A.; Howie, H.L.; Hay, A.M.; Dziewulska, K.H.; Brown, B.C.; Wither, M.J.; Karafin, M.; Stone, E.F.; Spitalnik, S.L.; Hod, E.A.; et al. Hematologic and systemic metabolic alterations due to Mediterranean class II G6PD deficiency in mice. *JCI Insight* **2021**, *6*, e147056. [CrossRef] [PubMed]
77. JAX Catalog: SLC22A15 Knockout Mouse. Available online: https://www.informatics.jax.org/allele/MGI:6392063 (accessed on 7 March 2024).
78. Pfeiffer, C.; Bach, M.; Bauer, T.; Campos da Ponte, J.; Schomig, E.; Grundemann, D. Knockout of the ergothioneine transporter ETT in zebrafish results in increased 8-oxoguanine levels. *Free Radic. Biol. Med.* **2015**, *83*, 178–185. [CrossRef] [PubMed]
79. Cheah, I.K.; Ong, R.L.; Gruber, J.; Yew, T.S.; Ng, L.F.; Chen, C.B.; Halliwell, B. Knockout of a putative ergothioneine transporter in Caenorhabditis elegans decreases lifespan and increases susceptibility to oxidative damage. *Free Radic. Res.* **2013**, *47*, 1036–1045. [CrossRef] [PubMed]
80. Toh, D.S.; Cheung, F.S.; Murray, M.; Pern, T.K.; Lee, E.J.; Zhou, F. Functional analysis of novel variants in the organic cation/ergothioneine transporter 1 identified in Singapore populations. *Mol. Pharm.* **2013**, *10*, 2509–2516. [CrossRef]
81. Urban, T.J.; Yang, C.; Lagpacan, L.L.; Brown, C.; Castro, R.A.; Taylor, T.R.; Huang, C.C.; Stryke, D.; Johns, S.J.; Kawamoto, M.; et al. Functional effects of protein sequence polymorphisms in the organic cation/ergothioneine transporter OCTN1 (SLC22A4). *Pharmacogenet. Genom.* **2007**, *17*, 773–782.
82. Dumaswala, U.J.; Zhuo, L.; Jacobsen, D.W.; Jain, S.K.; Sukalski, K.A. Protein and lipid oxidation of banked human erythrocytes: Role of glutathione. *Free Radic. Biol. Med.* **1999**, *27*, 1041–1049. [CrossRef]
83. Dumaswala, U.J.; Wilson, M.J.; Wu, Y.L.; Wykle, J.; Zhuo, L.; Douglass, L.M.; Daleke, D.L. Glutathione loading prevents free radical injury in red blood cells after storage. *Free Radic. Res.* **2000**, *33*, 517–529. [CrossRef] [PubMed]

84. Kim, C.Y.; Johnson, H.; Peltier, S.; Spitalnik, S.L.; Hod, E.A.; Francis, R.O.; Hudson, K.E.; Stone, E.F.; Gordy, D.E.; Fu, X.; et al. Deuterated Linoleic Acid Attenuates the RBC Storage Lesion in a Mouse Model of Poor RBC Storage. *Front. Physiol.* **2022**, *13*, 868578. [CrossRef] [PubMed]
85. Greenwalt, T.J.; Bryan, D.J.; Dumaswala, U.J. Erythrocyte membrane vesiculation and changes in membrane composition during storage in citrate-phosphate-dextrose-adenine-1. *Vox. Sang.* **1984**, *47*, 261–270. [CrossRef] [PubMed]
86. Blasi, B.; D'Alessandro, A.; Ramundo, N.; Zolla, L. Red blood cell storage and cell morphology. *Transfus. Med.* **2012**, *22*, 90–96. [CrossRef] [PubMed]
87. Roubinian, N.H.; Reese, S.E.; Qiao, H.; Plimier, C.; Fang, F.; Page, G.P.; Cable, R.G.; Custer, B.; Gladwin, M.T.; Goel, R.; et al. Donor genetic and nongenetic factors affecting red blood cell transfusion effectiveness. *JCI Insight* **2022**, *7*, e152598. [CrossRef] [PubMed]
88. Nemkov, T.; Stefanoni, D.; Bordbar, A.; Issaian, A.; Palsson, B.O.; Dumont, L.J.; Hay, A.; Song, A.; Xia, Y.; Redzic, J.S.; et al. Blood donor exposome and impact of common drugs on red blood cell metabolism. *JCI Insight* **2021**, *6*, e146175. [CrossRef] [PubMed]
89. Wang, L.Z.; Thuya, W.L.; Toh, D.S.; Lie, M.G.; Lau, J.Y.; Kong, L.R.; Wan, S.C.; Chua, K.N.; Lee, E.J.; Goh, B.C. Quantification of L-ergothioneine in human plasma and erythrocytes by liquid chromatography-tandem mass spectrometry. *J. Mass Spectrom.* **2013**, *48*, 406–412. [CrossRef] [PubMed]
90. McMenamy, R.H.; Lund, C.C.; Neville, G.J.; Wallach, D.F. Studies of Unbound Amino Acid Distributions in Plasma, Erythrocytes, Leukocytes and Urine of Normal Human Subjects. *J. Clin. Investig.* **1960**, *39*, 1675–1687. [CrossRef] [PubMed]
91. Jocelyn, P.C. The distribution of ergothioneine in blood as determined by a new method of estimation. *Biochem. J.* **1958**, *70*, 656–660. [CrossRef]
92. Mitsuyama, H.; May, J.M. Uptake and antioxidant effects of ergothioneine in human erythrocytes. *Clin. Sci.* **1999**, *97*, 407–411. [CrossRef]
93. Chaleckis, R.; Ebe, M.; Pluskal, T.; Murakami, I.; Kondoh, H.; Yanagida, M. Unexpected similarities between the Schizosaccharomyces and human blood metabolomes, and novel human metabolites. *Mol. Biosyst.* **2014**, *10*, 2538–2551. [CrossRef] [PubMed]
94. Fraser, R.S. Blood ergothioneine levels in disease. *J. Lab. Clin. Med.* **1951**, *37*, 199–206. [PubMed]
95. Touster, O.; Yarbro, M.C. The ergothioneine content of human erythrocytes; the effect of age, race, malignancy, and pregnancy. *J. Lab. Clin. Med.* **1952**, *39*, 720–724. [PubMed]
96. Chaleckis, R.; Murakami, I.; Takada, J.; Kondoh, H.; Yanagida, M. Individual variability in human blood metabolites identifies age-related differences. *Proc. Natl. Acad. Sci. USA* **2016**, *113*, 4252–4259. [CrossRef] [PubMed]
97. Kumosani, T.A. L-ergothioneine level in red blood cells of healthy human males in the Western province of Saudi Arabia. *Exp. Mol. Med.* **2001**, *33*, 20–22. [CrossRef] [PubMed]
98. Kumar, P.; Maurya, P.K. L-cysteine efflux in erythrocytes as a function of human age: Correlation with reduced glutathione and total anti-oxidant potential. *Rejuvenation Res.* **2013**, *16*, 179–184. [CrossRef] [PubMed]
99. Beatty, C.H. Levels of ergothioneine in the red blood cell in experimental diabetes. *J. Biol. Chem.* **1952**, *199*, 553–561. [CrossRef] [PubMed]
100. Jamshidi, N.; Xu, X.; von Lohneysen, K.; Soldau, K.; Mohney, R.P.; Karoly, E.D.; Scott, M.; Friedman, J.S. Metabolome Changes during In Vivo Red Cell Aging Reveal Disruption of Key Metabolic Pathways. *Iscience* **2020**, *23*, 101630. [CrossRef]
101. Zimring, J.C.; Smith, N.; Stowell, S.R.; Johnsen, J.M.; Bell, L.N.; Francis, R.O.; Hod, E.A.; Hendrickson, J.E.; Roback, J.D.; Spitalnik, S.L. Strain-specific red blood cell storage, metabolism, and eicosanoid generation in a mouse model. *Transfusion* **2014**, *54*, 137–148. [CrossRef]
102. Heath, H.; Rimington, C.; Searle, C.E.; Lawson, A. Some effects of administering ergothioneine to rats. *Biochem. J.* **1952**, *50*, 530–533. [CrossRef]
103. Spicer, S.S.; Wooley, J.G.; Kessler, V. Ergothioneine depletion in rabbit erythrocytes and its effect on methemoglobin formation and reversion. *Proc. Soc. Exp. Biol. Med.* **1951**, *77*, 418–420. [CrossRef]
104. Taubert, D.; Lazar, A.; Grimberg, G.; Jung, N.; Rubbert, A.; Delank, K.S.; Perniok, A.; Erdmann, E.; Schomig, E. Association of rheumatoid arthritis with ergothioneine levels in red blood cells: A case control study. *J. Rheumatol.* **2006**, *33*, 2139–2145. [PubMed]
105. Liu, H.; Zhang, Y.; Wu, H.; D'Alessandro, A.; Yegutkin, G.G.; Song, A.; Sun, K.; Li, J.; Cheng, N.Y.; Huang, A.; et al. Beneficial Role of Erythrocyte Adenosine A2B Receptor-Mediated AMP-Activated Protein Kinase Activation in High-Altitude Hypoxia. *Circulation* **2016**, *134*, 405–421. [CrossRef] [PubMed]
106. Seko, T.; Sato, Y.; Kuniyoshi, M.; Murata, Y.; Ishihara, K.; Yamashita, Y.; Fujiwara, S.; Ueda, T.; Yamashita, M. Distribution and Effects of Selenoneine by Ingestion of Extract from Mackerel Processing Residue in Mice. *Mar. Biotechnol.* **2023**, *25*, 1020–1030. [CrossRef] [PubMed]
107. Kobayashi, D.; Aizawa, S.; Maeda, T.; Tsuboi, I.; Yabuuchi, H.; Nezu, J.; Tsuji, A.; Tamai, I. Expression of organic cation transporter OCTN1 in hematopoietic cells during erythroid differentiation. *Exp. Hematol.* **2004**, *32*, 1156–1162. [CrossRef] [PubMed]
108. Tan, G.; Meier-Abt, F. Differential expression of hydroxyurea transporters in normal and polycythemia vera hematopoietic stem and progenitor cell subpopulations. *Exp. Hematol.* **2021**, *97*, 47–56.e45. [CrossRef] [PubMed]
109. Haiman, Z.B.; Palsson, B.; D'Alessandro, A. RBC-GEM: A Knowledge Base for Systems Biology of Human Red Blood Cell Metabolism. *bioRxiv* **2024**. [CrossRef]

110. Nakamura, T.; Sugiura, S.; Kobayashi, D.; Yoshida, K.; Yabuuchi, H.; Aizawa, S.; Maeda, T.; Tamai, I. Decreased proliferation and erythroid differentiation of K562 cells by siRNA-induced depression of OCTN1 (SLC22A4) transporter gene. *Pharm. Res.* **2007**, *24*, 1628–1635. [CrossRef]
111. Lee, S.J.; Jung, C.; Oh, J.E.; Kim, S.; Lee, S.; Lee, J.Y.; Yoon, Y.S. Generation of Red Blood Cells from Human Pluripotent Stem Cells-An Update. *Cells* **2023**, *12*, 1554. [CrossRef]
112. Mortensen, R.A. The effect of diet on methemoglobin levels of nitrite-injected rats. *Arch. Biochem. Biophys.* **1953**, *46*, 241–243 [CrossRef]
113. Rae, C.D.; Sweeney, K.J.; Kuchel, P.W. Stability and nonreactivity of ergothioneine in human erythrocytes studied by 1H NMR. *Magn. Reson. Med.* **1993**, *29*, 826–829. [CrossRef] [PubMed]
114. Brennan, F.M.; McInnes, I.B. Evidence that cytokines play a role in rheumatoid arthritis. *J. Clin. Investig.* **2008**, *118*, 3537–3545. [CrossRef] [PubMed]
115. Szarka, A.; Rigo, J., Jr.; Lazar, L.; Beko, G.; Molvarec, A. Circulating cytokines, chemokines and adhesion molecules in normal pregnancy and preeclampsia determined by multiplex suspension array. *BMC Immunol.* **2010**, *11*, 59. [CrossRef] [PubMed]
116. Alexandrova-Watanabe, A.; Abadjieva, E.; Giosheva, I.; Langari, A.; Tiankov, T.; Gartchev, E.; Komsa-Penkova, R.; Todinova, S. Assessment of Red Blood Cell Aggregation in Preeclampsia by Microfluidic Image Flow Analysis-Impact of Oxidative Stress on Disease Severity. *Int. J. Mol. Sci.* **2024**, *25*, 3732. [CrossRef] [PubMed]
117. Turner, E.; Brewster, J.A.; Simpson, N.A.; Walker, J.J.; Fisher, J. Imidazole-based erythrocyte markers of oxidative stress in preeclampsia--an NMR investigation. *Reprod. Sci.* **2009**, *16*, 1040–1051. [CrossRef] [PubMed]
118. Francis, R.O.; Jhang, J.S.; Pham, H.P.; Hod, E.A.; Zimring, J.C.; Spitalnik, S.L. Glucose-6-phosphate dehydrogenase deficiency in transfusion medicine: The unknown risks. *Vox Sang.* **2013**, *105*, 271–282. [CrossRef] [PubMed]
119. Darghouth, D.; Koehl, B.; Madalinski, G.; Heilier, J.F.; Bovee, P.; Xu, Y.; Olivier, M.F.; Bartolucci, P.; Benkerrou, M.; Pissard, S.; et al. Pathophysiology of sickle cell disease is mirrored by the red blood cell metabolome. *Blood* **2011**, *117*, e57–e66. [CrossRef] [PubMed]
120. Chaves, N.A.; Alegria, T.G.P.; Dantas, L.S.; Netto, L.E.S.; Miyamoto, S.; Bonini Domingos, C.R.; da Silva, D.G.H. Impaired antioxidant capacity causes a disruption of metabolic homeostasis in sickle erythrocytes. *Free Radic. Biol. Med.* **2019**, *141*, 34–46. [CrossRef] [PubMed]
121. Darghouth, D.; Koehl, B.; Heilier, J.F.; Madalinski, G.; Bovee, P.; Bosman, G.; Delaunay, J.; Junot, C.; Romeo, P.H. Alterations of red blood cell metabolome in overhydrated hereditary stomatocytosis. *Haematologica* **2011**, *96*, 1861–1865. [CrossRef]
122. McMenamy, R.H.; Lund, C.C.; Wallach, D.F. Unbound Amino Acid Concentrations in Plasma, Erythrocytes, Leukocytes and Urine of Patients with Leukemia. *J. Clin. Investig.* **1960**, *39*, 1688–1705. [CrossRef]
123. Moore, A.; Busch, M.P.; Dziewulska, K.; Francis, R.O.; Hod, E.A.; Zimring, J.C.; D'Alessandro, A.; Page, G.P. Genome-wide metabolite quantitative trait loci analysis (mQTL) in red blood cells from volunteer blood donors. *J. Biol. Chem.* **2022**, *298*, 102706. [CrossRef]

antioxidants

MDPI

Review

Targeting Cysteine Oxidation in Thrombotic Disorders

Moua Yang [1,*] and Roy L. Silverstein [2,3,*]

1 Division of Hemostasis and Thrombosis, Beth Israel Deaconess Medical Center, Harvard Medical School, 3 Blackfan Circle, CLS-924, Boston, MA 02115, USA
2 Department of Medicine, Medical College of Wisconsin, Hub 8745, 8701 W Watertown Plank Rd., Milwaukee, WI 53226, USA
3 Versiti Blood Research Institute, Milwaukee, WI 53226, USA
* Correspondence: myang4@bidmc.harvard.edu (M.Y.); rsilverstein@mcw.edu (R.L.S.); Tel.: +1-(617)-735-4017 (M.Y.); +1-(414)-955-0518 (R.L.S.)

Abstract: Oxidative stress increases the risk for clinically significant thrombotic events, yet the mechanisms by which oxidants become prothrombotic are unclear. In this review, we provide an overview of cysteine reactivity and oxidation. We then highlight recent findings on cysteine oxidation events in oxidative stress-related thrombosis. Special emphasis is on the signaling pathway induced by a platelet membrane protein, CD36, in dyslipidemia, and by protein disulfide isomerase (PDI), a member of the thiol oxidoreductase family of proteins. Antioxidative and chemical biology approaches to target cysteine are discussed. Lastly, the knowledge gaps in the field are highlighted as they relate to understanding how oxidative cysteine modification might be targeted to limit thrombosis.

Keywords: oxidative stress; cysteine; CD36; protein disulfide isomerase; thrombosis; platelets; nucleophiles

Citation: Yang, M.; Silverstein, R.L. Targeting Cysteine Oxidation in Thrombotic Disorders. *Antioxidants* **2024**, *13*, 83. https://doi.org/10.3390/antiox13010083

Academic Editor: Angelo D'Alessandro

Received: 5 December 2023
Revised: 29 December 2023
Accepted: 5 January 2024
Published: 9 January 2024

1. Introduction

Many pathologic conditions are associated with oxidative stress and have increased risk for clinically significant thrombotic events. These conditions include, but are not limited to, disorders of metabolism (e.g., dyslipidemia [1,2], diabetes mellitus [3], and obesity [4]), chronic systemic inflammation [1], aging [5], cancer [6], infection [7], and blood disorders including hemoglobinopathy [8,9], and antiphospholipid syndrome [10]. Clinically significant thrombotic events promote organ dysfunction, organ failure, and sometimes death [11]. Thrombus formation is a complex, multifaceted process, and oxidative events initiated by reactive species of oxygen, nitrogen, sulfur, and carbon play important roles. These oxidative species induce modifications of cellular constitutions, including lipids and proteins [12]. Oxidative modifications of proteins do not induce clotting on their own, nor are they the final effectors of clotting, but like phosphorylation events that sensitize downstream signaling pathways, oxidative modifications may enhance or limit thrombosis [13] and thus provide additional layers of control of the thrombotic process.

Essential to oxidative post-translational modification of proteins are the reactive species themselves. The chemical reduction or oxidation of molecular oxygen generates reactive oxygen species (ROS), and a detailed review of specific species and their generation mechanisms is provided in a recent review [14]. It is the two-electron oxidants, such as hydrogen peroxide, lipid hydroperoxide, and peroxynitrite, that oxidize thiol side chains of cysteines [15]. Other amino acids are also susceptible to oxidation by ROS; for example, peroxynitrite formed from interaction of superoxide radical and nitric oxide oxidizes tyrosine to form nitrotyrosine [16]. Hypochlorous acid (HOCl), generated by myeloperoxidases in white blood cells, oxidizes methionine and halogenates tyrosine to generate methionine sulfoxide [17] and halo-tyrosine [18], respectively. In the context of thrombosis and hemostasis, we recently provided a review on oxidative cysteine and methionine modification [13].

As we delve deeper into the redox biology of thrombosis, fundamental knowledge gaps remain pertinent to understanding how the thrombotic machinery is transformed by redox cues from a protective homeostatic process to a pathological vaso-occlusive thrombotic process. Essential to this knowledge gap is the question of whether it is one protein or a network of proteins that become oxidized or reduced to promote pathologic thrombosis. Also, could specific enzymes sense these redox cues and transform the promiscuous electrophilic nature of oxidants to an enzyme-driven process allowing for specificity in thrombotic signaling pathways? And lastly, could these oxidative events be targeted therapeutically to decrease pathologic thrombus without compromising hemostasis? In this review, we discuss cysteine biochemistry relating to its reactivity, the different types of oxidative cysteine modification, and recent advances in cysteine sulfenylation in oxidative stress related thrombosis. We also discuss potential therapeutic avenues to oxidative cysteine modification rooted in evidence from the use of a nucleophilic fragment library designed to target oxidized thiols.

2. The Biochemistry of Cysteines

2.1. Cysteine Reactivity: Not All Are Created Equal

Because of its unique redox properties, cysteine plays diverse roles in protein structure and function. Evolutionarily, cysteines are found at both highly conserved and non-conserved sites, and not surprisingly, mutations at these sites often result in pathologic conditions [19]. Cysteine pairs within proteins can form disulfide bonds, thereby contributing importantly to secondary and higher order structure. These residues are often buried from solvent exposure [20]. Cysteines can also be in their free thiol state, where they are exposed to solvent and play non-structural roles, including regulation of enzyme activity (e.g., in kinases) [21,22], metal binding (e.g., Zn-finger transcriptional factors) [23], catalytic redox reactions (e.g., thiol isomerases as described in Section 3) [24,25], and catalytic nucleophilic reactions (e.g., caspases and phosphatases) [26]. The nucleophilicity of the thiol is determined by the stabilization of the negatively charged thiolate [27], a physicochemical property of the sulfur atom.

Deprotonation of the cysteine thiol is a critical step in generating the nucleophilic thiolate and depends on the acid dissociation constant or pK_a of the cysteine and the local pH of the environment [28]. If pK_a is lower than the pH, the fraction of cysteine in the deprotonated form is higher. At physiologic pH ~7.4, the cysteines with pK_a of less than 7.4 should have a higher fraction in the thiolate form and are thus more "reactive". However, it should be noted that the pK_a of the thiol group of free cysteine is ~8.3 and can range substantially (from 3.5 to 13 depending on the microenvironment) [29,30]. The pH of the local environment is also not uniform [30]; lysosomes, for example, have much lower pH (pH 4–5) relative to the cytosol (pH 6.8–7.4) and mitochondria (pH~8.0 within mitochondrial matrix) [31]. As the pK_a is linked to the pH of the aqueous environment, solvent accessibility (e.g., buried vs. exposed) influences cysteine reactivity. Additional factors that influence thiolate formation include proximal structural elements of the protein that can stabilize the negative charge [32]. Hydrogen bond donors, such as the hydroxyl group of threonines as in thioredoxins [33], have a partial positive charge from the hydrogen allowing for stability in the thiolate through bonding with the sulfur atom [29]. Electropositive (basic) amino acids including arginine and histidine afford additional ionic interaction with the negative charge of the sulfur and stabilize the thiolate for redox events [34]. Similarly, positively charged macrodipoles of alpha helices afford electrostatic interactions with the thiolate for stability [35].

Cysteine reactivity is also controlled by the redox potential of the thiol. Electrons are poised to proceed from the more electronegative to the more electropositive redox potential, with the potential ranging substantially in biologic systems (from −480 to −80 mV) [36]. In an isolated system, these potentials could be calculated for a thiol using different pairs of redox couplers (e.g., reduced glutathione (GSH) with oxidized glutathione (GSSG); reduced dithiothreitol with oxidized dithiothreitol; cysteines and cystines); however, the dynamics of redox potential within a biological system are quite complex and difficult to calculate owing to the different redox couplers in any given location [30]. This suggests that cysteines could be more reactive within different sub-compartments of the cell, with a classic example being the endoplasmic reticulum where the environment is more oxidizing to facilitate oxidative protein folding.

Lastly, an environmental factor that influences cysteine reactivity is proximity to the source of oxidants. Many sources within a cell promote electron leakages (and transfers) that could generate reactive oxygen species [37]. A protein that is closer to a source of oxidant is more likely to encounter the reactive species prior to the oxidant degrading through an efficient antioxidative mechanism or through competing oxidizing targets. Proteins further away from a source of oxidant are protected as the oxidants have less of a chance to reach the target. Evidence connecting proximity of a protein to an oxidant source and oxidation of that protein were demonstrated in studies of an epidermal carcinoma cell line [38]. In these cells, Epidermal Growth Factor (EGF) binding to its receptor (EGFR) [38,39] promotes receptor dimerization and activation, allowing for downstream signaling for proliferation, differentiation, growth, and survival [40]. The proximity of EGFR to NADPH oxidase was shown to promote cysteine oxidation of the receptor, enhancing kinase activity and downstream signaling.

2.2. *Cysteine Oxoforms*

Oxidation of the cysteine thiolate generates many sulfur oxoforms. A depiction of these oxoforms is presented in Figure 1. The sulfur atom attains valences between −2 and +6 oxidation states [15]. Oxidation of the thiolate by peroxide generates a sulfenic acid, which is an intermediate oxoform that is labile and susceptible to convert to other oxoforms [15]. Further oxidation of the sulfenic acid could generate sulfinic and sulfonic acid. Oxidation of the sulfinic acid is reversible with sulfinic acid reductase [41], whereas the sulfonic acid is currently believed to be irreversible. Sulfenic acids could also lead to the formation of other oxidative cysteine modifications, including cysteine glutathionylation (by reaction with glutathione), sulfenamides (reacting with an amine) [42], thiolsulfenates (reacting with a nearby sulfenic acid) [15], and disulfides [15,43]. In some instances, other cysteine oxoforms could be converted over to a sulfenic acid (as is the case for nitrosothiols) [15]. These oxoforms emphasize the significance of the sulfenic acid as a "gateway" modification that may be able to be targeted pharmacologically.

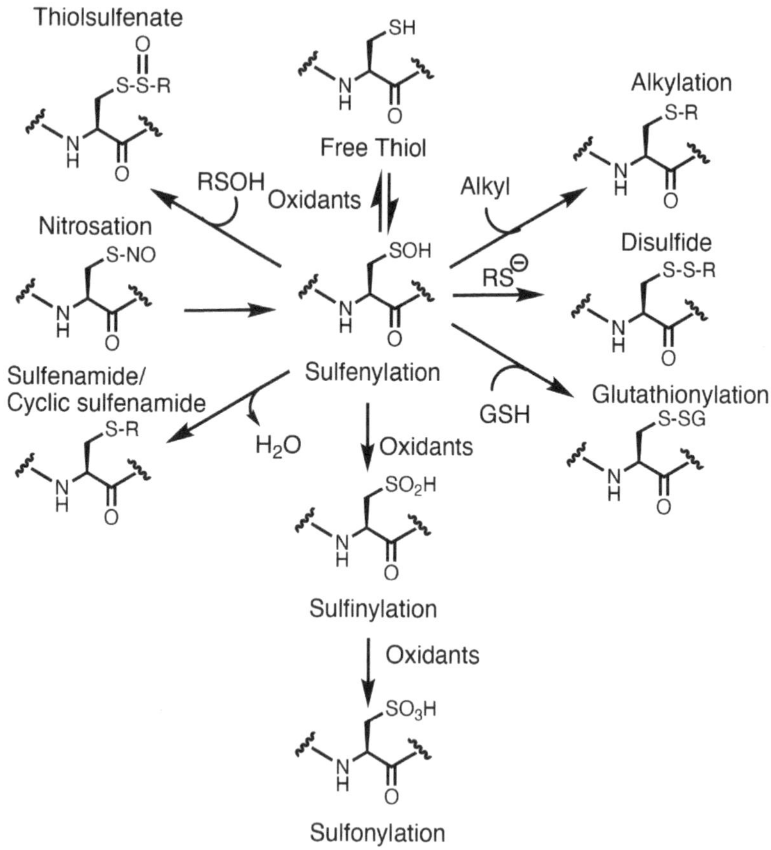

Figure 1. Oxidative modification of cysteine. The cysteine thiolate is nucleophilic and susceptible to oxidation. Oxidation of the thiol generates a transient and labile modification: the sulfenic acid (sulfenylation). Sulfur oxidation by nitric oxide or nitrosothiols could also lead to sulfenic acid formation. Sulfenylation is a cysteine oxoform that is at the crossroad of further oxidative cysteine modification, including sulfinylation, sulfonylation, glutathionylation, disulfide, alkylation, thiolsulfenate, and sulfenamides.

3. Oxidative Cysteine Modification in Oxidative Stress-Driven Thrombosis

3.1. Thrombosis Is a Multifaceted Process

Hemostasis, the physiological system that maintains blood fluidity and vascular integrity, is a highly regulated, tightly balanced process. The ability to prevent excessive bleeding after vascular injury involves many components of the vasculature, including cells and factors that are derived from both hematopoietic and non-hematopoietic lineages [44]. Endothelial cells that line the lumen of the vessel are quiescent and express antithrombotic entities [45], including prostacyclin [46] and nitric oxide [47], which are released to dampen platelet responses and inhibit thrombosis. These factors activate adenylyl and guanylyl cyclases within platelets, generating cAMP [48] and cGMP, respectively [47]. These cyclic nucleotides are secondary messengers promoting a cascade of phosphorylation events driven by protein kinase A and protein kinase G that limit platelet activation [48]. Cell surface expression of anticoagulant molecules including heparan sulfate proteoglycan and thrombomodulin also limit the thrombotic process [45] by interfering with thrombin generation and fibrin formation. Antiplatelet ectonucleotidases are associated with endothelial cells, including CD39 and CD73, and prevent nucleotide induced activation of platelets [49]. When the continuity of the endothelial layer is disrupted by injury (Figure 2, top), the underlining extracellular matrix is exposed to flowing blood leading to platelet recruitment, adhesion, and aggregation resulting in formation of the primary hemostatic plug [50]. In addition, exposure of the transmembrane protein tissue factor to blood at the site of injury activates the coagulation cascade through the prothrombinase and tenase serine protease complexes. These complexes ultimately lead to the generation of thrombin, a protease that cleaves plasma soluble fibrinogen to insoluble fibrin monomers which polymerize to form a fibrin mesh, stabilizing the clot [44]. Exposed phosphatidylserine on activated platelets further amplifies thrombin-generation potential by serving as a cofactor for the prothrombinase and tenase complexes. Dysregulation of this process by oxidants could contribute to the pro-thrombotic state associated with metabolic syndromes, chronic inflammation, and other pathogenic conditions where thrombus formation could occlude the lumen of the vasculature (Figure 2, bottom).

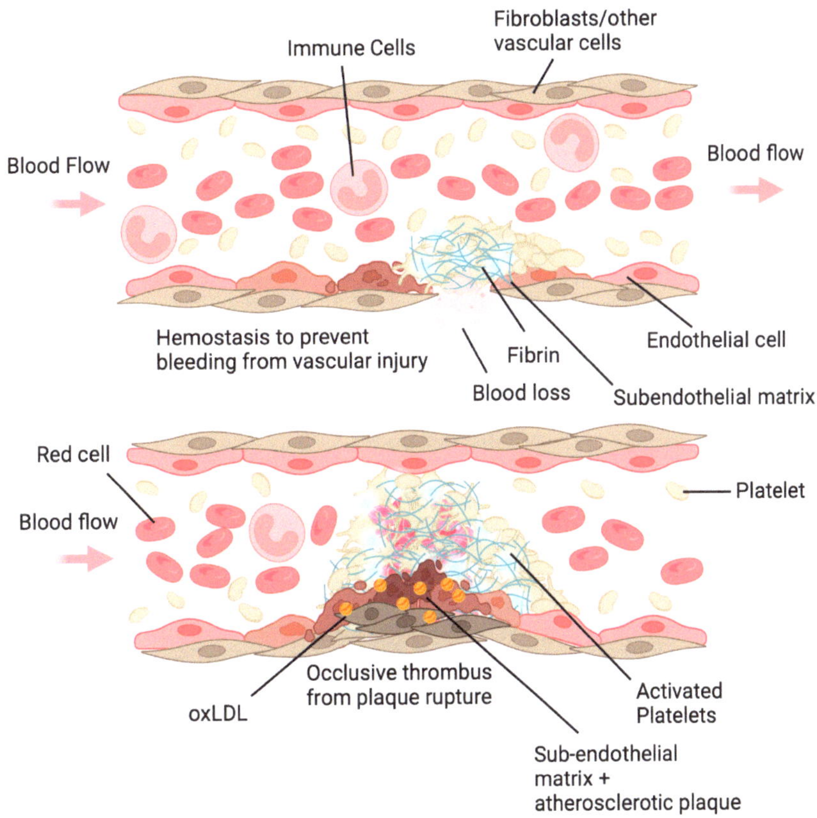

Figure 2. Hemostasis and thrombosis. (**Top**) In healthy states, a physical injury to a blood vessel causes extrusion of the blood, but the blood loss is limited by formation of a hemostatic plug generated by platelet adhesion and aggregation at the site of injury. Fibrin deposition generated from activation of the coagulation serine protease cascade at the site of injury creates a proteinaceous network that stabilizes the plug. (**Bottom**) In pathological conditions associated with oxidative stress, the normal hemostatic response is amplified and can lead to formation of an occlusive thrombus at the site of a minor injury, such as occurs in the setting of a small rupture of atherosclerotic plaque. Created with BioRender.com and accessed on 4 January 2024. oxLDL, oxidized low-density lipoprotein particle.

3.2. CD36 Links Oxidative Stress to Prothrombotic Cysteine Oxidative Signaling

Evidence for oxidative stress-driven thrombosis is derived from multiple independent studies using targeted interruption of genes involved in oxidant regulation [51–53], pharmacologic inhibition of oxidant sources [54,55], chemical degraders or generators of oxidants [56,57], cellular assays that involve oxidants (e.g., platelet activation and endothelial cell-mediated procoagulant activity) [58–60], and animal models of thrombosis [57,61]. We focus on atherothrombosis, which is arterial thrombosis in the context of atherosclerosis, based on evidence from our lab and others that oxidative stress is prominent in dyslipidemia and other chronic atherogenic states and that redox cues in these settings are converted to a pro-thrombotic response [2,62,63].

Dyslipidemia refers to abnormal levels and/or composition of lipids in the circulation that are mainly carried by lipoprotein particles [64,65]. In atherothrombosis, dyslipidemia links atherosclerotic plaque development, plaque instability, and ultimately plaque rupture with subsequent thrombosis [65]. Among the different types of lipoproteins present in the circulation are high-density lipoprotein (HDL) and low-density lipoprotein (LDL) [65]. These particles are susceptible to oxidation during the inflammatory process of plaque development, and oxidation generates oxidized lipid moieties on the particles that are recognized by specific proteins of the innate immune system [66,67]. Oxidation of LDL (oxLDL) creates a high-affinity ligand for the scavenger receptor CD36 [68,69], a transmembrane glycoprotein of the innate immune system that is expressed on multiple vascular cells and was first identified as the fourth-largest major membrane protein on platelets [70]. Recognition of oxLDL by platelet CD36 promotes multiple signaling events that link to desensitization of platelet inhibitory pathways [54,71,72] or sensitization of platelet activation pathways [55,71,73]. A comprehensive understanding of platelet CD36 signaling is still an active area of research and has recently been reviewed [71].

Generation of ROS in platelets by CD36 signaling contributes to its prothrombotic signaling (Figure 3). In addition to oxidized lipids, CD36 has many ligands including advanced glycated end products (AGEs) [74], apoptotic cells [75] and microparticles [76,77], thrombospondins and other proteins containing the thrombospondin type 1 repeat domain [78,79], long-chain fatty acids [80], *Plasmodium falciparum*-infected red cells [81], components of bacteria cell wall [82,83], beta amyloid [84], and the myeloid-related protein 14 (or S100A9) protein [85]. We and others have found that CD36 recognition of oxLDL promotes recruitment and activation of the Src family tyrosine kinases, Fyn and Lyn as signaling transducers [73]. Src kinases phosphorylate many downstream components [86], including a pathway that leads to the activation of the oxidant-sensitive mitogen-activated protein (MAP) kinase, Big MAP kinase/extracellular signal-regulated kinase 5 (ERK5) [55]. Src kinase signaling also cascades to the activation of the subunits of reduced nicotinamide adenine dinucleotide phosphate (NADPH) oxidase that promotes electron transfer from NADPH to molecular oxygen [87]. Aside from the mitochondrial respiratory chain, the electron transferring event induced by NADPH oxidase is a major pathway for generating superoxide radical anion [14,88]. Superoxide anion is an unstable ROS, however, and does not directly react with free thiols. In addition, superoxide anion also does not traverse the cell membrane due to the nature of its charge. Superoxide anion rapidly dismutates both spontaneously (10^5 M^{-1} s^{-1}) and catalytically via superoxide dismutase (2×10^9 M^{-1} s^{-1}) [89], converting the oxidant to the more stable hydrogen peroxide that is electrophilic, able to diffuse across the cellular membrane, and react with free thiols.

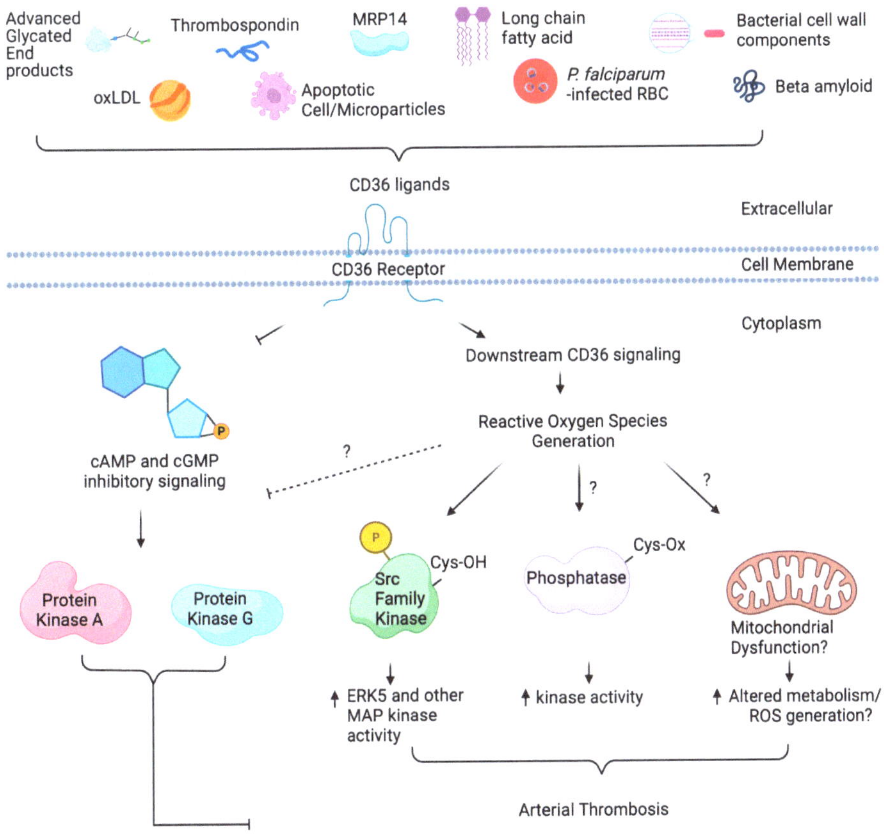

Figure 3. Platelet CD36 redox signaling in arterial thrombosis. Platelet CD36 has multiple ligands including oxidized lipids in lipoprotein particles, advanced glycated end products, thrombospondin, long chain fatty acids, *P. falciparum*-infected red cells, myeloid-related protein 14 (S100A9; MRP14), components of bacterial cell wall, apoptotic cells and microparticles, and beta amyloid. The best-characterized CD36 ligand is oxidized phospholipid present in oxLDL. Recognition of these ligands by CD36 at the surface of the cell promotes two downstream signaling pathway within platelets. One is through desensitization of the cAMP and cGMP signaling pathway that are the major inhibitory pathways within platelets, and the other is through downstream ROS signaling. The effects of ROS are multifaceted and include oxidative cysteine modification of Src family tyrosine kinases to enhance kinase activity and increase the activation of big MAP kinase ERK5 for a pro-aggregatory and pro-coagulant platelet phenotype. In macrophages and endothelial cells, CD36-mediated ROS generation inactivates specific mixed phosphatases including SHP-1 and SHP-2, resulting in prolonged activation of downstream kinases. In addition, CD36 alters mitochondrial metabolism within macrophages into a pro-inflammatory, pro-atherogenic, and ROS-generating organelle. We hypothesize similar mechanisms within platelets for phosphatase inactivation and mitochondrial dysfunction to enhance prothrombotic kinase activity and oxidative stress in dyslipidemia. It is also possible that ROS generated through CD36 activation participates in desensitizing the inhibitory pathways through oxidative cysteine modification. Created with BioRender.com and accessed on 4 January 2024. ?, hypothesized mechanism that requires investigation; ↑ with no faded tail, upregulates or increases arrow; ↓ with faded tail, cellular signaling arrow; ⊥ with faded tail, inhibitory signaling or inhibits.

We found that hydrogen peroxide generated from NADPH oxidase was able to sulfenylate exposed cysteines on Src kinases, converting the redox signals into autophosphorylation events. Although it was not directly tested, we hypothesized that Cys278 in Fyn is the key site of sulfenylation in platelets since this site is known to function as a negative regulator of kinase activity [90], and Lyn does not contain cysteine at its homologous site [91]. This supposition requires further investigation, for example, by mutating these cysteine residues to non-reactive alanines within platelets. Nonetheless, oxidation of Src family kinases promotes downstream platelet CD36 signaling, including activation of ERK5. A direct role for peroxides was further validated by inhibiting NADPH oxidase or directly scavenging hydrogen peroxide with catalase, both of which eliminated sulfenylation of Src family kinases and ERK5 activation. These approaches also prevented downstream functions of ERK5, including signaling integration with the glycoprotein VI (GPVI) pathway for apoptotic-like caspase activation and subsequent externalization of procoagulant phosphatidylserine (PSer) [92]. PSer provides the negatively charged surface for assembly of coagulation factors Xa and Va in an active prothrombinase complex, allowing for localized thrombin generation with ensuing fibrin deposition [44].

ERK5 also sensitizes platelets to integrin activation, lowering the threshold for platelets to aggregate for a full prothrombotic phenotype. In this context, the mechanism downstream of ERK5 is unclear. An unbiased approach to identifying phospho-targets of ERK5 by CD36 signaling would close the signaling gaps that link ERK5 to a procoagulant and pro-aggregatory platelet response. It is also unclear how Src activation by hydrogen peroxide promotes ERK5 phosphorylation. Given what is known of the mitogen-activated protein kinase kinase 5 (MEK5)-ERK5 pathway in other cell types [93], it is likely that Src family kinases directly phosphorylate upstream MAP kinase kinases upstream of ERK5, allowing for ERK5 to be activated. It could also be possible that ERK5 is directly sulfenylated. This supposition is supported by sulfenylation events with the related MAP kinase family member ERK2 [94]. ERK2 oxidation at Cys159 by peroxides influences substrate selectivity within its D-recruitment substrate recognition site (DRS). Sulfenylation at this site increases interaction between ERK2 and a model DRS-specific peptide substrate as well as the ribosomal S6 kinase A1 (RSK1) substrate. In this context, sulfenylation of ERK2 increases RSK1 kinase activity [94]. Although the presence of RSK1 and its function in platelets is yet to be studied, the oxidation of ERK2 suggests a likely sulfenylation site within ERK5 based on conservation of the cysteine and the 66% structural homology with ERK2 [95]. Sulfenylation of ERK5 could thus functionally promote platelet reactivity in dyslipidemia.

CD36 potentially links oxidants and cysteine oxidation to other cellular dysfunction that could impact the prothrombotic response during vessel injury as shown in Figure 3. Microvascular endothelial cells (MVECs) and macrophages express CD36 and ligation of the receptor with extracellular vesicles (EVs) [77], or oxLDL [96] enhances ROS generation from NADPH oxidase. ROS generation in MVECs was promoted through Fyn kinase activity, which blunts endothelial cell migration and apoptosis in response to angiogenic stimuli [77]. In macrophages, CD36-mediated ROS generation inhibits the activity of tyrosine phosphatase Src Homology Phosphatase-2 (SHP-2); inactivation of SHP-2 enhances phosphorylation of focal adhesion kinase to restrict macrophage migration while promoting macrophage spreading [96], with the net effect of trapping the cell in the neointima of a growing atherosclerotic plaque. The inactivation of SHP-2 is through oxidation of the catalytic cysteine within the active site [97]. Could similar downstream mechanisms for ROS in endothelial- and macrophage-CD36 signaling be present in platelets? While sulfenylation of platelet Src tyrosine kinases enhances kinase activity, coincident sulfenylation of the catalytic cysteines in phosphatases would serve to blunt tyrosine kinase inactivation and thereby prolong Src activity, further increasing NADPH oxidase-mediated ROS generation and platelet activation.

Additional evidence for oxidative stress-induced thrombosis by CD36 signaling was derived from studies of mice with deletion of the gene encoding the source of the oxidant. Genetic deletion of gp91phox, the catalytic subunit of NADPH oxidase, blunts the prothrombotic effect of oxLDL on murine platelets [54]. In addition, the NADPH oxidase inhibitor peptide gp91ds-tat, or the superoxide radical scavengers TEMPO or Manganese (III) tetrakis(1-methyl-4-pyridyl)porphyrin (MnTMPyP), inhibited oxLDL-mediated platelet accumulation in an ex vivo microfluidic thrombosis model [54]. It should be emphasized that deleting gp91phox or inhibition of NADPH oxidase had limited impact on platelet activation by classic agonists, such as ADP and collagen [98], suggesting that the systems required for oxidant-induced platelet sensitization are not necessary for supporting platelet activation under "normal" hemostatic conditions. This is further supported by the lack of bleeding disorder in CD36 deficient mice or humans. It would therefore be helpful to determine whether platelet NADPH oxidase activity supports CD36-mediated thrombosis in other oxidative stress-related conditions (e.g., in diabetes [74], infection [99]).

The mitochondria are additional sources of oxidants within hematopoietic and vascular cells and are important in thrombosis and hemostasis. Based on electron micrographs, there are approximately 10–12 mitochondria per platelet [100], whereas nucleated cells may have between 80 and 2000 mitochondria [101]. The mitochondria functions to maintain the metabolic demands of the cell through combined efforts of glycolysis and oxidative phosphorylation [102]. CD36 plays an essential role in fatty acid metabolism, and its metabolic involvement is intimately linked to the mitochondrial function. Recent studies by our lab and others in macrophages revealed that CD36 recognition of oxLDL promotes mitochondrial dysfunction by inducing a metabolic shift from oxidative phosphorylation to glycolysis and disrupting electron flow through the electron transport chain to generate ROS. This in effect converts mitochondrial from an energy producing factory to a pro-atherogenic, pro-inflammatory organelle [103]. We hypothesize that similar changes may occur in platelets with CD36 signaling where lipids from oxLDL particles could similarly alter metabolic pathways, repurposing them for platelet hyperreactivity, mitochondrial ROS generation, and a prothrombotic response.

The prothrombotic properties of oxidative stress in dyslipidemia are afforded by overcoming intrinsic antioxidant defense mechanisms. Multiple genes encode these mechanisms and are largely regulated by the interaction of transcription factor nrf2 with the antioxidant regulatory element (ARE) in their promotors [104]. Peroxiredoxin-2 (prdx2) is one such gene [51], and its deletion in hyperlipidemic mice was shown to promote thrombosis in a ferric chloride injury model of carotid artery thrombosis [51]. CD36 deletion in this context rescued the phenotype, supporting a role for CD36 in preventing antioxidant defenses from blunting thrombosis. Since 15 other antioxidant genes are controlled by nrf2, it is possible that deficiencies in these other genes could also promote thrombosis in the setting of dyslipidemia [104,105]; indeed, deficiencies in glutathione peroxidase and superoxide dismutase antioxidant genes have been related to thrombosis. The selenium-based glutathione peroxidase family of antioxidant enzymes degrade hydrogen peroxide by utilizing glutathione as a reducing agent, and the glutathionylated peroxidase is recycled with thioredoxin [106]. There are eight members of the glutathione peroxidase family, with some of the members directly implicated in the thrombotic machinery. Glutathione peroxidase-1 deficiency was found to be associated with age-related thrombosis risk, and Dayal et al. showed that aged mice over-expressing glutathione peroxidase-1 were protected from accelerated venous and arterial thrombosis in large vessels. When platelets were activated by thrombin, surface expression of the activation marker integrin αIIbβ3 was higher in old mice (18 months) compared to young mice (4 months) [53]. This was accompanied by an increase in platelet-derived hydrogen peroxide, and the addition of catalase to scavenge hydrogen peroxide or apocynin to inhibit NADPH oxidase decreased the hyperreactive platelet phenotype in old mice. Mutations in the promoter region of the gene for glutathione peroxidase-3 [107], an extracellular peroxidase, are associated with increased platelet reactivity [52], oxidative damage of lipids, and increased risk for myocardial infarction and stroke [108,109] in humans.

Another example of antioxidant deficiency promoting thrombosis is with the superoxide dismutase family of enzymes [110]. In the aging model of oxidative stress, Sonkar et al. found that platelet superoxide dismutase 2 (SOD2) functions to limit thrombin generation and arterial thrombosis [111]. Mice with platelet-specific SOD2 deficiency had increased in platelet oxidant generation, sustained calcium release with cellular activation, increased platelet granule secretion, and increased phosphatidylserine externalization after activation by thrombin relative to control aged mice [111]. The aged platelet-deficient SOD2 mice also had an increase in carotid artery thrombosis in the Rose Bengal phytochemical injury model [111]. The hyperreactive platelet phenotype from aged platelet-deficient SOD2 mice was rescued by a SOD2 mimetic compound GC4419 suggesting that this approach could be explored in other oxidant stress-induced thrombotic states.

3.3. *Thiol Isomerases Convert Redox Cues to a Thrombotic Response*

Src family kinases and phosphatases are not the only cysteine sensors converting the redox environment to a prothrombotic response. We recently found that thiol isomerases also participate in the thrombotic potential of oxLDL [112]. The thrombogenicity of intravenous injection of oxLDL was demonstrated by multiple independent laboratories [72,91,113]. Intravenous injection of oxLDL promotes thrombosis and increases pro-coagulant pathways through mechanisms that are yet to be defined [91,114]. OxLDL itself contains lipids and amino acid hydroperoxides [115] that we hypothesize could promote thrombosis through cysteine modification [24].

Thiol isomerases are a family of 21 oxidoreductase proteins that are found in the endoplasmic reticulum to promote disulfide formation and rearrangement. They are essential for protein folding and are thus essential to life. Thiol isomerases contain thioredoxin-like domains that contain a CXXC motif within their catalytic site for mediating oxidative protein folding [25]. Cysteines within the CXXC motif are among the most susceptible to electrophilic attack by oxidants and are some of the most reactive cysteines in the entire proteome [116]. In specific conditions where cells are activated, thiol isomerases escape the endoplasmic reticulum into the extracellular milieu and mediate disulfide chemistry in their non-native environment [117,118]. The mechanisms by which thiol isomerases escape the endoplasmic reticulum despite their KDEL endoplasmic reticulum retention sequence is currently an active area of research.

Protein disulfide isomerase (PDI) is the founding member of the thiol isomerase family. It contains four thioredoxin-like domains configured structurally in an **a-b-b′-a′** sequence. The active site CGHC motifs are within the **a** and **a′** domains, whereas the hydrophobic **b** and **b′** domains function in substrate recognition. A 15 amino acid linker flanked by the **b′** and **a′** domains affords structural flexibility for the protein's activity. An important role for PDI in thrombosis is supported by pharmacologic inhibitor studies [119,120] and targeted genetic deletion in platelets and endothelial cells [118,121,122]. Specifically, mice with conditional knockout of PDI in platelets showed a defect in thrombus formation in cremaster arterioles in the laser injury thrombosis model [117,122]. The defect was rescued by infusion of recombinant wildtype PDI but not a catalytically inactive PDI where the active site cysteines had been mutated to alanine [122]. Mutation studies of the active site cysteines in the **a** or **a′** domain revealed that the **a′** domain is important for controlling the thrombotic machinery, whereas the **a** domain is more important for cell survival [123]. The redox control of PDI's active site cysteines is multifaceted and governed by the structure of the protein, the local environment, and the redox potential [24]. Given the susceptibility of the active site cysteines to redox reactions, it is unclear whether PDI is reduced or oxidized in thrombosis and hemostasis.

Many substrates have been identified for PDI and are reviewed in [20,25]. Biochemical studies on these substrates revealed that PDI reduces allosteric disulfide bonds, thus influencing their functions [20]. In this regard, the active site cysteines of PDI are in the reduced redox state prior to its reductase activity (breaking disulfide bonds). Other evidence suggests that PDI oxidase activity promotes disulfide bond formation on a substrate, indicating that rather than being in a reduced form, the active cysteines are in an oxidized state prior to its oxidase activity [20]. Although there is a large body of evidence indicating that PDI is important for thrombosis and is a potential antithrombotic target, we still do not know how PDI functions within the thrombotic machinery.

Our recent findings indicate that extracellular PDI senses the redox environment and links oxidative stress to thrombosis through cysteine sulfenylation [112]. Using biochemical assays and in vitro cell biology approaches, we found that the active site cysteines of PDI are sulfenylated by peroxides (Figure 4A). Efficient sulfenylation of PDI is selective to the **a** domain and requires components of the **b** and **b′x** domains. Sulfenylation of the **a** domain is an intermediate cysteine oxoform to the disulfide. Mass spectrometry-based experiments revealed that the **a′** domain was already in an oxidized state (~80% oxidized), and that it is the **a** domain that is susceptible to oxidation. Disulfides on PDI are transferred to a ribonuclease (RNAse) that has been chemically reduced and denatured. The transfer of disulfides from PDI to RNAse accelerates the folding of RNAse into a functional enzyme in order to hydrolyze cCMP to CMP [124]. The enzymatic activity of RNAse is linked to PDI oxidase activity, where the dependency on sulfenic acid was tested with arsenite. Arsenite is a potentially selective but crude reagent to reduce sulfenic acid to free thiol [15,125,126], and using the agent prevented the enzyme-coupled refolding of RNAse by oxidized PDI.

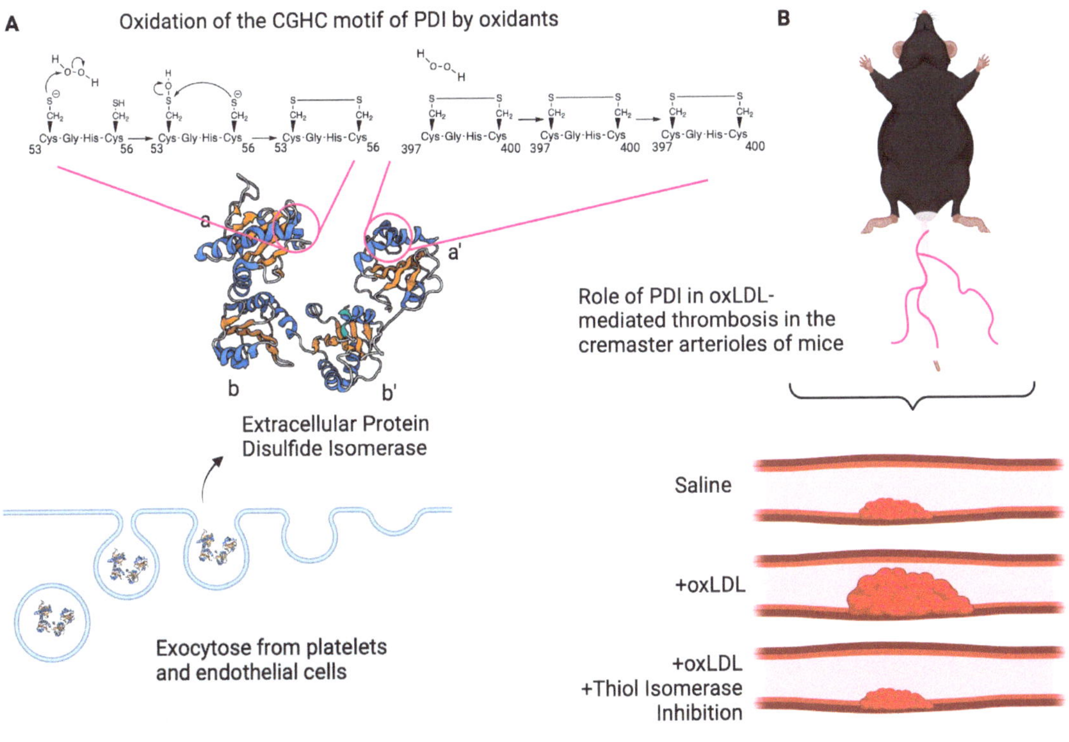

Figure 4. Thiol isomerases link oxidative stress to thrombosis. (**A**) Protein disulfide isomerase (PDI) contains redox-active cysteines within CGHC motifs of the catalytic **a** and **a′** domains. The CGHC motif of the **a** domain is sensitive to oxidation (sulfenylation) by peroxides, while the **a′** domain is in an oxidized state and is less sensitive. PDI secreted from activated human vein endothelial cells and platelets contains a fraction of PDI that is sulfenylated. The mechanism of sulfenylation in the CGHC motif is drawn based on the more nucleophilic N-terminal cysteine; it is not clear which cysteine is being sulfenylated by peroxides; and it is also not clear how much of the fraction of exocytosed PDI is in the reduced or oxidized state. (**B**) In a laser-injury mouse model of cremaster arteriolar thrombosis, oxLDL infusion results in increased platelet accumulation compared to saline infusion. Inhibition of thiol isomerases with a PDI blocking monoclonal antibody RL90 prevents the enhanced thrombosis observed with oxLDL. Created with BioRender.com and accessed on 4 January 2024.

Our hypothesis that PDI could become oxidized and contribute to thrombosis was tested ex vivo in cells and in vivo in mice. The first question was whether sulfenylated PDI is part of the pool of PDI secreted from activated cells. Platelets and endothelial cells were chosen because these are the cell types that have been shown to exocytose PDI [121,122] Consistent with our hypothesis, we found that activating human umbilical vein endothelial cells with thrombin or platelets with SFLLRN, a protease activated receptor agonist, increased sulfenylation of exocytosed PDI (Figure 4A), while the intracellular pool of PDI did not appear to be sulfenylated. Gaspar and Laurindo hypothesized that compartmentalization of PDI within the cells could insulate the protein from being oxidized [127]. In addition, structural components of the **b** and **b′x** domains may insulate the protein once it is sulfenylated, allowing for sulfenylation to be detected within the **a** domain. Furthermore, it is likely that sulfenylation of PDI may occur in only a subset of the fraction of oxidized PDI released from the cells. The stability of the sulfenic acid is competitive between the rate of cysteine oxidation by peroxides (~10–20 M^{-1} s^{-1}) [15], the rate of the probe's interaction with sulfenic acid (~10–1700 M^{-1} s^{-1} [128]; see Section 4), and the rate of the CXXC motif transitioning to the disulfide (~10^4–10^6 M^{-1} s^{-1}) [129]. These kinetic considerations within the context of PDI oxidation and sulfenic acid detection are yet to be tested. Nonetheless, the in vivo relevance of PDI sulfenylation in thrombosis was determined with a model of dyslipidemia, where wildtype C57Bl/6J mice were infused intravenously with oxLDL [72,113,114]. Direct infusion of oxLDL is an attempt to replicate conditions observed with atherosclerotic plaque rupture where micromolar levels of oxidized lipids are delivered to the local vascular environment. Intravenous infusion of recombinant PDI following oxLDL infusion caused sulfenylation of the enzyme, suggesting that sulfenylation of PDI is present in the pool of circulating PDI. Infusion of recombinant PDI was necessary in this study because endogenous circulating levels were below the level of detection. The thrombogenicity of intravenously injecting oxLDL was apparent by observing an increase in platelet accumulation after laser injury to the arteriole walls of the cremaster muscle. This increased platelet accumulation was prevented by limiting PDI oxidoreductase activity with RL90, an inhibitory monoclonal antibody [130,131]. These studies suggest that thiol isomerases participate in the prothrombotic phenotype in dyslipidemia possibly related to their oxidized state (Figure 4B). Although we have not yet linked PDI oxidation in vivo to the CD36 signaling pathway, such an experiment could suggest PDI as a therapeutic signaling node within the CD36 pathway for atherothrombosis. In addition, we do not know in vivo whether it is lipid hydroperoxide on oxLDL or peroxides that are generated from NADPH oxidase or other sources for PDI sulfenylation. Further sensitive methods to identify sulfenylation of endogenous circulating PDI is required. It is possible that the development of a sensitive enzyme-linked immunosorbent assay (ELISA) method or with mass spectrometry may help in determining whether endogenous circulating PDI is sulfenylated. Although typically used as a PDI oxidoreductase blocking antibody in vitro and in vivo, the RL90 antibody has been shown to inhibit other thiol isomerases (e.g., ERp57) [132]. It would be useful to further characterize existing and future thiol isomerase inhibitors to further dissect the mechanisms of each member in thrombosis.

4. Targeting Oxidative Cysteine Modification with Carbon Nucleophiles

Several molecules of natural or synthetic origin with antioxidative properties have shown potential therapeutic benefit in preventing oxidative stress-driven diseases. Yet interventions in humans have shown limited utility [105]. Many reasons may contribute to this lack of efficacy, including the challenges associated with increasing antioxidant defense mechanisms, the inability to properly scavenge specific oxidants, the bioavailability of natural antioxidants (e.g., flavonoids and other polyphenols), and the inability to target specific cells and tissues (all reviewed in [105]). One potential approach to overcoming these obstacles is to understand the targets of oxidants and the functional consequences of the oxidation. Oxidation of cysteine is one of the best models for targeted therapy given the available chemical biology tools for cysteine redox forms [133]. As discussed in Section 2, cysteine is oxidized to many sulfur species. The sulfenic acid oxoform is conveniently labeled by carbon nucleophiles [15,134]. Selective targeting with the nucleophiles serves two functions, as shown in Figure 5A: (1) it prevents further oxidation or reduction back to a free thiol as it covalently forms a thioether bond, allowing for the assaying of sulfenic acid function; and (2) coupled with azide–alkyne cycloaddition chemistry, commonly known as "click" chemistry, the nucleophiles provide a way to identify the protein target and site of oxidation [13] using fluorophores or other adducts. In this section, we discuss recent advances with a nucleophilic chemical library to characterize the sulfenome within cells and to show that the nucleophilic "tone" of the fragments could be leveraged to control the function of the oxidized protein.

In covalent ligand discovery, expanding a nucleophile library to target oxidized cysteines while not effecting the reduced state is at an early stage of development. Covalent targeting of oncogenic drivers (e.g., EGFR, KRAS) is an approach for the therapeutic treatment for cancer [135], although not in the context of oxidative modification. Oxidation of a cysteine to a sulfenic acid makes it electrophilic, a physicochemical property that can be leveraged for site directed labeling [15]. Using condensation reactions first studied in 1974 [136], the Carroll lab adapted the dimedone-based (1,3-dimethyl-1,3-cyclohexanedione; DYn-2) probe to label sulfenic acid (second order rate constants of 10 M^{-1} s^{-1}) [38,128]. Later, faster-reacting nucleophiles were identified that target different sulfenylated cysteines, with a benzothiazine-based probe having the fastest kinetics (1700 M^{-1} s^{-1}) [128]. Using this probe coupled to tandem mass spectrometry for global analysis, this team identified 622 sulfenic acid sites on 477 proteins in the proteome of adenocarcinoma cell lines [137]. They also synthesized 65 nucleophilic analogs [137] of an electrophilic cysteine-targetable library [138] and found that selective nucleophiles ligated reactive sulfenic acids differently when comparing the sulfenome versus the cysteinome [137]. Functionally, these nucleophilic covalent fragments impacted cellular activity. Specifically, oxidation of glyceraldehyde-3-phosphate dehydrogenase (GAPDH), glutathione S-transferase omega (GSTO1), and acetyl-CoA acyltransferase 1 (ACAT1) inhibited the enzymes, and the labeling with nucleophiles prevented the reversibility to a free thiol for enzyme reactivation. In another experiment, oxidation of the antioxidant peroxiredoxin-like 2A (PRXL2A) enzyme prevented its regulation of MAP kinase signaling; carbon nucleophiles targeting the oxidized cysteine of PRXL2A reactivated MAP kinase signaling. Protein–protein interaction was also investigated with heparin-binding growth factor (HDGF) interacting with nucleolin. Oxidation of HDGF and labeling with the nucleophile prevented the interaction. Lastly, BRCA2 and p21-interacting DNA repair protein (BCCIP)-mediated nucleotide damage was investigated linking oxidation events to the potential health of the cell through regulating genome instability. Covalent labeling of the functional cysteine that was oxidized in BCCIP decreased DNA damage repair by preventing p21 binding. These studies highlight specific approaches with a library of nucleophiles that can be used to identify sites of oxidation, determine the functional relevance of the oxidized sites, and suggest whether these sites could be targeted by specific nucleophiles for drug development [137,139].

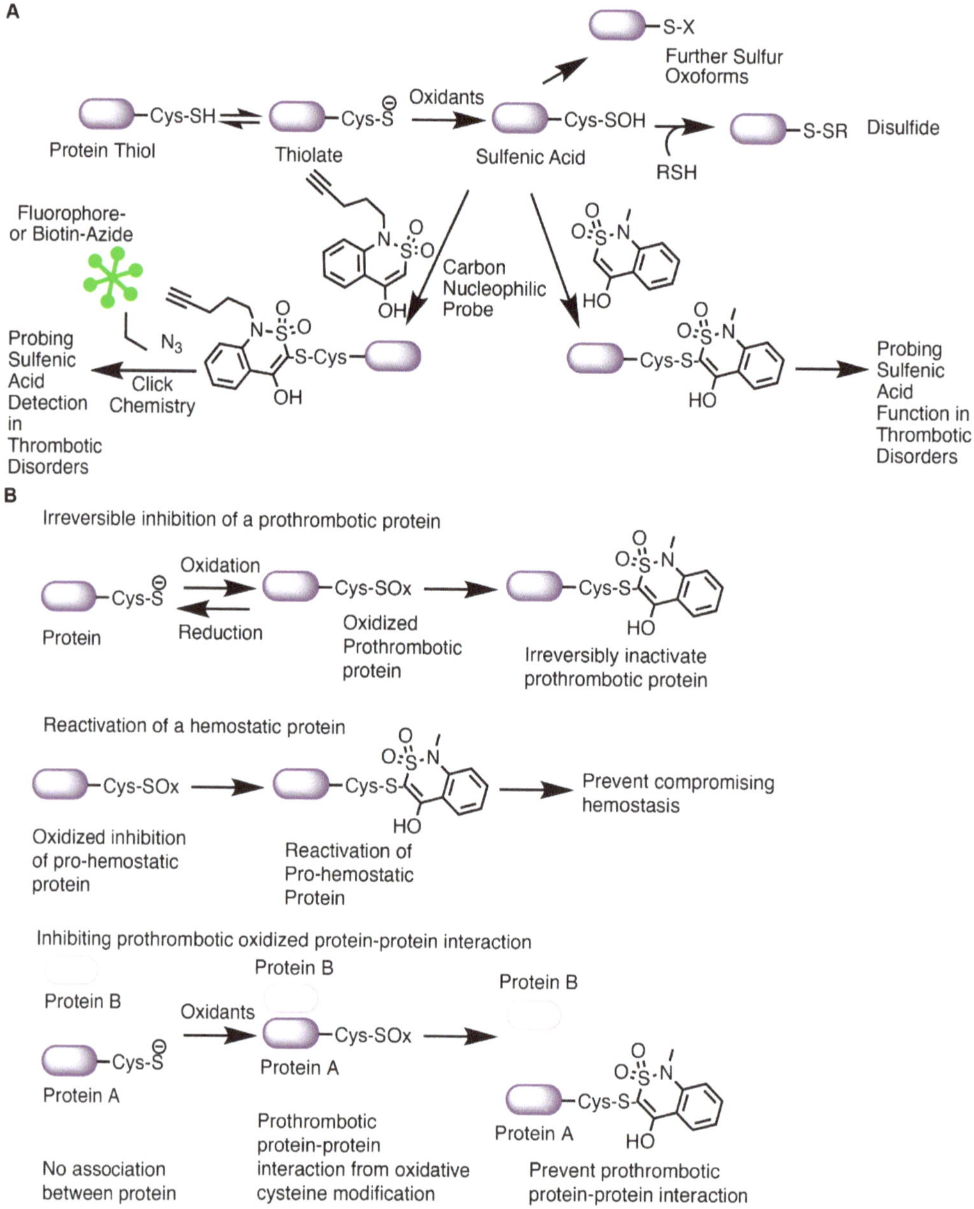

Figure 5. Carbon nucleophiles target oxidized cysteines. (**A**) Oxidation of the thiolate anion of cysteine promotes sulfenic acid formation, a precursor to other cysteine oxoforms. Sulfenic acids are conveniently labeled with carbon nucleophiles. In this figure, the benzothiazine-based carbon nucleophile was drawn as an example. The nucleophile modified with an alkyne arm can be adapted for click chemistry with biotin or a fluorophore to enable sulfenic acid detection by imaging or blotting. The nucleophile can also be used to assay the function of sulfenic acids in protein or cellular assays. (**B**) Covalent targeting of an oxidized cysteine could potentially serve several functions in thrombosis and hemostasis; thus, it is difficult to understand the mechanism of targeting oxidized

cysteines. These hypothesized mechanisms are based on the impact of sulfenic acid-selective nucleophiles recently published by Fu et al. [137] and highlighted in a commentary [139]. Oxidation of a protein could be prothrombotic. In this case, covalent targeting of an oxidized cysteine with a nucleophile would prevent reduction of the cysteine to the thiol, thus irreversibly inactivating the prothrombotic protein. If a protein is important for hemostasis and oxidation inactivates the pro-hemostatic function of the protein, covalent labeling of the oxidized cysteine by a nucleophile would potentially re-activate the protein thus alleviating issues with preventing hemostasis. Lastly, a prothrombotic protein–protein interaction (either allosterically or covalently through e.g., disulfides) could be prevented with nucleophiles, thus limiting thrombosis. These hypothesized mechanisms are yet to be tested in vivo and will require higher selectivity of the compounds to a specific protein as well as chemoproteomic approaches to understand what other proteins are ligated.

In our studies, we used carbon nucleophiles to detect sulfenylation (as described in Sections 3.2 and 3.3) and to assess the function of cysteine sulfenylation in oxidative stress-related thrombosis. Using probes [128] that differentially label sulfenic acids at different kinetic rates, we studied the effect of sulfenylation on platelet aggregatory and procoagulant activity induced by oxLDL/CD36 signaling [91,140]. We found that oxLDL enhanced platelet aggregation and phosphatidylserine externalization compared to control [91,92], and that a benzothiazine-based probe prevented oxLDL-induced platelet aggregation with an IC_{50} of 2 mM compared to 1,3-cyclohexanedione (5.9 mM), pyrrolidine (9.2 mM), and piperidine-based probes (10.2 mM) [91]. Importantly, the benzothiazine probe did not impact platelet aggregation mediated by the P2Y1/12 receptor agonist ADP or the GPVI receptor agonist collagen-related peptide [91]. In these experiments, platelet procoagulant activity mediated by the externalization of negatively charged phosphatidylserine (PS) [141] occurred after stimulating platelets with oxLDL in the presence of a GPVI receptor agonist [92,114] and was also impacted by protein sulfenylation as benzothiazine and 1,3-cyclohexanedione at the IC_{50} concentrations prevented the oxLDL-mediated increase in PS externalization [91]. These studies suggest that sulfenylation of proteins within platelet activation pathways in conditions simulating hyperlipidemic redox stress is relevant to thrombosis.

The connection between sulfenylation and oxidative stress-related thrombosis was further evaluated in vivo [91]. Wildtype C57Bl/6 mice fed a high fat/high cholesterol diet to elevate circulating lipoproteins to levels sufficient to induce platelet hypersensitivity [86,91,113] were injected intraperitoneally with increasing concentration of the benzothiazine-based probe (from 25 mg/kg up to 100 mg/kg) prior to initiating carotid artery injury by topical application of ferric chloride. The high fat diet-fed mice showed decreased time to cessation of arterial blood flow relative to control diet (indicative of a hyperreactive platelet phenotype) and this was normalized back to that seen in the control diet conditions with the 25 mg/kg dose of benzothiazine. Increasing the concentration of benzothiazine further increased the time to blood flow cessation. These studies were repeated in a second thrombosis model induced by laser injury to the cremaster arterioles after infusion of oxLDL. Similar to the findings with the ferric chloride model, platelet and fibrin deposition after laser injury were increased in the presence of oxLDL and this was prevented with 25 mg/kg benzothiazine. The probe had no impact in control mice infused with saline instead of oxLDL. These studies underscore the relevance of sulfenylation in oxidative stress thrombosis and suggest the possibility that targeted covalent inhibition with nucleophiles might be useful in disease conditions.

Covalent ligation of sulfenylated cysteines in thrombosis requires much more investigation. Ligating an oxidized cysteine could manifest into multiple mechanisms, as found in the study by Fu et al. [137]. Covalent ligation of an oxidized cysteine could inhibit a protein by irreversibly preventing reduction of the cysteine to a free thiol. In this context, it could be a way to inhibit a protein that is only activated in diseased conditions and thus does not impact hemostasis (Figure 5B, top). What would be the proteins that promote thrombosis in diseased conditions but are not essential for hemostasis? In another context, covalent ligation of the oxidized cysteine allows for re-activation of a protein (Figure 5B, middle). If a protein is pro-hemostatic and oxidation "kills" the protein, could covalent targeting of the oxidized cysteine re-activate the protein's hemostatic function during oxidative stress? This might be relevant in conditions where efficient antithrombotic regiments increase the risk for bleeding complication or in coagulopathy where bleeding is present. An example of this may be observed during the inflammation of disseminated intravascular coagulation. In this condition, clotting factors and platelets are "used up" in the later stage of the disease, increasing the risk for bleeding complications. Reactivation of an oxidized pro-hemostatic protein might alleviate the risk for bleeding. Lastly, targeting oxidized cysteines within a prothrombotic protein–protein interaction complex may prevent thrombosis while maintaining hemostasis (Figure 5B, bottom). CD36 is a great example of this; it is highly present in lipid rafts and coordinates with other membrane proteins, including tetraspanins [142], NaK-ATPase [143], toll like receptors [144], and integrins [145], for pro-atherogenic and pro-thrombotic signaling. Preventing such interactions allosterically through targeting oxidized cysteines important for protein–protein interaction could alleviate athero progression and subsequent thrombosis. The current limitation is the off-target effects observed with covalent ligation. The recent data from Julio et al. suggest that covalent targeting of cysteines could trigger a stress response within cells, causing proteins to aggregate, and increase proteasomal degradation [146], supporting the indispensable need to address the issue of selectivity. The structure of a lead nucleophilic fragment could be refined to target a specific protein of interest through traditional structure activity relationship coupled with biochemical validation (e.g., molecular modeling and simulations, in vitro protein assays). Further characterization of whether ligating the oxidized protein of interest impacts the protein's function and subsequent cellular activity would help in transitioning lead nucleophiles into animal thrombosis models and pre-clinical studies.

Overall, the proof-of-principles in the nucleophilic fragment screen from the Carroll lab, as well as our in vivo functional thrombosis studies in dyslipidemia with carbon nucleophiles, provides an opportunity for covalently targeting cysteine oxidation. A nucleophilic fragment screening in platelets, endothelial cells, immune cells, and red cells during oxidative stress coupled to global chemoproteomic platforms would be essential in understanding the sulfenome that controls the thrombotic machinery. Initial efforts have already been investigated for the sulfenome in pathogen-inactivated platelets [147], which identified many proteins of the cytoskeletal and integrin pathways that are cysteine oxidized. Although we found that the probe prevented thrombosis in dyslipidemia, further characterization of the compound in vivo is required.

5. Conclusions

Many antioxidative therapies have shown limited utility in decreasing the risk of cardiovascular events. One potential approach is to understand the target and the functional consequence of oxidation. Cysteine oxidation is a type of oxidative posttranslational modification present during oxidative stress. While pathologic conditions are associated with oxidative stress, strategies to target oxidative cysteine modification could be exploited to prevent thrombosis in diseases while not impacting hemostasis. This could be achieved by taking advantage of the physicochemistry of cysteines using chemical biology. The studies outlined in this review suggests that oxidative cysteine modification is important in regulating thrombosis. Specifically, the CD36 signaling pathway in atherothrombosis is a system where the redox environment is translated into a selective phosphorylation-driven prothrombotic response. In addition, thiol isomerases have highly reactive cysteines that are sulfenylated. Thiol isomerases thus convert the promiscuous redox cues into selective disulfide signaling, a different cysteine oxoform that is substrate-dependent for thrombosis. As carbon nucleophiles covalently label sulfenylation, they are useful tools for probing cysteine oxidation in model systems and could be used as future backbones for drug development against oxidized thiols.

Author Contributions: M.Y. and R.L.S. wrote and edited the manuscript. All authors have read and agreed to the published version of the manuscript.

Funding: This work was supported by the National Institute of Health (National Heart, Lung, and Blood grant numbers K99HL164888 to M.Y., R01 HL142152 to R.L.S., R01HL153397 to R.L.S., and R01HL164460 to R.L.S.), the American Society of Hematology Scholar Award (to M.Y.), the Eleanor Miles Shore Award (to M.Y.), and the Foundation for Women's Wellness (to M.Y.). These funding sources had no involvement in writing this review or submitting this review for publication.

Institutional Review Board Statement: Not applicable.

Informed Consent Statement: Not applicable.

Data Availability Statement: No new data were created or analyzed in this study. Data sharing is not applicable to this article.

Conflicts of Interest: The authors declare no financial conflicts of interest.

References

1. Masenga, S.K.; Kabwe, L.S.; Chakulya, M.; Kirabo, A. Mechanisms of Oxidative Stress in Metabolic Syndrome. *Int. J. Mol. Sci.* **2023**, *24*, 7898. [CrossRef]
2. Morotti, A.; Barale, C.; Melchionda, E.; Russo, I. Platelet Redox Imbalance in Hypercholesterolemia: A Big Problem for a Small Cell. *Int. J. Mol. Sci.* **2022**, *23*, 11446. [CrossRef]
3. Vaidya, A.R.; Wolska, N.; Vara, D.; Mailer, R.K.; Schröder, K.; Pula, G. Diabetes and Thrombosis: A Central Role for Vascular Oxidative Stress. *Antioxidants* **2021**, *10*, 706. [CrossRef] [PubMed]
4. Barale, C.; Russo, I. Influence of Cardiometabolic Risk Factors on Platelet Function. *Int. J. Mol. Sci.* **2020**, *21*, 623. [CrossRef] [PubMed]
5. Wang, Q.; Zennadi, R. Oxidative Stress and Thrombosis during Aging: The Roles of Oxidative Stress in RBCs in Venous Thrombosis. *Int. J. Mol. Sci.* **2020**, *21*, 4259. [CrossRef] [PubMed]
6. Zhang, X.; Yu, S.; Li, X.; Wen, X.; Liu, S.; Zu, R.; Ren, H.; Li, T.; Yang, C.; Luo, H. Research progress on the interaction between oxidative stress and platelets: Another avenue for cancer? *Pharmacol. Res.* **2023**, *191*, 106777. [CrossRef] [PubMed]
7. Alam, M.S.; Czajkowsky, D.M. SARS-CoV-2 infection and oxidative stress: Pathophysiological insight into thrombosis and therapeutic opportunities. *Cytokine Growth Factor Rev.* **2022**, *63*, 44–57. [CrossRef] [PubMed]
8. Beckman, J.D.; Sparkenbaugh, E.M. The invisible string of coagulation, complement, iron, and inflammation in sickle cell disease. *Curr. Opin. Hematol.* **2023**, *30*, 153–158. [CrossRef]
9. Bettiol, A.; Galora, S.; Argento, F.R.; Fini, E.; Emmi, G.; Mattioli, I.; Bagni, G.; Fiorillo, C.; Becatti, M. Erythrocyte oxidative stress and thrombosis. *Expert Rev. Mol. Med.* **2022**, *24*, e31. [CrossRef]
10. Nocella, C.; Bartimoccia, S.; Cammisotto, V.; D'amico, A.; Pastori, D.; Frati, G.; Sciarretta, S.; Rosa, P.; Felici, C.; Riggio, O.; et al. Oxidative Stress in the Pathogenesis of Antiphospholipid Syndrome: Implications for the Atherothrombotic Process. *Antioxidants* **2021**, *10*, 1790. [CrossRef]
11. Wendelboe, A.M.; Raskob, G.E. Global Burden of Thrombosis. *Circ. Res.* **2016**, *118*, 1340–1347. [CrossRef]
12. Demasi, M.; Augusto, O.; Bechara, E.J.; Bicev, R.N.; Cerqueira, F.M.; da Cunha, F.M.; Denicola, A.; Gomes, F.; Miyamoto, S.; Netto, L.E.; et al. Oxidative Modification of Proteins: From Damage to Catalysis, Signaling, and Beyond. *Antioxid. Redox Signal.* **2021**, *35*, 1016–1080. [CrossRef]
13. Yang, M.; Smith, B.C. Cysteine and methionine oxidation in thrombotic disorders. *Curr. Opin. Chem. Biol.* **2023**, *76*, 102350. [CrossRef] [PubMed]
14. Sies, H.; Belousov, V.V.; Chandel, N.S.; Davies, M.J.; Jones, D.P.; Mann, G.E.; Murphy, M.P.; Yamamoto, M.; Winterbourn, C. Defining roles of specific reactive oxygen species (ROS) in cell biology and physiology. *Nat. Rev. Mol. Cell Biol.* **2022**, *23*, 499–515. [CrossRef] [PubMed]
15. Gupta, V.; Carroll, K.S. Sulfenic acid chemistry, detection and cellular lifetime. *Biochim. Biophys. Acta (BBA) Gen. Subj.* **2014**, *1840*, 847–875. [CrossRef] [PubMed]
16. Yoon, S.; Eom, G.H.; Kang, G. Nitrosative Stress and Human Disease: Therapeutic Potential of Denitrosylation. *Int. J. Mol. Sci.* **2021**, *22*, 9794. [CrossRef] [PubMed]
17. Aussel, L.; Ezraty, B. Methionine Redox Homeostasis in Protein Quality Control. *Front. Mol. Biosci.* **2021**, *8*, 665492. [CrossRef]
18. Sun, H.; Jia, H.; Kendall, O.; Dragelj, J.; Kubyshkin, V.; Baumann, T.; Mroginski, M.-A.; Schwille, P.; Budisa, N. Halogenation of tyrosine perturbs large-scale protein self-organization. *Nat. Commun.* **2022**, *13*, 4843. [CrossRef]
19. Marino, S.M.; Gladyshev, V.N. Cysteine Function Governs Its Conservation and Degeneration and Restricts Its Utilization on Protein Surfaces. *J. Mol. Biol.* **2010**, *404*, 902–916. [CrossRef]
20. Chiu, J.; Hogg, P.J. Allosteric disulfides: Sophisticated molecular structures enabling flexible protein regulation. *J. Biol. Chem.* **2019**, *294*, 2949–5908. [CrossRef]

21. Awoonor-Williams, E.; Rowley, C.N. How Reactive are Druggable Cysteines in Protein Kinases? *J. Chem. Inf. Model.* **2018**, *58*, 1935–1946. [CrossRef]
22. Liu, R.; Zhan, S.; Che, Y.; Shen, J. Reactivities of the Front Pocket N-Terminal Cap Cysteines in Human Kinases. *J. Med. Chem.* **2022**, *65*, 1525–1535. [CrossRef] [PubMed]
23. Pace, N.J.; Weerapana, E. Zinc-Binding Cysteines: Diverse Functions and Structural Motifs. *Biomolecules* **2014**, *4*, 419–434. [CrossRef] [PubMed]
24. Yang, M.; Flaumenhaft, R. Oxidative Cysteine Modification of Thiol Isomerases in Thrombotic Disease: A Hypothesis. *Antioxid. Redox Signal.* **2021**, *35*, 1134–1155. [CrossRef] [PubMed]
25. Gaspar, R.S.; Gibbins, J.M. Thiol Isomerases Orchestrate Thrombosis and Hemostasis. *Antioxid. Redox Signal.* **2021**, *35*, 1116–1133. [CrossRef] [PubMed]
26. Ramos-Guzmán, C.A.; Ruiz-Pernía, J.J.; Zinovjev, K.; Tuñón, I. Unveiling the Mechanistic Singularities of Caspases: A Computational Analysis of the Reaction Mechanism in Human Caspase-1. *ACS Catal.* **2023**, *13*, 4348–4361. [CrossRef] [PubMed]
27. Paulsen, C.E.; Carroll, K.S. Cysteine-Mediated Redox Signaling: Chemistry, Biology, and Tools for Discovery. *Chem. Rev.* **2013**, *113*, 4633–4679. [CrossRef] [PubMed]
28. Marino, S.M.; Gladyshev, V.N. Analysis and Functional Prediction of Reactive Cysteine Residues. *J. Biol. Chem.* **2012**, *287*, 4419–4425. [CrossRef]
29. Roos, G.; Foloppe, N.; Messens, J. Understanding the pK_a of Redox Cysteines: The Key Role of Hydrogen Bonding. *Antioxid. Redox Signal.* **2013**, *18*, 94–127. [CrossRef]
30. Bak, D.W.; Bechtel, T.J.; Falco, J.A.; Weerapana, E. Cysteine reactivity across the subcellular universe. *Curr. Opin. Chem. Biol.* **2019**, *48*, 96–105. [CrossRef]
31. Yang, R.; Zhu, T.; Xu, J.; Zhao, Y.; Kuang, Y.; Sun, M.; Chen, Y.; He, W.; Wang, Z.; Jiang, T.; et al. Organic Fluorescent Probes for Monitoring Micro-Environments in Living Cells and Tissues. *Molecules* **2023**, *28*, 3455. [CrossRef] [PubMed]
32. Mazmanian, K.; Chen, T.; Sargsyan, K.; Lim, C. From quantum-derived principles underlying cysteine reactivity to combating the COVID-19 pandemic. *Wiley Interdiscip. Rev. Comput. Mol. Sci.* **2022**, *12*, e1607. [CrossRef] [PubMed]
33. Røhr, K.; Hammerstad, M.; Andersson, K.K. Tuning of Thioredoxin Redox Properties by Intramolecular Hydrogen Bonds. *PLoS ONE* **2013**, *8*, e69411. [CrossRef] [PubMed]
34. Karala, A.-R.; Lappi, A.-K.; Ruddock, L.W. Modulation of an Active-Site Cysteine pKa Allows PDI to Act as a Catalyst of both Disulfide Bond Formation and Isomerization. *J. Mol. Biol.* **2010**, *396*, 883–892. [CrossRef] [PubMed]
35. Kortemme, T.; Creighton, T.E. Ionisation of Cysteine Residues at the Termini of Model α-Helical Peptides. Relevance to Unusual Thiol pKaValues in Proteins of the Thioredoxin Family. *J. Mol. Biol.* **1995**, *253*, 799–812. [CrossRef] [PubMed]
36. Poole, L.B. The basics of thiols and cysteines in redox biology and chemistry. *Free Radic. Biol. Med.* **2015**, *80*, 148–157. [CrossRef] [PubMed]
37. Sies, H.; Jones, D.P. Reactive oxygen species (ROS) as pleiotropic physiological signalling agents. *Nat. Rev. Mol. Cell Biol.* **2020**, *21*, 363–383. [CrossRef] [PubMed]
38. Paulsen, C.E.; Truong, T.H.; Garcia, F.J.; Homann, A.; Gupta, V.; Leonard, S.E.; Carroll, K.S. Peroxide-dependent sulfenylation of the EGFR catalytic site enhances kinase activity. *Nat. Chem. Biol.* **2011**, *8*, 57–64. [CrossRef]
39. Truong, T.H.; Ung, P.M.-U.; Palde, P.B.; Paulsen, C.E.; Schlessinger, A.; Carroll, K.S. Molecular Basis for Redox Activation of Epidermal Growth Factor Receptor Kinase. *Cell Chem. Biol.* **2016**, *23*, 837–848. [CrossRef]
40. Rosenkranz, A.A.; Slastnikova, T.A. Epidermal Growth Factor Receptor: Key to Selective Intracellular Delivery. *Biochemistry* **2020**, *85*, 967–1092. [CrossRef]
41. Akter, S.; Fu, L.; Jung, Y.; Conte, M.L.; Lawson, J.R.; Lowther, W.T.; Sun, R.; Liu, K.; Yang, J.; Carroll, K.S. Chemical proteomics reveals new targets of cysteine sulfinic acid reductase. *Nat. Chem. Biol.* **2018**, *14*, 995–1004. [CrossRef]
42. Petroff, J.T.; Omlid, S.M.; Haloi, N.; Sith, L.; Johnson, S.; McCulla, R.D. Reactions of sulfenic acids with amines, thiols, and thiolates studied by quantum chemical calculations. *Comput. Theor. Chem.* **2020**, *1189*, 112979. [CrossRef]
43. Rehder, D.S.; Borges, C.R. Cysteine sulfenic Acid as an Intermediate in Disulfide Bond Formation and Nonenzymatic Protein Folding. *Biochemistry* **2010**, *49*, 7748–7755. [CrossRef] [PubMed]
44. Branchford, B.R.; Flood, V.H. 51—Bleeding and Thrombosis. In *Nelson Pediatric Symptom-Based Diagnosis: Common Diseases and Their Mimics*, 2nd ed.; Kliegman, R.M., Toth, H., Bordini, B.J., Basel, D., Eds.; Elsevier: Philadelphia, PA, USA, 2023; pp. 942–964.e1.
45. Neubauer, K.; Zieger, B. Endothelial cells and coagulation. *Cell Tissue Res.* **2022**, *387*, 391–398. [CrossRef] [PubMed]
46. Braune, S.; Küpper, J.-H.; Jung, F. Effect of Prostanoids on Human Platelet Function: An Overview. *Int. J. Mol. Sci.* **2020**, *21*, 9020. [CrossRef] [PubMed]
47. Gambaryan, S. The Role of NO/sGC/cGMP/PKG Signaling Pathway in Regulation of Platelet Function. *Cells* **2022**, *11*, 3704. [CrossRef] [PubMed]
48. Nagy, Z.; Smolenski, A. Cyclic nucleotide-dependent inhibitory signaling interweaves with activating pathways to determine platelet responses. *Res. Pract. Thromb. Haemost.* **2018**, *2*, 558–571. [CrossRef]
49. Wang, L.; Li, Y.-J.; Yang, X.; Yang, B.; Zhang, X.; Zhang, J.; Zhang, Q.; Cheng, X.-D.; Wang, J.-H.; Yu, N.-W. Purinergic signaling: A potential therapeutic target for ischemic stroke. *Purinergic Signal.* **2023**, *19*, 173–183. [CrossRef]
50. Scridon, A. Platelets and Their Role in Hemostasis and Thrombosis—From Physiology to Pathophysiology and Therapeutic Implications. *Int. J. Mol. Sci.* **2022**, *23*, 12772. [CrossRef]

51. Li, W.; Febbraio, M.; Reddy, S.P.; Yu, D.-Y.; Yamamoto, M.; Silverstein, R.L. CD36 participates in a signaling pathway that regulates ROS formation in murine VSMCs. *J. Clin. Investig.* **2010**, *120*, 3996–4006. [CrossRef]
52. Jin, R.C.; Mahoney, C.E.; Anderson, L.; Ottaviano, F.; Croce, K.; Leopold, J.A.; Zhang, Y.-Y.; Tang, S.-S.; Handy, D.E.; Loscalzo, J.; et al. Glutathione Peroxidase-3 Deficiency Promotes Platelet-Dependent Thrombosis In Vivo. *Circulation* **2011**, *123*, 1963–1973. [CrossRef] [PubMed]
53. Dayal, S.; Wilson, K.M.; Motto, D.G.; Miller, F.J., Jr.; Chauhan, A.K.; Lentz, S.R. Hydrogen Peroxide Promotes Aging-Related Platelet Hyperactivation and Thrombosis. *Circulation* **2013**, *127*, 1308–1316. [CrossRef] [PubMed]
54. Magwenzi, S.; Woodward, C.; Wraith, K.S.; Aburima, A.; Raslan, Z.; Jones, H.; McNeil, C.; Wheatcroft, S.; Yuldasheva, N.; Febbriao, M.; et al. Oxidized LDL activates blood platelets through CD36/NOX2–mediated inhibition of the cGMP/protein kinase G signaling cascade. *Blood* **2015**, *125*, 2693–2703. [CrossRef] [PubMed]
55. Yang, M.; Cooley, B.C.; Li, W.; Chen, Y.; Vasquez-Vivar, J.; Scoggins, N.O.; Cameron, S.J.; Morrell, C.N.; Silverstein, R.L. Platelet CD36 promotes thrombosis by activating redox sensor ERK5 in hyperlipidemic conditions. *Blood* **2017**, *129*, 2917–2927. [CrossRef] [PubMed]
56. Nishimura, S.; Manabe, I.; Nagasaki, M.; Kakuta, S.; Iwakura, Y.; Takayama, N.; Ooehara, J.; Otsu, M.; Kamiya, A.; Petrich, B.G.; et al. In vivo imaging visualizes discoid platelet aggregations without endothelium disruption and implicates contribution of inflammatory cytokine and integrin signaling. *Blood* **2012**, *119*, e45–e56. [CrossRef] [PubMed]
57. Li, W.; McIntyre, T.M.; Silverstein, R.L. Ferric chloride-induced murine carotid arterial injury: A model of redox pathology. *Redox Biol.* **2013**, *1*, 50–55. [CrossRef] [PubMed]
58. Trostchansky, A.; Alarcon, M. An Overview of Two Old Friends Associated with Platelet Redox Signaling, the Protein Disulfide Isomerase and NADPH Oxidase. *Biomolecules* **2023**, *13*, 848. [CrossRef]
59. Ansari, S.A.; Keshava, S.; Pendurthi, U.R.; Rao, L.V.M. Oxidative Stress Product, 4-Hydroxy-2-Nonenal, Induces the Release of Tissue Factor-Positive Microvesicles from Perivascular Cells into Circulation. *Arter. Thromb. Vasc. Biol.* **2021**, *41*, 250–265. [CrossRef]
60. Ansari, S.A.; Pendurthi, U.R.; Rao, L.V.M. The lipid peroxidation product 4-hydroxy-2-nonenal induces tissue factor decryption via ROS generation and the thioredoxin system. *Blood Adv.* **2017**, *1*, 2399–2413. [CrossRef]
61. Jourdan, A.; Aguejouf, O.; Imbault, P.; Doutremepuich, F.; Inamo, J.; Doutremepuich, C. Experimental thrombosis model induced by free radicals. Application to aspirin and other different substances. *Thromb. Res.* **1995**, *79*, 109–123. [CrossRef]
62. Berger, M.; Wraith, K.; Woodward, C.; Aburima, A.; Raslan, Z.; Hindle, M.S.; Moellmann, J.; Febbraio, M.; Naseem, K.M. Dyslipidemia-associated atherogenic oxidized lipids induce platelet hyperactivity through phospholipase Cγ2-dependent reactive oxygen species generation. *Platelets* **2018**, *30*, 467–472. [CrossRef] [PubMed]
63. Yang, M.; Silverstein, R.L. CD36 signaling in vascular redox stress. *Free. Radic. Biol. Med.* **2019**, *136*, 159–171. [CrossRef] [PubMed]
64. Berberich, A.J.; Hegele, R.A. A Modern Approach to Dyslipidemia. *Endocr. Rev.* **2022**, *43*, 611–653. [CrossRef] [PubMed]
65. Libby, P. 1—Overview of Lipids and Atherosclerosis. In *Clinical Lipidology*, 3rd ed.; Ballantyne, C.M., Ed.; Elsevier: New Delhi, India, 2024; pp. 1–15.e12.
66. Zhong, S.; Li, L.; Shen, X.; Li, Q.; Xu, W.; Wang, X.; Tao, Y.; Yin, H. An update on lipid oxidation and inflammation in cardiovascular diseases. *Free Radic. Biol. Med.* **2019**, *144*, 266–278. [CrossRef] [PubMed]
67. Berger, M.; Naseem, K.M. Oxidised Low-Density Lipoprotein-Induced Platelet Hyperactivity—Receptors and Signalling Mechanisms. *Int. J. Mol. Sci.* **2022**, *23*, 9199. [CrossRef] [PubMed]
68. Podrez, E.A.; Byzova, T.V.; Febbraio, M.; Salomon, R.G.; Ma, Y.; Valiyaveettil, M.; Poliakov, E.; Sun, M.; Finton, P.J.; Curtis, B.R.; et al. Platelet CD36 links hyperlipidemia, oxidant stress and a prothrombotic phenotype. *Nat. Med.* **2007**, *13*, 1086–1095. [CrossRef]
69. Podrez, E.A.; Poliakov, E.; Shen, Z.; Zhang, R.; Deng, Y.; Sun, M.; Finton, P.J.; Shan, L.; Gugiu, B.; Fox, P.L.; et al. Identification of a Novel Family of Oxidized Phospholipids That Serve as Ligands for the Macrophage Scavenger Receptor CD36. *J. Biol. Chem.* **2002**, *277*, 38503–38516. [CrossRef] [PubMed]
70. Okumura, T.; Jamieson, G.A. Platelet glycocalicin. I. Orientation of glycoproteins of the human platelet surface. *J. Biol. Chem.* **1976**, *251*, 5944–5949. [CrossRef]
71. Bendas, G.; Schlesinger, M. The Role of CD36/GPIV in Platelet Biology. In *Seminars in Thrombosis and Hemostasis*; Thieme Medical Publishers, Inc.: New York, NY, USA, 2023. [CrossRef]
72. Berger, M.; Raslan, Z.; Aburima, A.; Magwenzi, S.; Wraith, K.S.; Spurgeon, B.E.J.; Hindle, M.S.; Law, R.; Febbraio, M.; Naseem, K.M. Atherogenic lipid stress induces platelet hyperactivity through CD36-mediated hyposensitivity to prostacyclin: The role of phosphodiesterase 3A. *Haematologica* **2020**, *105*, 808–819. [CrossRef]
73. Chen, K.; Febbraio, M.; Li, W.; Silverstein, R.L. A Specific CD36-Dependent Signaling Pathway Is Required for Platelet Activation by Oxidized Low-Density Lipoprotein. *Circ. Res.* **2008**, *102*, 1512–1519. [CrossRef]
74. Zhu, W.; Li, W.; Silverstein, R.L. Advanced glycation end products induce a prothrombotic phenotype in mice via interaction with platelet CD36. *Blood* **2012**, *119*, 6136–6144. [CrossRef]
75. Banesh, S.; Ramakrishnan, V.; Trivedi, V. Mapping of phosphatidylserine recognition region on CD36 ectodomain. *Arch. Biochem. Biophys.* **2018**, *660*, 1–10. [CrossRef]

76. Ghosh, A.; Li, W.; Febbraio, M.; Espinola, R.G.; McCrae, K.R.; Cockrell, E.; Silverstein, R.L. Platelet CD36 mediates interactions with endothelial cell–derived microparticles and contributes to thrombosis in mice. *J. Clin. Investig.* **2008**, *118*, 1934–194[illegible] [CrossRef] [PubMed]
77. Ramakrishnan, D.P.; Hajj-Ali, R.A.; Chen, Y.; Silverstein, R.L. Extracellular Vesicles Activate a CD36-Dependent Signaling Pathway to Inhibit Microvascular Endothelial Cell Migration and Tube Formation. *Arterioscler. Thromb. Vasc. Biol.* **2016**, *36*, 534–544 [CrossRef] [PubMed]
78. Roberts, W.; Magwenzi, S.; Aburima, A.; Naseem, K.M. Thrombospondin-1 induces platelet activation through CD36-dependent inhibition of the cAMP/protein kinase A signaling cascade. *Blood* **2010**, *116*, 4297–4306. [CrossRef] [PubMed]
79. Chu, L.-Y.; Ramakrishnan, D.P.; Silverstein, R.L. Thrombospondin-1 modulates VEGF signaling via CD36 by recruiting SHP-1 to VEGFR2 complex in microvascular endothelial cells. *Blood* **2013**, *122*, 1822–1832. [CrossRef] [PubMed]
80. Coburn, C.T.; Knapp, F.F., Jr.; Febbraio, M.; Beets, A.L.; Silverstein, R.L.; Abumrad, N.A. Defective Uptake and Utilization of Long Chain Fatty Acids in Muscle and Adipose Tissues of CD36 Knockout Mice. *J. Biol. Chem.* **2000**, *275*, 32523–32529. [CrossRef]
81. Bachmann, A.; Metwally, N.G.; Allweier, J.; Cronshagen, J.; Tauler, M.d.P.M.; Murk, A.; Roth, L.K.; Torabi, H.; Wu, Y.; Gutsmann, T.; et al. CD36—A Host Receptor Necessary for Malaria Parasites to Establish and Maintain Infection. *Microorganisms* **2022**, *10*, 2356. [CrossRef] [PubMed]
82. Stuart, L.M.; Deng, J.; Silver, J.M.; Takahashi, K.; Tseng, A.A.; Hennessy, E.J.; Ezekowitz, R.A.B.; Moore, K.J. Response to *Staphylococcus aureus* requires CD36-mediated phagocytosis triggered by the COOH-terminal cytoplasmic domain. *J. Cell Biol.* **2005**, *170*, 477–485. [CrossRef] [PubMed]
83. Olonisakin, T.F.; Li, H.; Xiong, Z.; Kochman, E.J.K.; Yu, M.; Qu, Y.; Hulver, M.; Kolls, J.K.; Croix, C.S.; Doi, Y.; et al. CD36 Provides Host Protection against *Klebsiella pneumoniae* Intrapulmonary Infection by Enhancing Lipopolysaccharide Responsiveness and Macrophage Phagocytosis. *J. Infect. Dis.* **2016**, *214*, 1865–1875. [CrossRef]
84. Dobri, A.-M.; Dudău, M.; Enciu, A.-M.; Hinescu, M.E. CD36 in Alzheimer's Disease: An Overview of Molecular Mechanisms and Therapeutic Targeting. *Neuroscience* **2021**, *453*, 301–311. [CrossRef]
85. Wang, Y.; Fang, C.; Gao, H.; Bilodeau, M.L.; Zhang, Z.; Croce, K.; Liu, S.; Morooka, T.; Sakuma, M.; Nakajima, K.; et al. Platelet-derived S100 family member myeloid-related protein-14 regulates thrombosis. *J. Clin. Investig.* **2014**, *124*, 2160–2171. [CrossRef] [PubMed]
86. Chen, K.; Li, W.; Major, J.; Rahaman, S.O.; Febbraio, M.; Silverstein, R.L. Vav guanine nucleotide exchange factors link hyperlipidemia and a prothrombotic state. *Blood* **2011**, *117*, 5744–5750. [CrossRef] [PubMed]
87. Lee, H.-Y.; Lee, T.; Lee, N.; Yang, E.G.; Lee, C.; Lee, J.; Moon, E.-Y.; Ha, J.; Park, H. Src activates HIF-1α not through direct phosphorylation of HIF-1α-specific prolyl-4 hydroxylase 2 but through activation of the NADPH oxidase/Rac pathway. *Carcinogenesis* **2011**, *32*, 703–712. [CrossRef] [PubMed]
88. Hayyan, M.; Hashim, M.A.; AlNashef, I.M. Superoxide Ion: Generation and Chemical Implications. *Chem. Rev.* **2016**, *116*, 3029–3085. [CrossRef] [PubMed]
89. Winterbourn, C.C. Biological chemistry of superoxide radicals. *ChemTexts* **2020**, *6*, 7. [CrossRef]
90. Heppner, D.E. Structural insights into redox-active cysteine residues of the Src family kinases. *Redox Biol.* **2021**, *41*, 101934. [CrossRef] [PubMed]
91. Yang, M.; Li, W.; Harberg, C.; Chen, W.; Yue, H.; Ferreira, R.B.; Wynia-Smith, S.L.; Carroll, K.S.; Zielonka, J.; Flaumenhaft, R.; et al. Cysteine sulfenylation by CD36 signaling promotes arterial thrombosis in dyslipidemia. *Blood Adv.* **2020**, *4*, 4494–4507. [CrossRef] [PubMed]
92. Yang, M.; Kholmukhamedov, A.; Schulte, M.L.; Cooley, B.C.; Scoggins, N.O.; Wood, J.P.; Cameron, S.J.; Morrell, C.N.; Jobe, S.M.; Silverstein, R.L. Platelet CD36 signaling through ERK5 promotes caspase-dependent procoagulant activity and fibrin deposition in vivo. *Blood Adv.* **2018**, *2*, 2848–2861. [CrossRef]
93. Le, N.-T. The significance of ERK5 catalytic-independent functions in disease pathways. *Front. Cell Dev. Biol.* **2023**, *11*, 1235217. [CrossRef]
94. Postiglione, A.E.; Adams, L.L.; Ekhator, E.S.; Odelade, A.E.; Patwardhan, S.; Chaudhari, M.; Pardue, A.S.; Kumari, A.; LeFever, W.A.; Tornow, O.P.; et al. Hydrogen peroxide-dependent oxidation of ERK2 within its D-recruitment site alters its substrate selection. *iScience* **2023**, *26*, 107817. [CrossRef]
95. Nithianandarajah-Jones, G.N.; Wilm, B.; Goldring, C.E.; Müller, J.; Cross, M.J. ERK5: Structure, regulation and function. *Cell. Signal.* **2012**, *24*, 2187–2196. [CrossRef] [PubMed]
96. Park, Y.M.; Febbraio, M.; Silverstein, R.L. CD36 modulates migration of mouse and human macrophages in response to oxidized LDL and may contribute to macrophage trapping in the arterial intima. *J. Clin. Investig.* **2009**, *119*, 136–145. [CrossRef] [PubMed]
97. Netto, L.E.S.; Machado, L.E.S.F. Preferential redox regulation of cysteine-based protein tyrosine phosphatases: Structural and biochemical diversity. *FEBS J.* **2022**, *289*, 5480–5504. [CrossRef] [PubMed]
98. Sonkar, V.K.; Kumar, R.; Jensen, M.; Wagner, B.A.; Sharathkumar, A.A.; Miller, F.J., Jr.; Fasano, M.; Lentz, S.R.; Buettner, G.R.; Dayal, S. Nox2 NADPH oxidase is dispensable for platelet activation or arterial thrombosis in mice. *Blood Adv.* **2019**, *3*, 1272–1284. [CrossRef] [PubMed]
99. Tang, Z.; Xu, Y.; Tan, Y.; Shi, H.; Jin, P.; Li, Y.; Teng, J.; Liu, H.; Pan, H.; Hu, Q.; et al. CD36 mediates SARS-CoV-2-envelope-protein-induced platelet activation and thrombosis. *Nat. Commun.* **2023**, *14*, 5077. [CrossRef] [PubMed]

100. Pokrovskaya, I.D.; Yadav, S.; Rao, A.; McBride, E.; Kamykowski, J.A.; Zhang, G.; Aronova, M.A.; Leapman, R.D.; Storrie, B. 3D ultrastructural analysis of α-granule, dense granule, mitochondria, and canalicular system arrangement in resting human platelets. *Res. Pract. Thromb. Haemost.* **2020**, *4*, 72–85. [CrossRef]
101. Cole, L.W. The Evolution of Per-cell Organelle Number. *Front. Cell Dev. Biol.* **2016**, *4*, 85. [CrossRef]
102. Fuentes, E.; Araya-Maturana, R.; Urra, F.A. Regulation of mitochondrial function as a promising target in platelet activation-related diseases. *Free. Radic. Biol. Med.* **2019**, *136*, 172–182. [CrossRef]
103. Chen, Y.; Yang, M.; Huang, W.; Chen, W.; Zhao, Y.; Schulte, M.L.; Volberding, P.; Gerbec, Z.; Zimmermann, M.T.; Zeighami, A.; et al. Mitochondrial Metabolic Reprogramming by CD36 Signaling Drives Macrophage Inflammatory Responses. *Circ. Res.* **2019**, *125*, 1087–1102. [CrossRef]
104. He, F.; Ru, X.; Wen, T. NRF2, a Transcription Factor for Stress Response and Beyond. *Int. J. Mol. Sci.* **2020**, *21*, 4777. [CrossRef] [PubMed]
105. Forman, H.J.; Zhang, H. Targeting oxidative stress in disease: Promise and limitations of antioxidant therapy. *Nat. Rev. Drug Discov.* **2021**, *20*, 689–709. [CrossRef] [PubMed]
106. Brigelius-Flohé, R.; Maiorino, M. Glutathione peroxidases. *Biochim. Biophys. Acta (BBA) Gen. Subj.* **2013**, *1830*, 3289–3303. [CrossRef] [PubMed]
107. Bierl, C.; Voetsch, B.; Jin, R.C.; Handy, D.E.; Loscalzo, J. Determinants of Human Plasma Glutathione Peroxidase (GPx-3) Expression. *J. Biol. Chem.* **2004**, *279*, 26839–26845. [CrossRef] [PubMed]
108. Voetsch, B.; Jin, R.C.; Bierl, C.; Deus-Silva, L.; Camargo, E.C.; Annichino-Bizacchi, J.M.; Handy, D.E.; Loscalzo, J. Role of Promoter Polymorphisms in the Plasma Glutathione Peroxidase (GPx-3) Gene as a Risk Factor for Cerebral Venous Thrombosis. *Stroke* **2008**, *39*, 303–307. [CrossRef] [PubMed]
109. Voetsch, B.; Jin, R.C.; Bierl, C.; Benke, K.S.; Kenet, G.; Simioni, P.; Ottaviano, F.; Damasceno, B.P.; Annichino-Bizacchi, J.M.; Handy, D.E.; et al. Promoter polymorphisms in the plasma glutathione peroxidase (GPx-3) gene: A novel risk factor for arterial ischemic stroke among young adults and children. *Stroke* **2007**, *38*, 41–49. [CrossRef]
110. Dayal, S.; Gu, S.X.; Hutchins, R.D.; Wilson, K.M.; Wang, Y.; Fu, X.; Lentz, S.R. Deficiency of Superoxide Dismutase Impairs Protein C Activation and Enhances Susceptibility to Experimental Thrombosis. *Arter. Thromb. Vasc. Biol.* **2015**, *35*, 1798–1804. [CrossRef]
111. Sonkar, V.K.; Eustes, A.S.; Ahmed, A.; Jensen, M.; Solanki, M.V.; Swamy, J.; Kumar, R.; Fidler, T.P.; Houtman, J.C.; Allen, B.G.; et al. Endogenous SOD2 (Superoxide Dismutase) Regulates Platelet-Dependent Thrombin Generation and Thrombosis During Aging. *Arter. Thromb. Vasc. Biol.* **2023**, *43*, 79–91. [CrossRef]
112. Yang, M.; Chiu, J.; Scartelli, C.; Ponzar, N.; Patel, S.; Patel, A.; Ferreira, R.B.; Keyes, R.F.; Carroll, K.S.; Pozzi, N.; et al. Sulfenylation links oxidative stress to protein disulfide isomerase oxidase activity and thrombus formation. *J. Thromb. Haemost.* **2023**, *21*, 2137–2150. [CrossRef]
113. Badrnya, S.; Schrottmaier, W.C.; Kral, J.B.; Yaiw, K.-C.; Volf, I.; Schabbauer, G.; Söderberg-Nauclér, C.; Assinger, A. Platelets Mediate Oxidized Low-Density Lipoprotein–Induced Monocyte Extravasation and Foam Cell Formation. *Arter. Thromb. Vasc. Biol.* **2014**, *34*, 571–580. [CrossRef]
114. Zheng, T.J.; Kohs, T.C.; Mueller, P.A.; Pang, J.; Reitsma, S.E.; Parra-Izquierdo, I.; Melrose, A.; Yang, L.; Choi, J.; Zientek, K.D.; et al. Effect of antiplatelet agents and tyrosine kinase inhibitors on oxLDL-mediated procoagulant platelet activity. *Blood Adv.* **2022**, *7*, 1366–1378. [CrossRef] [PubMed]
115. Parthasarathy, S.; Raghavamenon, A.; Garelnabi, M.O.; Santanam, N. Oxidized low-density lipoprotein. *Methods Mol. Biol.* **2010**, *610*, 403–417. [CrossRef] [PubMed]
116. Quan, S.; Schneider, I.; Pan, J.; Von Hacht, A.; Bardwell, J.C.A. The CXXC Motif Is More than a Redox Rheostat. *J. Biol. Chem.* **2007**, *282*, 28823–28833. [CrossRef] [PubMed]
117. Cho, J.; Furie, B.C.; Coughlin, S.R.; Furie, B. A critical role for extracellular protein disulfide isomerase during thrombus formation in mice. *J. Clin. Investig.* **2008**, *118*, 1123–1131. [CrossRef] [PubMed]
118. Reinhardt, C.; von Brühl, M.-L.; Manukyan, D.; Grahl, L.; Lorenz, M.; Altmann, B.; Dlugai, S.; Hess, S.; Konrad, I.; Orschiedt, L.; et al. Protein disulfide isomerase acts as an injury response signal that enhances fibrin generation via tissue factor activation. *J. Clin. Investig.* **2008**, *118*, 1110–1122. [CrossRef] [PubMed]
119. Jasuja, R.; Passam, F.H.; Kennedy, D.R.; Kim, S.H.; van Hessem, L.; Lin, L.; Bowley, S.R.; Joshi, S.S.; Dilks, J.R.; Furie, B.; et al. Protein disulfide isomerase inhibitors constitute a new class of antithrombotic agents. *J. Clin. Investig.* **2012**, *122*, 2104–2113. [CrossRef] [PubMed]
120. Zwicker, J.I.; Schlechter, B.L.; Stopa, J.D.; Liebman, H.A.; Aggarwal, A.; Puligandla, M.; Caughey, T.; Bauer, K.A.; Kuemmerle, N.; Wong, E.; et al. Targeting protein disulfide isomerase with the flavonoid isoquercetin to improve hypercoagulability in advanced cancer. *JCI Insight* **2019**, *4*, e125851. [CrossRef]
121. Jasuja, R.; Furie, B.; Furie, B.C. Endothelium-derived but not platelet-derived protein disulfide isomerase is required for thrombus formation in vivo. *Blood* **2010**, *116*, 4665–4674. [CrossRef]
122. Kim, K.; Hahm, E.; Li, J.; Holbrook, L.-M.; Sasikumar, P.; Stanley, R.G.; Ushio-Fukai, M.; Gibbins, J.M.; Cho, J. Platelet protein disulfide isomerase is required for thrombus formation but not for hemostasis in mice. *Blood* **2013**, *122*, 1052–1061. [CrossRef]
123. Zhou, J.; Wu, Y.; Wang, L.; Rauova, L.; Hayes, V.M.; Poncz, M.; Essex, D.W. The C-terminal CGHC motif of protein disulfide isomerase supports thrombosis. *J. Clin. Investig.* **2015**, *125*, 4391–4406. [CrossRef]

124. Xiao, R.; Lundström-Ljung, J.; Holmgren, A.; Gilbert, H.F. Catalysis of thiol/disulfide exchange. Glutaredoxin 1 and protein-disulfide isomerase use different mechanisms to enhance oxidase and reductase activities. *J. Biol. Chem.* **2005**, *280*, 21099–21106. [CrossRef]
125. Saurin, A.T.; Neubert, H.; Brennan, J.P.; Eaton, P. Widespread sulfenic acid formation in tissues in response to hydrogen peroxide. *Proc. Natl. Acad. Sci. USA* **2004**, *101*, 17982–17987. [CrossRef] [PubMed]
126. Radi, R.; Beckman, J.S.; Bush, K.M.; Freeman, B.A. Peroxynitrite oxidation of sulfhydryls. The cytotoxic potential of superoxide and nitric oxide. *J. Biol. Chem.* **1991**, *266*, 4244–4250. [CrossRef]
127. Gaspar, R.S.; Laurindo, F.R.M. Sulfenylation: An emerging element of the protein disulfide isomerase code for thrombosis. *J. Thromb. Haemost.* **2023**, *21*, 2054–2057. [CrossRef] [PubMed]
128. Gupta, V.; Yang, J.; Liebler, D.C.; Carroll, K.S. Diverse Redoxome Reactivity Profiles of Carbon Nucleophiles. *J. Am. Chem. Soc.* **2017**, *139*, 5588–5595. [CrossRef] [PubMed]
129. Nagy, P. Kinetics and Mechanisms of Thiol–Disulfide Exchange Covering Direct Substitution and Thiol Oxidation-Mediated Pathways. *Antioxid. Redox Signal.* **2013**, *18*, 1623–1641. [CrossRef]
130. Kaetzel, C.S.; Rao, C.K.; Lamm, M.E. Protein disulphide-isomerase from human placenta and rat liver. Purification and immunological characterization with monoclonal antibodies. *Biochem. J.* **1987**, *241*, 39–47. [CrossRef] [PubMed]
131. Mandel, R.; Ryser, H.J.; Ghani, F.; Wu, M.; Peak, D. Inhibition of a reductive function of the plasma membrane by bacitracin and antibodies against protein disulfide-isomerase. *Proc. Natl. Acad. Sci. USA* **1993**, *90*, 4112–4116. [CrossRef]
132. Wu, Y.; Ahmad, S.S.; Zhou, J.; Wang, L.; Cully, M.P.; Essex, D.W. The disulfide isomerase ERp57 mediates platelet aggregation, hemostasis, and thrombosis. *Blood* **2012**, *119*, 1737–1746. [CrossRef]
133. Shi, Y.; Carroll, K.S. Activity-Based Sensing for Site-Specific Proteomic Analysis of Cysteine Oxidation. *Acc Chem. Res.* **2020**, *53*, 20–31. [CrossRef]
134. Gupta, V.; Carroll, K.S. Profiling the reactivity of cyclic C-nucleophiles towards electrophilic sulfur in cysteine sulfenic acid. *Chem. Sci.* **2016**, *7*, 400–415. [CrossRef]
135. Boike, L.; Henning, N.J.; Nomura, D.K. Advances in covalent drug discovery. *Nat. Rev. Drug Discov.* **2022**, *21*, 881–898. [CrossRef] [PubMed]
136. Benitez, L.V.; Allison, W.S. The inactivation of the acyl phosphatase activity catalyzed by the sulfenic acid form of glycer-aldehyde 3-phosphate dehydrogenase by dimedone and olefins. *J. Biol. Chem.* **1974**, *249*, 6234–6243. [CrossRef] [PubMed]
137. Fu, L.; Jung, Y.; Tian, C.; Ferreira, R.B.; Cheng, R.; He, F.; Yang, J.; Carroll, K.S. Nucleophilic covalent ligand discovery for the cysteine redoxome. *Nat. Chem. Biol.* **2023**, *19*, 1309–1319. [CrossRef] [PubMed]
138. Backus, K.M.; Correia, B.E.; Lum, K.M.; Forli, S.; Horning, B.D.; González-Páez, G.E.; Chatterjee, S.; Lanning, B.R.; Teijaro, J.R.; Olson, A.J.; et al. Proteome-wide covalent ligand discovery in native biological systems. *Nature* **2016**, *534*, 570–574. [CrossRef]
139. Bak, D.W. Flipping the polarity switch. *Nat. Chem. Biol.* **2023**, *19*, 1292–1293. [CrossRef] [PubMed]
140. Tsoupras, A.; Zabetakis, I.; Lordan, R. Platelet aggregometry assay for evaluating the effects of platelet agonists and antiplatelet compounds on platelet function in vitro. *MethodsX* **2019**, *6*, 63–70. [CrossRef]
141. Kholmukhamedov, A.; Jobe, S. Mitochondria and Platelet Cell Death. *Thromb. Haemost.* **2017**, *117*, 2207–2208. [CrossRef]
142. Huang, W.; Li, R.; Zhang, J.; Cheng, Y.; Ramakrishnan, D.P.; Silverstein, R.L. A CD36 transmembrane domain peptide interrupts CD36 interactions with membrane partners on macrophages and inhibits atherogenic functions. *Transl. Res.* **2023**, *254*, 68–76. [CrossRef]
143. Chen, Y.; Kennedy, D.J.; Ramakrishnan, D.P.; Yang, M.; Huang, W.; Li, Z.; Xie, Z.; Chadwick, A.C.; Sahoo, D.; Silverstein, R.L. Oxidized LDL–bound CD36 recruits an Na^+/K^+-ATPase–Lyn complex in macrophages that promotes atherosclerosis. *Sci. Signal.* **2015**, *8*, ra91. [CrossRef]
144. Chen, Y.; Huang, W.; Yang, M.; Xin, G.; Cui, W.; Xie, Z.; Silverstein, R.L. Cardiotonic Steroids Stimulate Macrophage Inflammatory Responses Through a Pathway Involving CD36, TLR4, and Na/K-ATPase. *Arter. Thromb. Vasc. Biol.* **2017**, *37*, 1462–1469. [CrossRef]
145. Miao, W.-M.; Vasile, E.; Lane, W.S.; Lawler, J. CD36 associates with CD9 and integrins on human blood platelets. *Blood* **2001**, *97*, 1689–1696. [CrossRef]
146. Julio, A.R.; Shikwana, F.; Truong, C.; Burton, N.R.; Dominguez, E.; Turmon, A.; Cao, J.; Backus, K. Pervasive aggregation and depletion of host and viral proteins in response to cysteine-reactive electrophilic compounds. *bioRxiv* **2023**. [CrossRef]
147. Sonego, G.; Le, T.M.; Crettaz, D.; Abonnenc, M.; Tissot, J.; Prudent, M. Sulfenylome analysis of pathogen-inactivated platelets reveals the presence of cysteine oxidation in integrin signaling pathway and cytoskeleton regulation. *J. Thromb. Haemost.* **2021**, *19*, 233–247. [CrossRef] [PubMed]

MDPI AG
Grosspeteranlage 5
4052 Basel
Switzerland
Tel.: +41 61 683 77 34

Antioxidants Editorial Office
E-mail: antioxidants@mdpi.com
www.mdpi.com/journal/antioxidants

www.ingramcontent.com/pod-product-compliance
Lightning Source LLC
Chambersburg PA
CBHW041348210926
51850CB00013B/3019